SIXTH EDITION

# THE SKILLED HELPER

# Gerard Egan

Gerard Egan, Ph.D., is Professor Emeritus of Organization Development and Psychology in the Center for Organization Development of Loyola University of Chicago. He has written over a dozen books, some in the field of counseling and communication, including *The Skilled Helper, Interpersonal Living,* and *People in Systems. The Skilled Helper,* translated into both European and Asian languages, is currently the most widely used counseling text in the world.

His other books, dealing with business and management, include *Change Agent Skills in Helping and Human Service Settings, Change Agent Skills A: Designing and Assessing Excellence, Change Agent Skills B: Managing Innovation and Change, Adding Value: A Systematic Guide to Business-Based Management and Leadership,* and *Working the Shadow Side: A Guide to Positive Behind-the-Scenes Management.* Through these writings, complemented by extensive consulting, he has created a comprehensive business-based system of management focusing on strategy, operations, structure, human resource management, the managerial role itself, and leadership. The management system includes a framework for initiating and managing change and a framework for managing such "shadow-side" complexities as organizational culture and politics and resistance to change. He has lectured, consulted, and given workshops in Africa, Asia, Australia, Europe, and North America. In China he has worked with university- and community-based professionals on counselor-training systems.

He consults to a variety of companies and institutions worldwide. Most of the counseling he does takes place within these organizations as part of his fourfold role — consultant, coach, counselor, and confidant. He specializes in working with senior managers, often on a longer-term basis, on board relationships, strategy, business and organization effectiveness, human-resource development, management development, leadership, challenging and redesigning corporate culture, and the design and management of change.

SIXTH EDITION

# THE SKILLED HELPER

## A PROBLEM-MANAGEMENT
## APPROACH TO HELPING

### Gerard Egan

Professor Emeritus
Loyola University of Chicago

BROOKS/COLE PUBLISHING COMPANY

I(T)P® *An International Thomson Publishing Company*

Pacific Grove • Albany • Belmont • Bonn • Boston • Cincinnati • Detroit • Johannesburg
London • Madrid • Melbourne • Mexico City • Paris • Singapore

Sponsoring Editor: *Eileen Murphy*
Marketing Representative: *Thomas L. Braden*
Editorial Assistant: *Susan Carlson*
Production Service: *Greg Hubit Bookworks*
Production Coordinator: *Marjorie Z. Sanders*
Marketing Team: *Romy Taormina and
  Jean Thompson*
Manuscript Editor: *Carole R. Crouse*

Interior Design: *John Edeen*
Interior Illustration: *Lotus Art*
Cover Design: *Katherine Minerva*
Typesetting: *ColorType, San Diego*
Cover Printing: *Phoenix Color Corporation*
Printing and Binding:
  R. R. Donnelley/Crawfordsville

*For more information, contact:*

BROOKS/COLE PUBLISHING COMPANY
511 Forest Lodge Road
Pacific Grove, CA 93950
USA

International Thomson Publishing Europe
Berkshire House 168–173
High Holborn
London WC1V 7AA
England

Thomas Nelson Australia
102 Dodds Street
South Melbourne, 3205
Victoria, Australia

Nelson Canada
1120 Birchmount Road
Scarborough, Ontario
Canada MIK 5G4

International Thomson Editores
Seneca 53
Col. Polanco
11560 México, D. F., México

International Thomson Publishing GmbH
Königswinterer Strasse 418
53227 Bonn
Germany

International Thomson Publishing Asia
221 Henderson Road
#05–10 Henderson Building
Singapore 0315

International Thomson Publishing Japan
Hirakawacho Kyowa Building, 3F
2-2-1 Hirakawacho
Chiyoda-ku, Tokyo 102
Japan

**Library of Congress Cataloging-in-Publication Data**
Egan, Gerard.
    The skilled helper : a problem-management approach to helping /
Gerard Egan. — 6th ed.
        p.  cm.
    Includes bibliographical references and indexes.
    ISBN 0-534-34948-X  (alk. paper)
    1. Counseling.   2. Helping behavior.   I. Title.
BF637.C6E39   1998
158'.3 — dc21

                                    97-17270
                                      CIP

# CONTENTS

PART TWO

# BASIC COMMUNICATION SKILLS FOR HELPING  59

CHAPTER 11

# STEP I-B: III. THE WISDOM OF CHALLENGING    188

CHAPTER 12

# STEP I-C: LEVERAGE: HELPING CLIENTS WORK ON THE RIGHT THINGS    201

## CHAPTER 15

# STEP II-C: COMMITMENT—WHAT ARE YOU WILLING TO PAY FOR WHAT YOU WANT?   257

## PART FIVE

# STAGE III: HELPING CLIENTS WORK FOR WHAT THEY NEED AND WANT   269

## CHAPTER 16

# STEP III-A: STRATEGIES FOR ACTION — WHAT DO I NEED TO DO TO GET WHAT I NEED AND WANT?   273

## CHAPTER 17

# STEP III-B: BEST-FIT STRATEGIES — WHAT STRATEGIES ARE BEST FOR ME? 286

## CHAPTER 18

# STEP III-C: HELPING CLIENTS MAKE PLANS — WHAT KIND OF PLAN WILL HELP ME GET WHAT I NEED AND WANT? 299

# PREFACE

Those intending to enter one of the helping professions will find a professional landscape that has been drastically altered over the past decade. Managed care for both physical and behavioral problems has become the norm. The number of psychologists in private practice has steadily declined as managed-care group practices have grown. Some helpers readily embrace managed care, others see it as a political and economic reality, and still others condemn it as "immoral" (see Broskowski, 1995; Fox, 1995; Fraser, 1996; "Managed Care," 1996, 323–325; Miller, 1996; Patricelli & Lee, 1996; Rosenberg, 1996). Furthermore, third-party payers are demanding that helpers improve the quality of their psychological services and make themselves more accountable for results (Steenbarger & Smith, 1996).

The practical helping model outlined in the pages of this book — problem management and opportunity development — can help human-service providers meet these challenges. There are many reasons for using a helping model based on problem solving:

- Problem solving is one of the most highly researched paradigms in psychology. It is not fad. It is not based on an unsupported theory.

- Problem management and opportunity development constitute key dynamics underlying every form of helping. The reason for this is that the process focuses on the client's needs, not the assumptions of a theory.

- The basic problem-solving process or model is universal; therefore, it crosses cultures easily and is easily adapted to cultural differences.

- It is practical. Since it deals with individuals and focuses on results, it is useful in managed-care environments. Each case is a study in itself; goals are set and progress toward those goals is plotted.

- Clients recognize, understand, and can use the problem-management process, once it is shared with them. A mutual understanding of the helping process allows clients and helpers to become partners.

- Once the helping process is mutually owned, then helping sessions become "labs" in which clients learn how to become better problem managers and opportunity developers in their everyday lives.

- Helpers can use the problem-management framework to organize a wide range of helping methods to serve their clients. The framework becomes a tool in helping them spot and integrate the best of research and the best of new methods into their practice.

Therefore, no matter what approach to helping a helper ultimately adopts, the problem-managing and opportunity-developing framework together with the communication skills and methods that make it work provides a valuable foundation and adds substance to the practice of any approach to helping.

As you will see, the approach espoused here focuses not just on problems but also on unused opportunities. At its best, it is about strengths rather than weaknesses. In helping clients, helpers (and I include myself) tend to focus most of their attention on problem situations rather than missed opportunities. Problems are center stage; they grab our attention. Missed opportunities languish in the shadows; they are too easily overlooked. Certainly problems cannot be ignored because of their "in your face" character. Nor should they. However, clients can often manage or even transcend problems by focusing on opportunities.

In our society, prevention, like unused opportunities, gets short shrift. The problem-management and opportunity-development framework would provide much greater benefit to individuals, indeed to society itself, if it were more widely used in the interest of prevention. The economics of prevention are well known. Many studies demonstrate that prevention "works," even though there is disagreement on how to go about doing it (Albee & Ryan-Finn, 1993; Cole, Watt, West, Hawkins, Asarnow, Markman, Ramey, Shure, & Long, 1993; Heller, 1993; Humphreys, 1996; Landsman, 1994; Lieberman, 1997; Perry & Albee, 1994). Resources spent on preventing social-emotional, physical, family, workplace, and societal problems provide a "return" that is much greater than the return from resources spent on cure. Despite this, the institutions of society continue to underspend on prevention. It isn't sexy. The helping professions keep talking about prevention, but in the end it is not personally, socially, or politically compelling. Indeed, the world's indifference to prevention is one of the reasons for the current crisis in both physical and social-emotional health care.

The problem-management process and skills outlined in this book are even more valuable when used preventively. When I ask parents how important — on a scale from 1 to 100 — interpersonal communication skills and competency in problem-management are for their children, they rate both near 100. But when I ask them how their children learn these skills, the hemming and hawing begin. After a while, I summarize what I am hearing. "We live in a society," I say, "in which these basic skills are extremely important, but we leave their development to chance." The development of these skills is too important to be left to any one social setting. Ideally, they would be taught, modeled, and reinforced in all social settings — family, peer groups, school, church, and community. We have the belief that these skills are essential for our children, but we do not have the social or political will to do much about it.

When and how are we to find the common sense, courage, and political will to put such things as competency in interpersonal communication and problem management and other forms of prevention on the national agenda? Certainly pockets of excellence already exist — this or that individual, this or that family, this or that school, this or that church, this or that community, this or that government program. But these do not add up to a national commitment. We don't need prevention and skill development shoved down our throats the way managed care is today. We do need a movement that is so appealing that a critical mass of citizens would like to participate in it because of the benefits. Do we live in a society that can market almost anything except things that count the most?

For offering their comments and suggestions for this edition of *The Skilled Helper*, I would like to thank the following reviewers: Jean F. Ayers, Towson State University; Dale Blumen, University of Rhode Island; John Bowers, Northwest Missouri State University; Diane H. Coursol, Mankato State University; and Tracey Manning, The College of Notre Dame of Maryland.

*Gerard Egan*

## PART ONE

# LAYING THE GROUNDWORK

Although the centerpiece of this book is the helping
model itself and the methods and communication
skills that make it work, there is some groundwork to
be laid. This includes outlining the nature and goals of
helping (Chapter 1), outlining the helping process
(Chapter 2), and crafting the preferred culture of
helping from the assumptions, beliefs, values, and
norms that should drive helping behavior (Chapter 3).

## CHAPTER 1

# INTRODUCTION

## FORMAL AND INFORMAL HELPERS — A VERY BRIEF HISTORY

Throughout history there has been a deeply embedded conviction that, under the proper conditions, some people are capable of helping others come to grips with problems in living. This conviction, of course, plays itself out differently in different cultures, but it is still a cross-cultural phenomenon. Today this conviction is often institutionalized in a variety of formal helping professions. Counselors, psychiatrists, psychologists, social workers, and ministers of religion are counted among those whose formal role is to help people manage the distressing problems of life.

There is also a second set of professionals who, although they are not helpers in the formal sense, often deal with people in times of crisis and distress. Included here are organizational consultants, dentists, doctors, lawyers, nurses, probation officers, teachers, managers, supervisors, police officers, and practitioners in other service industries. Although these people are specialists in their own professions, there is still some expectation that they will help those they serve manage a variety of problem situations. For instance, teachers teach English, history, and science to students who are growing physically, intellectually, socially, and emotionally and struggling with developmental tasks and crises. Teachers are, therefore, in a position to help their students, in direct and indirect ways, explore, understand, and deal with the problems of growing up. Managers and supervisors help workers cope with problems related to work performance, career development, interpersonal relationships in the workplace, and a variety of personal problems that affect their ability to do their jobs. This book is addressed directly to the first set of professionals and indirectly to the second.

To these professional helpers can be added any and all who try to help relatives, friends, acquaintances, strangers (on buses and planes), and themselves come to grips with problems in living. Indeed, only a small fraction of the help provided on any given day comes from helping professionals. Informal helpers — bartenders and hairdressers are often mentioned — abound in the social settings of life. Friends often help one another through troubled times. Parents need to manage their own marital problems while helping their children grow and develop. Finally, all of us must help ourselves cope with the problems and crises of life. In short, the world is filled with informal helpers. Since helping is such a common human experience, training in both solving one's own problems and helping others solve theirs should be as common as training in reading, writing, and math. But that is not the case.

## WHAT HELPING IS ABOUT

To determine what helping is about, it is useful to consider (1) why people seek — or are sent to get — help in the first place and (2) what the principal goals of the helping process are.

Why do people come?

# Clients with Problem Situations
## and Unused Opportunities

Many clients become clients because, either in their own eyes or in the eyes of others, they are involved in problem situations that they are not handling well. Others come because they feel they are not living as fully as they might. Therefore, clients' problem situations and unused opportunities constitute the starting point of the helping process. In this book, client-centered helping means that the needs of the client, not the models and methods of the helper, are central. The ultimate focus is, of course, client-enhancing outcomes.

<u>Problem situations.</u> Clients come for help <u>because they have crises, troubles, doubts, difficulties, frustrations, or concerns.</u> These are often called, generically, "problems," but they are not problems in a mathematical sense, because often emotions run high and often there are no clear-cut solutions. It is probably better to say that clients come not with problems but with *problem situations* — that is, with complex and messy problems in living that they are not handling well. These problem situations are not academic puzzles to be solved. Rather, they are more like problems people face in their everyday lives — that is, "(a) unformulated or in need of reformulation, (b) of personal interest, (c) lacking in information necessary for solution, (d) related to everyday experience, (e) poorly defined, (f) characterized by multiple 'correct' solutions, each with liabilities as well as assets, and (g) characterized by multiple methods for picking a problem solution" (Sternberg, Wagner, Williams, & Horvath, 1995).

Even devastating problem situations, when worked through, can often be handled more effectively. Consider the following example.

> Fred L. was thunderstruck when the doctor told him that he had terminal cancer. He was only 52; death couldn't be imminent. He felt confused, bitter, angry, and depressed. After a period of angry confrontations with doctors and members of his family, in his despair he finally talked to a clergyman who had, on a few occasions, gently and nonintrusively offered his support. They decided to have some sessions together, but the clergyman also referred him to a counselor who worked in a hospice for the dying. With their help Fred gradually learned how to manage the ultimate problem situation of his life. He came to grips with his religious convictions, put his affairs in order, began to learn how to say goodbye to his family and the world he loved so intensely, and, with the help of the hospice in which the counselor worked, set about the process of managing physical decline. There were some outbursts of anger and some brief periods of depression and despair, but generally he managed the process of dying much better than he would have done without the help of his family and his counselors.

This case demonstrates in a dramatic way that the <u>goal of helping is not to "solve" everything</u>. Fred did die. Some problem situations are simply more unmanageable than others. <u>Helping clients review their problems and the</u>

options they have for dealing with them is, as we shall see, a central part of the helping process.

Problem situations arise in our interactions with ourselves, with others, and with the organizations, institutions, and communities of life. Clients — whether they are hounded by self-doubt, tortured by unreasonable fears, grappling with cancer, addicted to alcohol or drugs, involved in failing marriages, fired from jobs because they do not have the skills needed in the "new economy," suffering from a catastrophic loss, jailed because of child abuse, wallowing in a midlife crisis, lonely and out of community with no family or friends, battered by their spouses, or victimized by racism — all face problem situations that move them to seek help or move others to send them for help.

**Missed opportunities and unused potential.** Some clients come for help not because they are dogged by problems like those listed above but because they are not as effective as they would like to be. They have resources they are not using or opportunities they are not developing. People who feel locked in dead-end jobs or bland marriages, who are frustrated because they lack challenging goals, who feel guilty because they are failing to live up to their own values and ideals, who want to do something more constructive with their lives, or who are disappointed with their uneventful interpersonal lives — such clients come to helpers not to manage their problems but to live more fully.

And so clients' missed opportunities and unused potential constitute a second starting point for helping. In this case, it is a question not of what is going wrong but of what could be better. It has often been suggested that most of us use only a small fraction of our potential. Most of us are capable of dealing much more creatively with ourselves, with our relationships with others, with our work life, and, generally, with the ways in which we involve ourselves with the social settings of our lives. Consider the following case.

> After ten years as a helper in several mental health centers, Carol was experiencing burnout. In the opening interview with a counselor, she berated herself for not being dedicated enough. Asked when she felt best about herself, she said that it was on those relatively infrequent occasions when she was asked to help another mental health center to get started or reorganize itself. The counseling sessions helped her explore her potential as a consultant to human-service organizations and make a career adjustment. She enrolled in courses in the Center for Organization Development at a local university. Carol stayed in the helping field, but with a new focus and a new set of skills.

In this case, the counselor helped the client manage her problems (burnout, guilt) more effectively through the development of an opportunity that made enormous sense at this stage of her life and of her career.

## The Two Principal Goals of Helping

Since this book discusses helping as a formal process, those being helped will be referred to as *clients*. However, most of what is said applies also to helping as an informal process. The goals of helping must be based on the needs of clients. There are two basic goals — one relating to clients' managing specific problems in living more effectively and the other relating to their general ability to manage problems and develop opportunities.

### GOAL ONE: *Help clients manage their problems in living more effectively and develop unused or underused opportunities more fully.*

Let's look at the helping process from both the helper's and the client's point of view. *Helpers* are successful to the degree to which their clients — because of client-helper interactions — are in a better position to manage specific problem situations and develop specific unused resources and opportunities more effectively. Notice that I stop short of saying that clients actually end up managing both problems and opportunities better. Although counselors help clients achieve valued outcomes, they do not control those outcomes. In the end, clients can choose to live more effectively or not.

Since helping is a two-way, collaborative process, clients, too, have a primary goal. *Clients* are successful to the degree that they commit themselves to the helping process and capitalize on what they learn from the helping sessions by using those learnings to manage problem situations more effectively *results* and develop opportunities more fully. Of course, many clients, because of their *Outcomes* interactions with helpers, not only are in a better position to manage the ups *accomplish-* and downs of their lives more effectively but also actually do so. *ment*

A corollary to Goal One is that **helping is about results, outcomes, ac-** *Impact* **complishments, impact.** Helping is an "–ing" word: It includes a series of activities in which helpers and clients engage. These activities, however, have value only to the degree that they lead to valued outcomes in clients' lives. Ultimately, statements such as "We had a good session," whether spoken by the helper or by the client, must translate into more effective living on the part of the client. If a helper and a client engage in the counseling process effectively, something of value will be in place that was not in place before the helping sessions: Unreasonable fears will disappear or diminish to manageable levels, self-confidence will replace self-doubt, addictions will be conquered, an operation will be faced with a degree of equanimity, a better job will be found, a woman and man will breathe new life into their marriage, a battered wife will find the courage to leave her husband, and a person embittered by institutional racism will regain his self-respect. In a word, helping should make a substantive *difference* in the life of the client. Helping is about *constructive change.* Skilled counselors help clients develop programs for constructive change.

The need for results is seen clearly in the case of a battered woman, Andrea N., outlined by Driscoll (1984, p. 64).

The mistreatment had caused her to feel that she was worthless even as she developed a secret superiority to those who mistreated her. These attitudes contributed in turn to her continuing passivity and had to be challenged if she was to become assertive about her own rights. Through the helping interactions, she developed a sense of worth and self-confidence. This was the first outcome of the helping process. As she gained confidence, she became more assertive; she realized that she had the right to take stands, and she chose to challenge those who took advantage of her. She stopped merely resenting them and did something about it. The second outcome was a pattern of assertiveness, however tentative in the beginning, that took the place of·a pattern of passivity. When her assertive stands were successful, her rights became established, her social relationships improved, and her confidence in herself increased, thus further altering the original self-defeating pattern. This was a third set of outcomes. As she saw herself becoming more and more an "agent" rather than a "patient" in her everyday life, she found it easier to put aside her resentment and the self-limiting satisfactions of the passive-victim role and to continue asserting herself. This constituted a fourth set of outcomes. The activities in which she engaged, either within the helping sessions or in her day-to-day life, were valuable because they led to these valued outcomes.

Andrea needed much more than "good sessions" with a helper. She needed to work for outcomes that made a difference in her life. Today there is another reason for focusing on outcomes. Many psychological services are offered in managed-care settings. In these settings, more and more third-party payments depend on meaningful treatment plans and the delivery of problem-managing outcomes. But economics should not force helpers to do what they should be doing anyway in the service of their clients.

### GOAL TWO: Help clients become better at helping themselves in their everyday lives.

Clients often are poor problem solvers, or whatever problem-solving ability they have tends to disappear in times of crisis. What Miller, Galanter, and Pribram (1960, pp. 171, 174) said many years ago is, unfortunately, probably just as true today.

> In ordinary affairs we usually muddle about, doing what is habitual and customary, being slightly puzzled when it sometimes fails to give the intended outcome, but not stopping to worry much about the failures because there are still too many other things still to do. Then circumstances conspire against us and we find ourselves caught failing where we must succeed—where we cannot withdraw from the field, or lower our self-imposed standards, or ask for help, or throw a tantrum. Then we may begin to suspect that we face a problem. . . . An ordinary person almost never approaches a problem systematically and exhaustively unless he or she has been specifically educated to do so.

Most people in our society are not "educated to do so." We have yet to find ways of making sure that our children develop what most consider to be essential "life skills" such as problem solving and the skills of interpersonal relating.

And so the second goal of helping deals with clients' need (1) to participate actively in the problem-management process during the helping sessions themselves and (2) to continue to manage their lives more effectively after the period of formal helping is over. Helpers are effective to the degree that clients, through the helping process, learn how to help themselves more effectively. Just as doctors want their patients to learn how to prevent illness through exercise, good nutritional habits, and the avoidance of toxic substances and activities, just as dentists want their patients to engage in effective prevention activities, so skilled helpers want to see their clients not just managing *this* problem situation more effectively but also becoming more capable of managing subsequent problems in living more effectively. That is, helping at its best empowers clients to become more effective self-helpers — better problem managers and opportunity developers. Therefore, although this book is about a process helpers can use to help clients, more fundamentally, it is about a problem-management and opportunity-development process that clients can use to help themselves.

A problem-management and opportunity-development process designed to help clients achieve these two goals is outlined and illustrated in Chapter 2.

## DOES HELPING HELP? THE GOOD, THE CAUTIONARY, AND THE BAD NEWS

The answer to the question "Does helping help?" contains some very good news, some cautionary news, and some bad news ("Outcome Assessment," 1996; Lambert & Cattani-Thompson, 1996).

**The good news.** There is a great deal of evidence to show that helping, indeed, does help many people in many different situations. There is a convergence of positive evidence from different kinds of outcome studies. A word is in order about efficacy studies, meta-analytic studies, and a recent *Consumer Reports* survey study.

*Efficacy studies* focus on the usefulness of a specific helping methodology for a particular kind of problem — for instance, cognitive therapy with clients suffering from panic disorders. Comparisons are made between the methodology in question and some other methodology or between clients with some disorder who do receive the treatment and those who do not or between two different methodologies for treating the same disorder. These studies are carried out under carefully controlled conditions (see Luborsky, 1993; Seligman, 1995). Seligman has laid down eight rigorous conditions for the ideal efficacy study. Because of the rigor demanded, these studies are expensive and time-consuming. There are hundreds of efficacy studies, many of them very well designed, that demonstrate the efficacy of a particular therapy for a particular psychological disorder.

Next, although conventional reviews of groups of helping-outcome studies have produced mixed and ambiguous results (see Rossi & Wright, 1984; Schmidt, 1992), a relatively recent approach called *meta-analysis* — a kind of study of studies and a reinterpretation of their findings — has strongly demonstrated that helping helps (see Smith, Glass, & Miller, 1980; Lipsey & Wilson, 1993; Shapiro & Shapiro, 1982). In the words of Lipsey and Wilson, "Meta-analytic reviews [of helping outcomes] show a strong, dramatic pattern of positive overall effects that cannot readily be explained as artifacts of meta-analytic technique or generalized placebo effects" (p. 1181). Hundreds of meta-analytical studies have been done over the past 20 years, and although some are, admittedly, quite crude, they still add up to convincing evidence of the overall efficacy of helping.

Finally, *Consumer Reports* (1994; 1995) published the results of a sophisticated large-scale *survey* project on helping. The findings indicated that

- clients believed that they had benefited very substantially from psychotherapy;
- psychotherapy alone did not differ in effectiveness from psychotherapy plus medication;
- no specific form of helping did better than any other for any particular kind of problem;
- psychiatrists, psychologists, and social workers did not differ in their effectiveness as helpers;
- long-term treatment produced appreciably better results than did short-term treatment;
- clients whose choice of helper or length of therapy was limited by insurance or managed-care systems did not benefit as much as clients without those restrictions.

This study deals with the responses of real clients to questions about themselves, their helpers, processes used, and benefit received (see Seligman, 1995, for a discussion and critique of this study).

**The cautionary news.** First, all researchers discuss the limitations of their research methodologies. For instance, although Seligman praises the *Consumers Reports* survey as one of the best of its kind, he also takes pains to point out and discuss its serious methodological flaws. He also points out that many efficacy studies do not follow the rigorous requirements he lays down for such studies.

Second, different research methodologies sometimes lead to different conclusions. To the point, both meta-analytic studies and the *Consumers Reports* survey tend to suggest that no specific form of helping does better than any other for any particular kind of problem. This seems to contradict the efficacy study finding that there are specific techniques for specific disorders.

On the other hand, some researchers claim that that finding would be upheld if meta-analytic studies were to focus on the right kinds of variables (see Shadish & Sweeney, 1991).

Third, many outcome studies on helping are done in the lab or under lab-like conditions. But lab results cannot be automatically compared to clinical-setting results. Real helping does not take place in a lab (see Henggeler, Schoenwald, & Pickrel, 1995; Weisz, Donenberg, Han, & Weiss, 1995). Seligman (1995) finds fault with efficacy studies because they "omit too many crucial elements of what is done in the field" (p. 966). Getting help is not like taking a pill, the results of which are somewhat automatic (see Stiles, 1994; Stiles & Shapiro, 1994). In the end, all clients are individuals with problem situations to be managed, and the main question is, Is *this* client being helped by *this* helper, using *this* approach?

Furthermore, most of the help people get in the United States comes from friends, relatives, self-help and other kinds of support groups (the self-help group is such an important form of helping that a brief overview of it is given in the Appendix), and ministers of religion, not from professional helpers. There are also many paraprofessional helpers — that is, practitioners who have learned helping techniques but who do not have some professional qualification. Although studies have demonstrated that paraprofessional helpers can be as effective as professional helpers and in some cases even more helpful (Durlak, 1979; Hattie, Sharpley, & Rogers, 1984), most helping-outcome research focuses on professional helpers. That is, it excludes those who do most of the helping.

**The bad news.** Some critics express grave doubts about the legitimacy of the helping professions, even claiming that helping is a fraudulent process, a manipulative and malicious enterprise (see, for example, Cowen, 1982; Eysenck, 1984, 1994; Masson, 1988). Masson went so far as to claim that in the United States, helping is a multibillion dollar business that does no more than profit from people's misery. He also maintained that devaluing people is part and parcel of all therapy and that the helper's values and needs are inevitably imposed on the client. Although such criticisms are extreme, they should not be dismissed out of hand. What they say may be true of some forms of helping and of some helpers.

There is indeed some evidence that therapy sometimes not only does not help but also actually makes things worse. That is, some helping leads to negative outcomes (see Mohr, 1995, and Strupp, Hadley, & Gomes-Schwartz, 1977, for reviews of the negative-outcome literature). Research shows that some of the factors associated with negative outcomes in helping are associated with clients, others with helpers. Sometimes clients who have severe interpersonal problems and severe symptomatology, who are poorly motivated, or who expect helping to be painless become more dysfunctional through therapy. Helpers who underestimate the severity of clients' problems, experience interpersonal difficulties with clients, use poor techniques, overuse any given technique, or disagree with clients over helping methodology can make things worse rather than better.

Finally, some helpers are outright incompetent. And even the competent and committed have their lapses. As noted by Luborsky and his associates (1986):

- There are considerable differences between therapists in their average success rates.
- There is considerable variability in outcome within the caseload of individual therapists.
- Variations in success rates typically have more to do with the therapist than with the type of treatment.

Although helping can and often does work, there is plenty of evidence that ineffective helping also abounds. Helping is a powerful process that is all too easy to mismanage. It is no secret that because of inept helpers some clients get worse from treatment. Helping is not neutral; it is "for better or for worse." Ellis (1984) claimed that inept helpers are either ineffective or inefficient. Even though the inefficient may ultimately help their clients, they use "methods that are often distinctly inept and that consequently lead these clients to achieve weak and unlasting results, frequently at the expense of enormous amounts of wasted time and money" (p. 24). Since studies on the efficacy of counseling and psychotherapy do not usually make a distinction between high-level and low-level helpers, and since the research on deterioration effects in therapy suggests that there is a large number of low-level or inadequate helpers, the negative results found in many studies are predictable.

As Mohr (1995) pointed out, however, there is good news even in the bad news: Helping failures, if examined open-mindedly, are sources of learning. An entire issue of the *Journal of Psychotherapy Integration* was dedicated to failure in psychotherapy (see Stricker, 1995, for an introduction to this issue; see also Kottler & Blau, 1989). Six failed cases were explored at length, and suggestions for improving the psychotherapy process were offered. The process outlined in Chapter 2 and developed in this book offers both frameworks for understanding the source of failure and ways of preventing it.

**Therefore.** In the hands of skilled and socially intelligent helpers, helping can do a great deal of good. Norman Kagan (1973) long ago suggested that the basic issue confronting the helping professions is not validity — that is, whether helping helps or not — but reliability: "Not, can counseling and psychotherapy work, but does it work consistently? Not, can we educate people who are able to help others, but can we develop methods which will increase the likelihood that most of our graduates will become as effective mental health workers as only a rare few do?" (p. 44) The question, then, is not

"Does helping work?" but rather "How and under what conditions does it work?" The answer to the first question is easier than the answer to the second (see Bergin & Garfield, 1994).

A third question is, "What's the best way to train helpers, whether professional or paraprofessional?" Effective training programs can help improve the reliability of helping. The model of helping presented in this book, together with the skills and techniques that make it work, is designed precisely to increase both the validity and the reliability of the helping process. This book is aimed at enabling helpers of all sorts "deliver the goods" consistently and ethically to clients who seek out or accept the services of helpers to manage their lives more effectively.

Until we have better answers to the questions posed here, *caveat emptor* — that is, "Let the buyer [the client] beware." The problem is that the typical client is not aware of the issues that are being discussed here. Often clients' problems are so pressing that they just want help. Therefore, "Let the practitioner beware." Become competent. Don't overpromise. Remain professionally self-critical. Keep your eye on results — that is, problem-managing and opportunity-developing outcomes for clients. Because helping in some generic sense "works," do not assume that it will always work with everyone. Skilled helpers do not confuse difficult cases with impossible cases.

You are encouraged to acquaint yourself with the ongoing debate concerning the efficacy of helping. Study of this debate is meant not to discourage you but to help you (1) appreciate the complexity of the helping process, (2) acquaint yourself with the issues involved in evaluating the outcomes of helping, (3) appreciate that, poorly done, helping can actually harm others, (4) make you reasonably cautious as a helper, and (5) motivate you to become a high-level helper, learning and using practical models, methods, skills, and guidelines for helping.

## IS HELPING FOR EVERYONE?

Just because, in the main, helping works, that does not mean that it is for everyone. Most people muddle through without professional help. I often show a popular series of videotaped counseling sessions in which the client receives help from a number of helpers, each with a different approach to helping. Once the course participants have seen the tapes, I ask, "Does this client need counseling?" They all practically yell, "Yes!" Then I ask, "Well, if this client needs counseling, how many people in the world need counseling?" The question proves to be very sobering because the client in question is struggling with issues we all struggle with. Just because a person might well benefit from counseling, that does not mean that he or she "needs" counseling. Furthermore, many clients can "get better" in a variety of ways without help. Professional help is not meant for everyone. Working with clients who, for whatever reason, don't want to grapple with their problem situations and develop their unused resources is a waste of time and money. Of course, it doesn't hurt to try, but the helper should know when to quit. In addition, clients with certain kinds of problems seem to be beyond help, at least at this stage of the development of the helping professions. Kierulff (1988, p. 436) presented the following case.

> A young man (whom I will call John) attempted to rob a bar. He and his older partner carried loaded shotguns, but the bartender was armed and quick on the draw, and John was shot. The . . . bullet severed John's spine; his legs collapsed under him, and he was left paraplegic. When he was taken to the hospital, he was assigned to me for psychotherapy.
>
> John and I talked for hours, day after day, as he lay prone in his hospital bed. This 20-year-old armed robber fascinated me. He lied to me straightfaced. He portrayed himself as a victim. He changed his story whenever he thought he could gain an advantage. . . . He displayed no loyalty, no honor, no compassion. He trusted no one, and he displayed not even a crumb of trustworthiness. He used everyone, including me.
>
> I had not dealt with any sociopaths before interacting with John. . . . I believed that Rogerian unconditional positive regard would eventually work its wonders and soften John's tough shell, allowing me to connect with him on a level of mutual empathy, caring, and respect.
>
> "You can't change these people," my supervisor admonished. . . . I hoped that my supervisor was wrong. I kept on trying, but when the internship was over and I said goodbye to John, I could tell from the look in his eyes that in spite of the extra concern I had shown him, in spite of my warmth and care and effort, I was just another in the long list of people he had coldly manipulated and discarded.

In the ensuing article, Kierulff mused on the meaning of and the relationship between free will and determinism and what kind of treatment could get through to "guiltless, manipulative people." He certainly learned that helpers who allow themselves to be conned are doing no good. He also learned that the ultimate success of counseling is never in the hands of the helper.

Knowing when to help is important. So is knowing how much to help. Even though the studies mentioned above show that clients who spend a longer time in therapy benefit more, that does not mean that spending a great deal of time in therapy is feasible. Helping is an expensive proposition, both monetarily and psychologically. Even when it is "free," someone is paying for it through tax dollars, insurance premiums, or free-will offerings. Therefore, without rushing nature or your clients, get to the point. Do not assume that you have a client for life. Helping can be "lean and mean" and still be most human. A colleague of mine experimented, quite successfully, with shortening the counseling "hour." He arrived at the point where he would begin a session by saying, "We have five minutes together. Let's see what we can get done." He was very respectful, and it was amazing how much he and his clients could get done in a short time. The helping industry, driven by politics and the financial dynamics of "managed care," is focusing more and more on results-oriented "brief psychotherapy." But even if that were not the case, helpers would still owe their clients value for money. Helping that achieves only partial results may, at times, be the best that we can do.

# WHAT THIS BOOK IS — AND WHAT IT IS NOT

"Beware the person of one book," we are told. For some people the central message of a book becomes a cause; religious books such as the Bible or the Koran are examples. Although causes empower people, we should beware when a person believes that all he or she needs to know is in one book. Such a person remains closed to new ideas and growth. Certainly all truth about helping cannot be found in one book. Therefore, *The Skilled Helper* is not and cannot be "all that you've ever wanted to know about helping."

**It is a practical model of helping.** Since this book cannot do everything, it is important to state what it is meant to do. Its purpose is to provide helpers — whether novices or those with experience — a practical framework or model of helping and some of the methods and skills that make the model work. It is designed to enable helpers to engage in activities that will help their clients manage their lives more effectively. In short, this book is part of the helping curriculum — an extremely important part — but it is not the whole.

**It is not the total curriculum.** Clients are the "customers" of helpers and have every right to expect the best of service from them. Beyond a helping model and the skills that make it work, what kind of training enables helpers to "deliver the goods" to their clients? A practical curriculum is one that enables helpers to understand and work with their clients in the service of problem management and opportunity development. The curriculum includes both working knowledge and skills. "Working knowledge" is the translation of theory and research into the kind of applied understandings that enable helpers to work with clients. "Skill" refers to the actual ability to deliver services.

A fuller curriculum for training professional helpers might include, besides a working model of helping, many of the following:

- A working knowledge of *applied developmental psychology*, how people develop or create their lives across the life span and the impact of environmental factors such as culture and socioeconomic status on development
- An understanding of the principles of *cognitive psychology* as applied to helping, since the way people think and construct their worlds has a great deal to do with both getting into and getting out of trouble
- The ability to apply *the principles of human behavior* — what we know about incentives, rewards, and punishment — to the helping process,

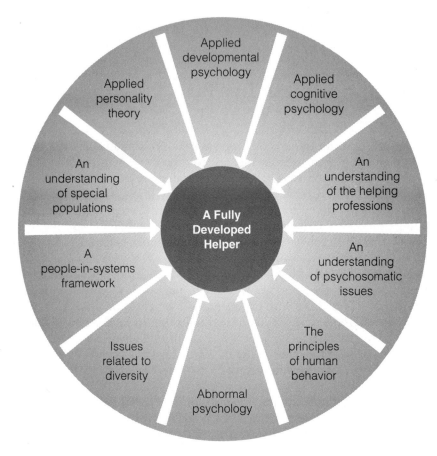

**FIGURE 1-1**
**A Sample Curriculum for Helpers**

since wrestling with problem situations and undeveloped opportunities always involves incentives and rewards

- *Applied personality theory*, since this area of psychology helps us understand in very practical ways what makes people "tick" and many of the ways in which individuals differ from one another

- An understanding of clients as *psychosomatic beings* and the interaction between physical and psychological states

- *Abnormal psychology*, a systematic understanding of the ways in which individuals get into psychological trouble

- An understanding of the ways in which people act when they are in social settings — that is, the *principles of social psychology* — together with an understanding of clients in context — that is, an understanding of clients in the *social settings* of their lives

- An understanding of the *diversity* of age, race, ethnicity, religion, sexual orientation, culture, social standing, economic standing, and the like among clients
- An understanding of the needs and problems of *special populations* with which one works, such as the physically challenged, substance abusers, the homeless
- An understanding of the *dynamics of the helping professions* themselves as they are currently practiced in our society together with the challenges they face

In the end there is no such thing as the perfect professional curriculum to which all helpers would subscribe. Although much of this curriculum, summarized in Figure 1-1, is touched on indirectly in the pages that follow, especially through the many examples offered, this book is not a substitute for such a curriculum. However, the fact that paraprofessional and informal helpers do not go through such a rigorous curriculum does not mean that they cannot be effective helpers.

## MOVING FROM SMART TO WISE: MANAGING THE SHADOW SIDE OF HELPING

This book outlines a model of helping that is rational, linear, and systematic. What good is that, you well might ask, in a world that is often irrational, nonlinear, and chaotic? One answer is that rational models help clients bring much-needed discipline and order into their chaotic lives. Effective helpers do not apologize for using such models. Rather, they make sure that their humanity permeates their models.

More than intelligence is needed to apply the model well — smart is not enough. The helper who understands and uses the model together with the skills and techniques that make it work might well be smart, but he or she must also be wise. Effective helpers understand the limitations not only of helping models but also of helpers, the helping profession, clients, and the environments that affect the helping process. One dimension of wisdom is the ability to understand and manage these limitations, which in sum constitute what I call the arational dimensions, or the "shadow side," of life. Helping models are flawed; helpers are sometimes selfish, lazy, and even predatory, and they are prone to burnout. Clients are sometimes selfish, lazy, and predatory even in the helping relationship.

Indeed, if the world were completely rational, the pool of clients would soon dry up, since many clients cause their own problems. But there is no danger of that because as Cross and Guyer (1980) note, people knowingly head down paths that lead to trouble. As Cross and Guyer go on to suggest, however, the popular view that people lacking foresight deserve what they get in life is itself based on the optimistic premise that human behavior is (or should be) determined by a goal-oriented intellectual process of evaluating alternative destinations and then following the path to the best one, rather than on the simpler and more direct procedure of permitting immediate rewards and punishments to dictate direction. (p. 41)

Life is not a straight road; often it is more like a maze. It often seems to be a contradictory process in which good and evil, the comic and the tragic, cowardice and heroism are inextricably intermingled.

In the pages of this book the helping relationship and process are described and illustrated in very positive terms. They are described as they should be, not as they always are. The shadow side of helping can be defined as:

**All those things that adversely affect the helping relationship,**
**process, outcomes, and impact in substantive ways but**
**that are not identified and explored by helper or client**
**or even the profession itself.**

Understanding and learning how to use helping models can make you smart, but using them well demands wisdom. Much of the stuff in the shadows contributes to the downside of helping, but the shadow side also has its upside (see Kottler, 1992, 1993, 1997).

## The Downside: The Messiness of Helping

The ability to understand how helper, client, the relationship, and the helping process itself can go wrong is the first step toward managing the shadow side. For instance, helpers' motives are not always as pure as they are portrayed in this book. Incompetent helpers pass themselves off as professionals. Some helpers are not very committed even though they are in a profession in which success demands a high degree of commitment. Some helpers are competent, others are not.

Clients often play games with themselves, their helpers, and the helping process. Helpers seduce their clients, clients seduce their helpers, not necessarily sexually. Hidden agendas are pursued by both helper and client. The helping relationship itself can end up as a conspiracy to do nothing. Clients, aided and abetted by their helpers, work on the wrong issues. Helping methods that are not likely to benefit *this* client are still used. Though decisions are made throughout the helping process, decision making itself remains murky. Helping is continued even though it is going nowhere. The list goes on.

Managing the shadow side is an exercise in social intelligence and competence, not in cynicism. Clearly, not all these things happen all the time. To assume that they do would be cynical. The shadow side is not usually an exercise in ill will. Few helpers set out to seduce their clients. Few clients set out to seduce their helpers. Helpers don't realize that they are incompetent.

Clients don't realize that they are playing games. Most clients most of the time have what Schein (1990) called a "constructive intent," and most helpers do their best to put their own concerns aside to help their clients as best they can. But this does not mean that any given helper is always giving his or her best. Both clients and helpers have their temptations. Clients, as we shall see, have their blind spots, but helpers also have theirs.

All human endeavors have their shadow side. Companies and institutions are plagued with internal politics and are often guided by covert or vaguely understood beliefs, values, and norms that do not serve their best interests. If helpers don't know what's in the shadows, they are naive. If they believe that shadow-side realities win out more often than not, they are cynical. Helpers should be neither naive nor cynical. Rather, they should pursue a course of upbeat realism.

## The Upside: Common Sense and Wisdom in the Helping Professions

Although helping is sometimes referred to as an art, the emphasis in the journals is on theory and research. In a way, the emphasis is on "smart" rather than "wise." Some writers and researchers, however, have begun to focus on such things as wisdom, sagacity, street smarts, practical intelligence, and common sense in helping (for instance, Hanna, 1994; Hanna & Ottens, 1995; Schmidt & Hunter, 1993; Sternberg, 1990; Sternberg, Wagner, Williams, & Horvath, 1995). For instance, many clients would benefit from self-control (who of us wouldn't?), but the wisdom of self-control has been with us for ages.

> In truth, most of the techniques used to facilitate self-control were not derived from any formal theory. The Bible, Koran, Talmud, and other old books contain much of the essential wisdom that occasionally appears in the pages of our learned journals. Common sense, as exemplified in *Poor Richard's Almanac* or the pages of newspaper advice columns, is another ready source of ideas (both good and bad). (Karoly, 1995, p. 273)

Whereas some smart helpers might see the client, in Hanna's (1994) words, as a "clinical entity subject to the templates and tools of the psychotherapeutic trade," wise helpers would see each client as a "vital, dynamic personage" (p. 132) who needs to be helped to take advantage of the wisdom and common sense that is already with us.

What is it that characterizes helping wisdom? Here are some possibilities (see Sternberg, 1990):

- Self-knowledge, maturity; the guts to admit mistakes and the sense to learn from them
- A psychological and a human understanding of others; insight into human interactions

- The ability to "see through" situations; the ability to understand the meaning of events
- Tolerance for ambiguity and the ability to work with it; being comfortable with messy and ill-structured cases and, in general, the messiness of human beings; openness to events that don't fit comfortably into logical or traditional categories
- The ability to frame a problem so that it is workable; the ability to reframe information
- Avoidance of stereotypes; holistic thinking; open-mindedness; open-endedness; contextual thinking; "meta-thinking," or the ability to think about thinking and become aware about being aware
- The ability to see relationships; the ability to spot flaws in reasoning; intuition; the ability to synthesize
- The refusal to let experience become a liability through blind spots; the ability to take the long view of problems
- The ability to blend seemingly antithetical helping roles — being one who cares and understands together with being one who challenges and "frustrates" (Levin & Shepherd, 1974)

We are all aware of the person who is brilliant in academic endeavors but incompetent in social interaction. But the underlying phenomenon goes much further than the "bumbling" professor. Although intelligence as measured by intelligence test scores makes a difference in everyday life, most of the difference comes from other forms of intelligence or other abilities.

The wisdom just described does not send the helper spinning off into space. On the contrary, it serves as the basis for practical intelligence, common sense, or street smarts. Knowing "how" rather than merely knowing is critical in helping others. If social intelligence is the ability to "read" the dynamics of a relationship or a social setting, then social competence is the ability to respond creatively to what one finds. In helping, this means, among other things, the ability to understand what you can change and what you can't (see Seligman, 1994). People with common sense may have a great deal of formal academic knowledge, but they are in a special way beneficiaries of "tacit knowledge" (see Schmidt & Hunter, 1993) — that is, "action-oriented knowledge, acquired without direct help from others, that allows individuals to achieve goals they personally value" (Sternberg, Wagner, Williams, & Horvath, 1995, p. 916). To my way of thinking, such things as wisdom and common sense are part of the upside shadows because they do not receive a great deal of attention in the helping literature, they do not form part of the curriculum in helper-training programs, and their cause is not always advanced through the internships in which helpers engage.

Despite the downside problems reviewed earlier, the tone of this book is unabashedly upbeat. However, helpers-to-be must not ignore the less palatable dimensions of the helping professions, including the less palatable dimensions of themselves. Therefore, throughout this book some of the common shadow-side realities that plague client, helper, and the profession itself are noted at the service of managing them. This book is by no means a treatise on the shadow side of helping. Rather its intent is to get helpers to begin to think about the shadow side of the profession. Wise helpers are idealistic without being naive. They also know the difference between realism and cynicism and opt for the former. They see the journey "from smart to wise" as a never-ending one.

## CHAPTER 2

# OVERVIEW OF
# THE HELPING MODEL

A NATURAL PROBLEM-MANAGEMENT PROCESS

THE SKILLED-HELPER MODEL

STAGE I: THE CURRENT STATE OF AFFAIRS—
CLARIFICATION OF THE KEY ISSUES CALLING
FOR CHANGE

    Identifying and Clarifying Problem Situations and
    Unused Opportunities

    The Three "Steps" of Stage I

STAGE II: THE PREFERRED SCENARIO—HELPING
CLIENTS DETERMINE WHAT THEY NEED AND WANT

    Developing a Preferred Scenario

    The Three "Steps" of Stage II

STAGE III: STRATEGIES FOR ACTION—HELPING
CLIENTS DISCOVER HOW TO GET WHAT THEY
NEED AND WANT

    Developing Action Strategies

    The Three "Steps" of Stage III

ACTION: MAKING IT ALL HAPPEN—HELPING CLIENTS
TURN DECISIONS INTO PROBLEM-MANAGING ACTION

ONGOING EVALUATION OF THE HELPING PROCESS:
HOW ARE WE DOING?

FLEXIBILITY IN THE USE OF THE MODEL

DEVELOPING A WHOLE-PROCESS MENTALITY:
MINIVERSIONS OF THE ENTIRE MODEL

UNDERSTANDING AND DEALING WITH THE SHADOW
SIDE OF HELPING MODELS

# A NATURAL
# PROBLEM-MANAGEMENT PROCESS

There is a natural decision-making process that helping processes can assist, modify, and accelerate but not replace (see Yankelovich, 1992). Understanding and respecting this natural process and its variations is part of the wisdom of helping. Applied to problem management and opportunity development, the steps of this natural process look something like this:

**1. _Initial awareness._** First, clients become _aware_ of an issue or a set of issues — for instance, not just this or that marital problem but also vague dissatisfaction with the relationship itself.

**2. _Urgency._** Second, a _sense of urgency_ develops, especially as the problem situation becomes more annoying or painful. Subsequent annoyances in the marriage are now seen in light of overall dissatisfaction. These two steps constitute the consciousness-raising stage of the process.

**3. _Initial search for remedies._** Third, clients begin to _look for remedies_. However perfunctorily, different strategies for managing the problem situation are explored. Clients in difficult marriages begin thinking about complaining openly, separating, getting a divorce, instituting subtle acts of revenge, having an affair, going to a marriage counselor, seeing a minister, unilaterally "withdrawing" from the marriage in one way or another, and so forth.

**4. _Estimation of costs._** Fourth, the _costs of pursuing different solutions_ begin to emerge. "If I confront my partner openly, I'll have to go through the agony of confrontation, denial, argument, counteraccusations, and who knows what else." Or, "What would I do if I were to go out on my own?" Or, "What would happen to the kids?" The possibilities here are endless, and clients often retreat because there is no cost-free or painless way of dealing with the problem situation.

**5. _Deliberation._** Fifth, since the problem situation is now seen for what it is, it is impossible to retreat completely and so _a more serious weighing of choices_ takes place. For instance, the costs of confronting the situation are weighed against the costs of merely withdrawing. Often, a kind of dialogue goes on in the client's mind between steps 4 and 5. "I might have to go through the agony of a separation for the kids' sake. Maybe time apart is what we need."

**6. _Rational decision._** Sixth, an _intellectual decision_ is made to accept some choice and pursue a certain course of action. "I'm going bring all of this up to my spouse and suggest we see a marriage counselor." Or, "I'm going to get on with my life, find other things to do, and let the marriage go where it will." A merely intellectual decision, however, is often not enough to drive action.

**7. _Rational-emotional decision._** Finally, the _heart joins the head_ in the decision. One spouse might finally say, "I've had enough of this!" That is, the decision is permeated by values and emotion. This fuller decision is more likely to drive action. A spouse might say, "It is unfair to both of us to go on like this; it just isn't right," and this drives the decision to seek help, even if it means going alone.

Two things should be noted. First, these steps, however logically sequenced on paper, are often jumbled and intermingled in real-life problem-management situations. Second, this natural process can be derailed at almost any point along the way. For instance, the costs of managing the problem (step 4) might seem too high and so the process is put on the back burner.

The problem-management process outlined in this chapter and developed in the rest of the book borrows from this natural process, complements it with other steps and techniques, provides ways of challenging backsliding, and, ideally, speeds it up. Moreover, since the natural process evolves differently in different individuals, individual differences must also be respected.

 # THE SKILLED-HELPER MODEL

All worthwhile helping models (or call them frameworks or processes) ultimately help clients ask and answer for themselves four fundamental questions:

- **Current scenario.** What are the problems (issues, concerns, undeveloped opportunities) I should be working on? The answers to this question constitute the client's *current state of affairs* or *current scenario*.
- **Preferred scenario.** What do I need or want in place of what I have? Answers to this question constitute the *preferred state of affairs* or *preferred scenario*.
- **Strategies.** What do I have to do to get what I need or want? Answers to this question produce *strategies for goal-accomplishing action*.
- **Action.** How do I make all this happen? Answers to this question help clients move from planning mode to *action, getting-it-done,* or *accomplishment* mode.

These four questions, illustrated in Figure 2-1, provide the basic framework for the helping process. Note that in the figure the term "stage" is used. The term "stage" is placed in quotation marks because it has sequential overtones that are somewhat misleading. In practice the three stages overlap and interact with one another as clients struggle through the natural process of constructive change outlined in the rest of this chapter. The term "stage" is appropriate if we focus exclusively on the *logic* of the process — moving from assessment to goal setting to strategy development to implementation. But helping, like life itself, is not as logical as the models used to describe it. More will be said about the flexible and "messy" nature of the helping model after a review of its logical stages. A problem-management model in counseling and therapy has the advantage of the vast amount of research that has been done on the problem-solving process itself. The model, techniques, and skills outlined in this book tap that research base.

The extended example that follows serves to bring this process to life. The case, though real, has been disguised and simplified. It is not a session-by-session presentation. Rather, it illustrates ways in which one client asked

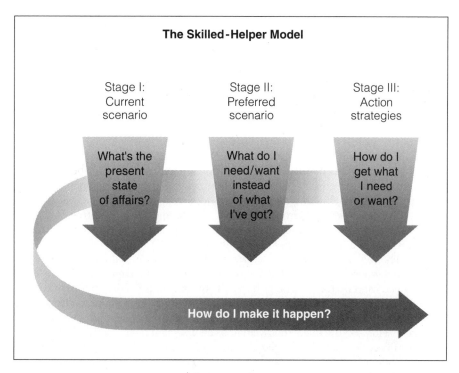

**FIGURE 2-1**
Overview of the Helping Model

and answered the four questions outlined earlier. The client, Maria, is voluntary, verbal, and, for the most part, cooperative. In actual practice, cases do not always flow as easily as this one. The simplification of the case, however, will help you see the main features of the model in action.

## STAGE I: THE CURRENT STATE OF AFFAIRS — CLARIFICATION OF THE KEY ISSUES CALLING FOR CHANGE

### Identifying and Clarifying Problem Situations and Unused Opportunities

The current scenario spells out the range of difficulties the client is facing. What are the problems, issues, concerns, and undeveloped opportunities with which Maria needs to grapple?

> Maria, a 37-year-old single parent, has two children in the upper grades of primary school. Since her husband has disappeared, she works two rather menial jobs, one full-time and one part-time, to make ends meet. Her problem is that she finds herself more and more exhausted and angry. Because on occasion she

blows up at work, she runs the risk of being fired from her full-time job. She has also begun to take her anger out on her kids. At one juncture, nervous neighbors reported her to the police as a possible child abuser. An investigation revealed no real signs of child abuse, but the incident left her even more angry. The investigator suggests that she might get some help from a local church and leaves the name of a contact person and a phone number to call. Although Maria hasn't been in church for years, in desperation she calls the number.

Two people from the church visit her, one to baby-sit and one to take Maria out for a cup of coffee and listen to her story. The visitors are not professional counselors. The one who talks with Maria is a paraprofessional counselor who works with the church-based self-help program. The participants in this program, whether church members or not, join small groups. There is a "facilitator" in each group. Group members are people who are struggling with problems like Maria's and who have found help and support both from the facilitators and from fellow group members.

The self-help counselor—a person who provides initial counseling for potential group members and facilitates two of the self-help groups—listens carefully to Maria's story and helps her spell out some of the more important details. For instance, by probing a bit, she discovers that Maria has done nothing to try to track down her husband with a view to getting some child support. The counselor also discovers that Maria has an associate of arts degree from a community college. Maria has talents that are currently not being used and this adds to her anger. She is grossly underemployed. Furthermore, given her work schedule and the demands of raising two kids, Maria has no social life to speak of.

During the first couple of sessions, the counselor notices some hesitancies on Maria's part. She skillfully and sensitively helps Maria talk about her fears of being "sucked into" membership in the church. Her early experiences with church and religion were quite negative. The counselor assures her that no one will push her into anything and that the groups are self-help groups, not theology or prayer groups.

After a couple of sessions with the counselor, Maria notices two things. First, she realizes that she herself is "hearing" her own story for the first time. Merely naming the issues she faces every day seems to clear the air a bit. Second, though there are no dramatic changes in her life, she finds herself less stressed and less angry. She also begins to realize that, although she sees herself as a "doer"—after all, she has two jobs and is trying, with mixed success, to raise two kids—she has done little to manage the most vexing problem situations of her life. Instead, she has allowed herself to develop a "victim" mentality.

This is a thumbnail sketch of Maria's current scenario. It is filled with a range of problem situations and undeveloped opportunities, the main features of which become clearer to her through her interaction with the counselor. The counselor, who has been trained in the helping model described in this book, shares its main features with Maria so that Maria can become an active collaborator in the process. Indeed, since all the members of the self-help groups learn the basic features of the model before they enter the groups, they use it to provide focus and direction and to guide their interactions with one another.

# The Three "Steps" of Stage I

As we shall see in the chapters that follow, each "stage" is divided into three "steps." Like the stages, the "steps" are not steps in a mechanistic, "now do this" sense. Rather, they are activities that help clients develop answers to the questions outlined earlier.

**Step I-A: The story.** Help clients tell their stories with the kind of detail that enables them to move on to the goal-setting ("What do I really want?") stage of the helping process. Maria's counselor needs to help her tell her "story" with the kind of detail that enables Maria to understand her problem situation and begin thinking about what she might do to manage it. Once Maria gets a clear picture of what's going on in her life, she has a better chance to do something about it.

**Step I-B: Blind spots.** Help clients break through blind spots that prevent them from seeing themselves, their problem situations, and their unused opportunities as they really are and that prevent them from successfully carrying out the work of the helping process. Maria's counselor can add great value if she can help Maria identify significant blind spots and develop the kind of new perspectives that will help her move forward.

**Step I-C: Choosing the right problems/opportunities to work on.** If a client has a range of issues, help him or her to work on things that will make a difference. Or, if a client wants to work only on trivial things or does not want to work at all, then it might be better to defer counseling. Maria, like other clients, may have a whole range of problems. The counselor adds value by helping Maria work on problems and unused opportunities that will make a substantial difference in her life.

These concepts are explained in detail and illustrated in later chapters together with ways of making sure that clients' needs rather than the helper's theories drive the helping process.

# STAGE II: THE PREFERRED SCENARIO — HELPING CLIENTS DETERMINE WHAT THEY NEED AND WANT

## Developing a Preferred Scenario

The preferred scenario spells out possibilities for a better future and culminates in the client's *change agenda* fashioned from those possibilities. Stage II deals with answers to goal- or accomplishment-oriented questions. What does Maria want? What does she need, whether she currently wants it or not? What would her life look like if it were more tolerable or, even better, more fulfilling?

Encouraged to look into the future by the counselor, Maria soon discovers that there are many things she wants — a better-paying job, a job she actually enjoys, one that requires her to think and make decisions; better relationships with her kids; some time for herself away from both the job and the kids; a better place to live; some discretionary money for a few simple luxuries such as some decent clothes, a meal out once in a while, a movie; some male companionship, some social life; possibly a car to get to work in; occasional help with the kids; less guilt, less anger; financial help from the husband who has disappeared; greater peace with herself. These are some of the options that come tumbling out during the course of her conversations with the counselor.

Further encouraged by the counselor to separate pressing needs from more idealized wants, Maria sorts out her needs and wants and sets some priorities. Overall she needs some relief from the pressure-cooker lifestyle she is living. This in her mind means three things: a better-paying job that would give her more time at home; a restructured relationship with her kids; and some kind of social life, even though at this time it might not include a special companion or partner. She notices that the very act of setting a few priorities makes her begin thinking of how she might accomplish them.

The counselor helps her explore and clarify these priorities. For instance, she helps Maria spell out in more detail what "a better relationship with her kids" would look like. And, instead of encouraging Maria just to go out and look for a new job, she helps her explore what kind of job would best fit her needs and wants. For instance, a better-paying menial job would do little to tap Maria's unused talents.

Unfortunately, some approaches to problem solving or management skip Stage II. They move from a "What's wrong?" stage (Stage I) to a "What do I do about it?" stage (Stage III). As we shall see, however, helping clients discover what they want has a profound impact on the entire helping process.

## The Three "Steps" of Stage II

Stage II also has three "steps" — that is, three ways of helping clients answer as creatively as possible the question "What do I need and want?"

**Step II-A: Possibilities for a better future.** Help clients use their imaginations to spell out elements of a better future. This often helps clients move beyond the problem-and-misery mind-set they bring with them and develop a bit of hope. Brainstorming possibilities for a better future often helps clients understand their problem situations better — "Now that I am beginning to know what I want, I can more easily identify the most important issues in my life."

**Step II-B: The change agenda.** Help clients choose realistic and challenging goals designed to manage the key problems and unused opportunities that have been identified in Stage I. The change agenda is nothing more than the priorities clients set for themselves. Maria's initial change agenda includes getting a better job, reconstructing her relationship with her kids, and getting "into community" in some way. If she accomplishes these goals, her life will be much more fulfilling than it is today and her children will be better off.

**Step II-C: Commitment.** Pursuing substantial priorities or goals such as Maria's demands a great deal of work and often takes courage. Setting unrealistic goals prevents real commitment from the start. Counselors need to help clients find the incentives that will help them persist until they get what they need and want. Maria's real love for her children can become a powerful incentive for constructive change.

In real-life counseling sessions, these steps are intermingled with one another and with the steps of other stages. Ways of helping clients discover possibilities for a better future, fashion a change agenda that will make a difference in their lives, and discover within themselves and in their daily lives the incentives that make them want to work at constructive change are all outlined and illustrated in the chapters dealing with Stage II.

# STAGE III: STRATEGIES FOR ACTION— HELPING CLIENTS DISCOVER HOW TO GET WHAT THEY NEED AND WANT

## Developing Action Strategies

Stage III defines the *work* that needs to be done to translate priorities into problem-managing accomplishments. Maria needs to discover ways of bridging the gap between the current scenario—what she has—and the preferred scenario—what she needs and wants. She has to ask herself, "What do I have to do to get what I need or want?"

> Along the way, Maria, at the counselor's suggestion, joins one of the self-help groups, one facilitated by her counselor. Here she finds people grappling with problem situations similar to hers. Through her interactions with her fellow group members, she gets a better understanding of her main problems, discovers and breaks through some blind spots, and gets a clearer understanding of her priorities. She also begins to explore what she needs to do to get what she wants.
>
> In the group Maria learns a great deal about searching for jobs. She talks about the kind of job she wants and then explores what she has to do to get the kind of job she wants. For instance, she soon learns that many job openings are passed along by word of mouth. Therefore, she needs some kind of "network" to find out about job openings. The self-help program provides a ready-made network of people who have at least limited access to job possibilities. She complements networking with a systematic do-it-yourself job-search program developed by the self-help program. In the group she discusses ways of tailoring the program to her specific needs.
>
> In the group Maria finds not cult but community. In the ensuing months she finds the weekly meetings both work and a source of social satisfaction, makes some friends, and even has a date or two. All of this helps relieve that stress she had been experiencing from being "out of community." It also enables her to concentrate on her two other priorities, getting a better-paying and more satisfying job and renewing her relationship with her children.

Outcome-producing action is at the heart of constructive change. Stage III helps clients discover the kinds of actions that will make a difference.

## The Three "Steps" of Stage III

Stage III, too, has three "steps" that, in practice, intermingle with one another and with the steps of the other stages.

**Step III-A: Possible actions.** Help clients see that there are many different ways of achieving their goals. Hasty and disorganized action is often self-defeating. "I tried this and I did that and nothing worked!" is usually a sign of poor planning rather than of the impossibility of the task. Stimulating clients to think of different ways of achieving their goals is usually an excellent investment of time. If Maria wants an effectively "reconstructed relationship" with her children, she needs both a clear picture of what this means and a broad view of the different possible routes to this goal.

**Step III-B: Choosing best-fit strategies.** Help clients choose the action strategies that best fit their talents, resources, style, temperament, and timetable. For instance, introverted clients that choose extrovert-oriented strategies to accomplish social goals might well be courting failure. To rebuild her relationship with her children, Maria needs to choose means that make sense both for herself and for her children.

**Step III-C: Crafting a plan.** Help clients organize the actions they are going to take to accomplish their goals. Plans are simply maps clients use to get where they want to go. A plan can be quite simple. Indeed, overly sophisticated plans are usually self-defeating. Maria soon realizes that she needs to be systematic and consistent in redeveloping her relationships with her kids. A simple plan helps her to do so.

Ways of helping clients brainstorm strategies to achieve their goals, tailoring those strategies to their resources and the conditions in which they have to work, and organizing actions into a coherent goal-accomplishment plan are outlined and illustrated in the sections dealing with Stage III.

## ACTION: MAKING IT ALL HAPPEN— HELPING CLIENTS TURN DECISIONS INTO PROBLEM-MANAGING ACTION

In many helping or problem-management models, action or implementation is tacked on at the end. In this model all three stages sit on the "action arrow," indicating that clients need to act in their own behalf right from the beginning of the counseling process. As we shall see, clients need to act both within the helping sessions themselves and in their real day-to-day worlds.

In the self-help group, Maria spends little time talking about her priority of re-constructing her relationship with her children. She is too embarrassed about the way she let the relationship disintegrate and certainly does not want to talk about the "investigation." She assumes that she can take care of this issue on her own. Unfortunately, she underestimates both the work involved in improving deteriorated relationships and the time needed to do so. The fact that she is less stressed and angry and therefore more civil at home on a day-to-day basis constitutes the heart of her undiscussed and rather simplistic "plan."

Her children are unresponsive to her newfound civility. They remain sullen and continue to take every opportunity to engage in activities outside the apartment. Although Maria feels hurt, she persists in her campaign of civility. Since this gets her nowhere, resentment builds up. It all comes to a head one evening when she loses her temper and screams at her kids about their ingratitude and rants and rages about the things they have been doing to hurt her. This brings stunned silence from her kids and a knock on the wall from one of her neighbors. Shocked, she breaks down crying and retreats to her bedroom. The kids remain dazed in the kitchen.

She continues to be so upset that she "comes clean" in the next group meeting. The group facilitator and her fellow group members help her retrace her steps and discuss openly the ways in which the relationship deteriorated. Acutely aware of Maria's shame, embarrassment, and sense of failure, they provide a great deal of support but don't let her off the hook. They help her take a more detailed and more realistic look at her previous, almost cavalierly announced, priority of redoing her relationship with her kids.

In the end, Maria apologizes to her children for her outbursts and, in a few separate family meetings with the counselor, talks through with them what has happened and what they can all do to turn the apartment into a home. This gives the children an opportunity to discuss their own feelings and views of what has happened and to "own" the process of reconciliation. This produces no miracles, but the climate at home steadily improves, even though there are some predictable setbacks.

Helping is ultimately about the client's working toward constructive change. Talking about problems and opportunities, discussing goals, and figuring out strategies for accomplishing goals is just so much blah, blah, blah without action. There is nothing magic about change; it is hard work. If clients do not act in their own behalf, nothing happens.

Action, like evaluation, does not come at the end of the helping process. A spirit of constructive change needs to permeate the entire process. As we shall see, each stage and step of the process can promote problem-managing and opportunity-developing action right from the beginning. Figure 2-2 adds the "steps" outlined briefly earlier and includes two-way arrows between both stages and steps to suggest the kind of flexibility needed to serve the interests of clients (more on flexibility later).

In the end, Maria did get a better, but not a perfect, job and was able to enjoy more free time and more of life's amenities; her relationship with her children did improve and the apartment became more of a home; she did develop a more fulfilling social life through both the self-help group and her new

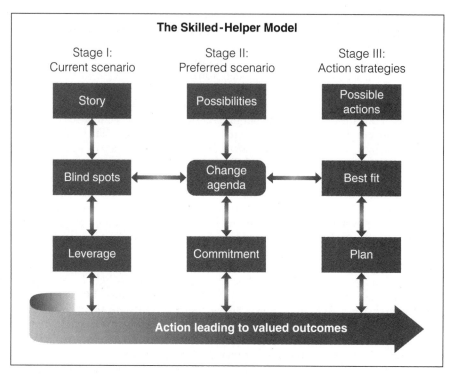

**FIGURE 2-2**
**The Helping Model Showing Interactive Stages and Steps**

job, but she struggled to find a male companion to her liking. Although she worked hard to accomplish all of this, she knew she could have done more.

## ONGOING EVALUATION OF THE HELPING PROCESS: HOW ARE WE DOING?

In many helping models, evaluation is presented as the last step in the model. However, if evaluation occurs only at the end, it is too late. As Mash and Hunsley (1993) noted, early detection of what is going wrong in the helping process is needed to prevent failure. They claimed that an early-detection framework should be theory-based, ongoing, practical, and sensitive to whatever new perspectives might emerge from the helping process. The model presented in this book can help achieve all these goals. In sum, quality assurance needs to be built into the helping process, not tacked on to it after it is too late. Helpers and clients need to collaborate in this ongoing evaluation process. In later chapters questions designed to help with the evaluation process will be provided.

The evaluation process can also be used to determine when the helping process has served its purpose and should be terminated. As Waehler and Lenox (1994) have noted, termination issues can arise anytime, even in early helping sessions. Some helpers and clients contract for a fixed number of sessions to begin with and by that very fact build a termination dimension into the process.

## FLEXIBILITY IN THE USE OF THE MODEL

A helping model is like a map that helps you know, at any given moment, "where you are" with clients and what kinds of interventions would be most useful. At its best, it is a *shared* map, because it outlines both what *clients* need to do to manage a problem situation or develop an opportunity and what helpers can do to *assist* clients as they move through these messy and intermingled stages toward problem-managing action. In the map metaphor, the stages and steps of the model are orientation devices.

**Flexibility.** The helping model, while providing guidance, must remain flexible to the needs of clients. One form of rigidity is to drag clients mechanically through the stages and steps of the model. This is not how clients work.

First, clients start differently. For instance, Client A might start with something that he tried to do to solve a problem but that did not work—"I threatened to quit if they didn't give me a leave of absence, but it didn't work." Client B might start with what she believes she wants but does not have—"I need a boyfriend who will take me as I am." Client C might start with the roots of his problem situation—"I don't think I've ever gotten over being abused by my uncle." Client D might announce that she really has no problems but is still vaguely dissatisfied with her life—"I don't know. Everyone tells me I've got a great life, but *something's* missing." Other clients may have multiple starting points.

Second, clients tell their stories differently. Since clients do not always present all their problems at once in neat packages or all the relevant elements of a named problem, it is impossible to work through Stage I completely before moving on to Stages II and III. It is not even advisable to do so. Naming goals often demands a return to the presenting problem. Naming actions to be taken may require revisiting a goal. New problems and new dimensions of a named problem often need to be explored and understood as they are presented. A client might actually try out some action strategy even before the problem situation is adequately defined and some goal has been set. That is, action sometimes *precedes* understanding. If the action is not successful, then the counselor helps the client learn from it and return to the tasks of clarifying the problem situation and setting some realistic goals. Here is an example of how one client did it.

Woody, a sophomore in college, came to the student counseling services with a variety of interpersonal and somatic complaints. He felt attracted to a number of women on campus but did very little to become involved with them. After exploring this issue briefly, he said to the counselor, "Well, I just have to go out and do it." Two months later he returned and said that his experiment had been a disaster. He had gone out with a few women, but none of them really interested him. Then he did meet someone he liked quite a bit. They went out a couple of times, but the third time he called, she said that she didn't want to see him any more. When he asked why, she muttered vaguely about his being too preoccupied with himself and ended the conversation. He felt so miserable he returned to the counseling center. He and the counselor took another look at his social life. This time, however, he had some experiences to probe. He wanted to explore his being "too preoccupied with himself."

This student put into practice Weick's (1979) dictum that chaotic action is sometimes preferable to orderly inactivity. Some of his learnings were painful, but now there is a chance of examining his interpersonal style much more concretely. One of the assumptions of Alcoholics Anonymous is that people sometimes need to act themselves into new ways of thinking rather than think themselves into new ways of acting.

Third, clients sometimes deal with parts of different stages and steps in an intermingled way in the same session. Let's say that the client says this:

> Every time I try to be nice to her, she throws it back in my face. So who says being more considerate is the answer? Maybe my problem is that I'm a wimp, not the self-centered bastard she makes me out to be. Maybe I'm being a wimp with you and you're letting me do it. Maybe it's time for me to start looking out for my own interests. A lot of 'maybes.'

In those few sentences there can be found a failed action strategy, the questioning of a previously set goal, a possible new problem, a reference to a possible difficulty with the helping relationship, and a reference to opportunity development.

Since the stages and steps of the model intermingle, helpers will often find themselves moving back and forth in the model. Often two or more steps or even two stages of the process merge into one another. For instance, clients can name parts of a problem situation, set goals, and develop strategies to achieve them in the same session. New and more substantial concerns arise while goals are being set, and the process moves back to an earlier, exploratory stage.

**Direction.** Flexibility, however, is not chaos. Focus and direction in helping are also essential. Letting clients wander around in the morass of problem situations under the guise of flexibility is of no help at all. The structure of the helping model is the very foundation for flexibility; it is the underlying "system" that keeps helping from being a set of random events. The model-as-map tells you where clients are in the helping process and what you might do to add value. The model-as-topological-map — that is, a map dealing with different layers — helps you understand and help clients when they are engaged in more than one stage or step at the same time. The helping model provides "channels" for problem management and opportunity development. These channels can be used, as we shall see, to "prod" clients — or, better, to help them prod themselves — to move intelligently to problem-managing action and outcomes.

**Cultural sensitivity.** Client diversity is addressed in Chapter 3. Given the diversity of clients, helping models should be vehicles of personal growth, not cultural domination. The advantage of problem-management and opportunity-development models of helping is that they are easily recognized across the world. At least, that is my experience. Many years ago, before presenting an earlier version of the helping process outlined in this book to some 300 college students and faculty members in Tanzania, I said, "All I can do is present to you the helping process I teach and use. You have to decide whether it makes sense in your own culture." At the end they said two things. First, the communication skills used in the helping process would have to be modified somewhat to fit their culture. Second, the problem-management helping process itself was very useful.

Since then, this scene has been repeated — in conferences and training events I and others have presented — over and over again on every continent. The model presented here spells out, in a flexible, step-by-step fashion, the way human beings think about constructive change. The reason this process crosses cultures so easily is *that in essence it is already there*. Its basic structure is recognized around the world. It is, to use Orlinsky and Howard's (1987) term, a "generic" model of helping. Of course, the process as outlined in these pages together with the skills and techniques that make it work still has to be adapted both to different cultural settings and to different individuals within those settings. This demands cultural sensitivity on the part of helpers.

Waehler and Lenox (1994) rightfully maintained that the intermingling of stages and steps in any given session must be presented to prospective helpers as part of the training program and that this intermingling must be depicted in some way. The logic is this: (1) The stages and steps of the *helping model* provide both direction and flexibility for the issues clients talk about; (2) the stages and steps of the helping framework are delivered and filtered through *the ever-developing relationship* between helper and client; (3) the *helping sessions* are the "melting pot" in which the sometimes messy work of helping takes place in flexible but still directionful ways; and (4) all this takes place at the service of real-life problem-managing and opportunity-developing outcomes. Ongoing evaluation deals with all four dimensions of the process.

One systematic way of using the stages and steps of the helping process flexibly is outlined in the Appendix (see "A Future-Centered Approach to Helping"). Also outlined in the Appendix is a way of using the model outlined in this chapter as an instrument for "mining" value from other approaches to helping (see "Using the Model as a 'Browser': The Search for Best Practice"). Finally, the training process for learning both the helping model and the skills that make it work are outlined in the Appendix (see "Becoming a Skilled Helper: The Basic Steps of the Training Process").

## Developing a Whole-Process Mentality: Miniversions of the Entire Model

Helpers, whether professional or paraprofessional, and ordinary people engage in miniversions of the helping process more often than they realize. That is, the whole process, which can take weeks or months or even years, can also take place within literally minutes. No matter how long it takes, however, helpers need to adopt a whole-process mentality. Consider the following scenario.

> Lara, a pastoral-care worker in an urban hospital, gets a call from a nurse who says that a patient who is to undergo surgery the next day wants to see someone from the pastoral-care team. She visits the patient, a man who is going to have surgery for colon cancer. She listens to his concerns, sees his anxiety, and realizes that he is seeking some modicum of peace in the face of adversity. She gently asks whether praying together might help. He says that although he has not been very "religious," he would like to face the "Almighty" in a better frame of mind. She chooses the Psalm that begins with the words "The Lord is my Shepherd." She emphasizes the themes that are most relevant to his current plight and state of mind. Between phrases of the prayer, they talk about his concerns. At the end, he says that he is still frightened but feels he is in a much better state to face the next day. Lara says that she will have the nurse stop by to see whether he might want something for his anxiety. Later the nurse drops by and gives him some medication to help reduce his anxiety and get some sleep.

Lara listens to the patient's concerns and helps him give expression to them (Stage I), realizes that he wants some relief from his anxiety and to feel more at peace, from a religious point of view, about facing surgery (Stage II), and uses prayer and dialogue about the themes within the prayer to help him find some of the peace he is looking for (Stage III). Even though they have only one meeting, it would be a mistake to assume that only Stage I things could happen. Helpers need to develop a whole-process mentality—any part of any stage or step can be invoked in any session if it proves beneficial for the client.

Think of a hologram—that laser-generated three-dimensional image that seems to float in space. The technologically gifted tell us that the whole is found in each of its parts. The helping model is more like a hologram than a tool kit—it works best when the whole is found in each of its parts.

## Understanding and Dealing with the Shadow Side of Helping Models

Besides the broad shadow-side themes mentioned in Chapter 1, there are a number of shadow-side pitfalls in the use of any helping model, including the constructive-change model used here.

**No model.** Some helpers "wing it." They have no consistent, integrated model that has a track record of benefiting clients. Professional training programs often offer a wide variety of approaches to helping drawn from the "major brands" listed above. Helpers leave such programs knowing a great deal about different approaches but lacking an integrated approach for themselves.

**Misuse of parts of the model.** As previously mentioned, the model is a hologram rather than a tool kit. Its organic nature is destroyed when helpers break it up into fragments and deliver what they are interested in or good at rather than what the client needs. Even when, in successful cases, helpers seem to deliver just parts of it—say, problem identification or goal setting—the other stages and steps are being done by the client.

**Fads.** The helping profession is not immune to fads. A fad very often is an insight or a technique that would have some merit were it to be integrated into some overriding model or framework of helping. Instead it is marketed more or less on its own as the central, if not the only meaningful, concept or intervention needed. A fad need not be something new; it can be the "rediscovery" of a truth or a technique that had not found its proper place in the helping tool kit. The profession is enamored of these ideas and techniques for a while and then abandons them. In the process, sometimes the baby—some useful insight or methodology—is thrown out with the bath water. Constructionism and multiculturalism currently run the fad risk.

**Rigid models rigidly applied.** Some helpers use decent models but do so ineptly. One form of ineptness is rigidity. Once they do adopt a particular approach, they use it rigidly. The purity of the model becomes more important than the needs of clients. Beginning helpers, or even experienced but unskilled ones, can apply the helping model too rigidly. They drag clients in a linear way through the model even though that is not what clients need. Helping is for the client; the model exists only to aid the helping process. Flexibility is essential.

**Overcontrol.** A second form of ineptness is overcontrol. Helping models can provide focus and direction, thus helping clients develop self-discipline in the management of their problems in living. On the other hand, since helping models are powerful tools, they can be used to control clients rather than helping them liberate themselves.

**Virtuosity.** A third form of ineptness is virtuosity. Some helpers tend to specialize in certain techniques and skills—exploring the past, assessment, goal setting, empathy, challenging, and the like. Helpers who specialize not only run the risk of ignoring client needs but also often are not very effective even in their chosen specialties. For example, the counselor whose specialty is challenging clients is often an ineffective challenger. The reason is obvious:

Challenge must be based solidly on an understanding of clients and must serve a wider process. Effective counselors have a wide repertory of skills and use them in a socially intelligent way. These skills enable them to respond spontaneously to a wide variety of client needs. In sum, counseling is for the client; it is not a virtuoso performance by the helper.

**Overly cognitive training programs.** The training of helpers is often too academic, too cognitive in nature. It focuses too narrowly on step 1 of the training process outlined in the Appendix for Chapter 2 (see p. 337). Theory and research findings in, say, developmental psychology are not translated into models and techniques that helpers can use to serve clients across the life span. Helpers-to-be don't get their "hands dirty" enough trying things out right from the beginning. Most graduates from professional training programs become practitioners, not theoreticians and researchers. This is not surprising since they are inundated with theories and research in school and the teaching of research methodologies is often aversive. Yet the academic culture of helper-trainer institutions prevails over the professional culture.

## CHAPTER 3

# THE HELPING RELATIONSHIP: VALUES IN ACTION

THE HELPING RELATIONSHIP

THE WORKING ALLIANCE

THE CULTURE OF HELPING: VALUES IN ACTION
Putting Values into the Broader Context of Culture
The Importance and Practicality of Values in Helping

THE VALUE OF RESPECT
Norms That Translate Respect into Behaviors
Norms Related to Diversity and Multiculturalism
Guidelines Related to Diversity and Multiculturalism

THE VALUE OF GENUINENESS: BEYOND PROFESSIONALISM
AND PHONINESS

THE VALUE OF CLIENT EMPOWERMENT: HELPING
CLIENTS DEVELOP SELF-RESPONSIBILITY
Helping as a Social-Influence Process
Norms for Empowerment and Self-Responsibility

A WORKING CHARTER: THE CLIENT-HELPER
CONTRACT

SHADOW-SIDE REALITIES IN THE RELATIONSHIP

# THE HELPING RELATIONSHIP

Although theoreticians, researchers, and practitioners alike, not to mention clients, agree that the relationship between client and helper is important, there are significant differences as to how this relationship is to be characterized and played out in the helping process (Gaston, Goldfried, Greenberg, Horvath, Raue, & Watson, 1995; Hill, 1994; Safran & Muran, 1995b; Sexton & Whiston, 1994; Weinberger, 1995). Some stress the relationship in itself (see Bailey, Wood, & Nava, 1992; Kahn, 1990; Kelly, 1994; Patterson, 1985), and others stress the work that is done through the relationship (Reandeau & Wampold, 1991) or the outcomes achieved through the relationship (Horvath & Symonds, 1991).

**The relationship in itself.** Patterson (1985) made the relationship itself central to helping. He claimed that counseling or psychotherapy does not merely *involve* an interpersonal relationship; rather, it *is* an interpersonal relationship. Some schools of psychotherapy emphasize the centrality of the helping relationship. For instance, in psychoanalytic or psychodynamic approaches, "transference" — the complex and often unconscious interpersonal dynamics between helper and client that are rooted in the client's and even the helper's past — is central (Carter, 1996; Gelso & Carter, 1994; Greenberg, 1994; Multon, Patton, & Kivlighan, 1996a, 1996b). Resolving these often murky dynamics is seen as intrinsic to successful therapeutic outcomes.

In a different mode, Carl Rogers (1951, 1957), representing the humanistic-experiential approach to helping, claimed that the quality of the relationship with respect to the unconditional positive regard, accurate empathy, and genuineness offered by the helper and perceived by the client was both necessary and sufficient for therapeutic progress. He claimed that such "client-centered" counselors helped clients understand themselves, liberate their resources, and manage their lives more effectively. Rogers's work spawned the widely discussed client-centered approach to helping (Rogers, 1965). Unlike psychodynamic approaches, however, the helping relationship was considered a facilitative condition, not a "problem" in itself to be explored and resolved.

**The relationship as a means to an end.** Others see the helping relationship as very important but still as a means to an end. In this view, a good relationship is practical because it enables client and counselor to do the work called for by whatever helping process is being used. The relationship is instrumental in achieving the goals of the helping process. Practitioners using cognitive and behavioral approaches to helping, although sensitive to relationship issues (Arnkoff, 1995), tend toward the means-to-end view. Overstressing the relationship is a mistake because it obscures the ultimate goal of helping: clients' managing their lives better. This goal won't be achieved if the relationship is poor, but if too much focus is placed on the relationship itself, both client and helper can be distracted from the real work to be done.

# THE WORKING ALLIANCE

The term *working alliance*, first coined by Greenson (1967) and now used by advocates of different schools of helping, can be used to bring together the best of the relationship-in-itself and relationship-as-means approaches. Bordin (1979) defined the working alliance as the collaboration between the client and the helper based on their agreement on the goals and tasks of counseling and on the development of an attachment bond. Although there is, predictably, considerable disagreement among practitioners as to what the critical dimensions of the working alliance are, how it operates, and what results it is to produce (see Weinberger, 1995), it is relatively simple to outline what it means in the context of the problem-management and opportunity-development process, which, of course, deals with the goals and tasks of counseling.

**The collaborative nature of helping.** In the working alliance, helpers and clients are collaborators. Helping is not something that helpers do to clients; rather, it is a process that helpers and clients work through together. Helpers do not "cure" their patients. Both have work to do in the problem-management and opportunity-development stages and steps, and both have responsibilities related to outcomes. Outcomes depend on the competence and motivation of the helper, on the competence and motivation of the client, and on the quality of their interactions. Helping is a two-person *team* effort in which helpers need to do their part and clients theirs. If either party refuses to play or plays incompetently, then the enterprise can fail. Of course, when all is said and done, clients must act on what they learn through the alliance.

**The relationship as a forum for relearning.** Even though helpers don't cure their clients, the relationship itself can be therapeutic. In the working alliance, the relationship itself is often a forum or a vehicle for social-emotional relearning (Mallinckrodt, 1996). Effective helpers model attitudes and behavior that help clients challenge and change their own attitudes and behavior. It is as if a client were to say to himself, "She [the helper] obviously cares for and trusts me, so perhaps it is all right for me to care for and trust myself." Or, "He takes the risk of challenging, so what's so bad about challenge when it's done well?" Or, "I came here frightened to death by relationships and now I'm experiencing a nonexploitative relationship that I cherish." Furthermore, protected by the safety of the helping relationship, clients can experiment with different behaviors during the sessions themselves. The shy person can speak up, the reclusive person can open up, the aggressive person can back off, the overly sensitive person can ask to be challenged, and so forth.

These learnings can then be transferred to other social settings. It is as if a client might say to himself, "He [the helper] listens to me so carefully and makes sure that he understands my point of view even when he thinks I should reconsider it. My relationships outside would be a lot different if I were to do the same." Or, "I do a lot of stuff in the sessions that would make anyone angry. But she doesn't let herself become a victim of emotions, either her own or mine. And her control doesn't diminish her humanity at all. That would make a big difference in my life." Clients may not use such explicit words, but the relearning dynamic, however subtle or covert, is often powerful. In sum, needed changes in both attitudes and behavior often take place within the sessions themselves through the relationship.

**Relationship flexibility.** The idea that one kind of perfect relationship or alliance fits all clients is a myth. Different clients have different needs, and those needs are best met through different kinds of relationships and different modulations within the same relationship. One client may work best with a helper who expresses a great deal of warmth, whereas another may work best with a helper who is more objective and businesslike. Some clients come to counseling with a fear of intimacy. If helpers, right from the beginning, communicate a great deal of empathy and warmth, these clients might be put off. Once the client learns to trust the helper, stronger interventions can be used. Effective helpers use a mix of styles, skills, and techniques tailored to the kind of relationship that is right for each client (Lazarus, 1993; Mahrer, 1993). And they remain themselves while they do so.

# THE CULTURE OF HELPING: VALUES IN ACTION

One of the best ways to characterize a helping relationship is through the values that should permeate and drive it. The relationship is the vehicle through which values come alive. Values, expressed concretely through working-alliance behaviors, play a critical role in the helping process (Bergin, 1991; Beutler & Bergan, 1991; Kerr & Erb, 1991; Norcross & Wogan, 1987; Vachon & Agresti, 1992). Since it has become increasingly clear that helpers' values influence clients' values over the course of the helping process, it is essential to build a value orientation into the process itself.

## Putting Values into the Broader Context of Culture

Values are part of a broader concept. Assumptions, beliefs, values, and norms constitute the *culture* of helping.

> The *assumptions and beliefs* held by helpers about themselves, their clients, the helping process, and the world around them interact with their *values* to generate *norms* that drive *patterns of behavior* in the helping relationship.

For instance, if a helper believes in the dignity of the person and prizes caring as a value, he or she will develop norms or standards that will be expressed in his or her behavior in the helping sessions.

> The *assumptions and beliefs* held by clients about themselves, the people in their social settings, the world around them, and the helping process itself interact with their *values* to generate *norms* that drive *patterns of behavior* both in their daily lives and in the helping relationship.

For instance, if a client assumes that she is a "bad person" and values security, these internal states will be expressed in a variety of external behaviors, including behavior in the helping sessions. She may be both hesitant and deferent.

Patterns of behavior constitute the "bottom line" of culture. Therefore, a popular definition of institutional culture is "the way we do things here." A popular definition of a client's personal culture is "the way I do things in my life." This definition applied to the helper is "the way I do helping." Inevitably, the helper's personal culture interacts with the client's for better or for worse. The focus in this chapter is on, directly, the values of helping and the norms they generate. Indirectly, this entire book is on the culture of helping—that is, the beliefs, values, and norms that can and should drive the helping process.

## The Importance and Practicality of Values in Helping

Values are not just ideals. They are also a set of practical criteria for making decisions. Ideally, they are drivers of behavior. For instance, a helper might say to himself or herself during a session with a difficult client something like this:

> The arrogant, I'm-always-right attitude of this client needs to be challenged. How to challenge her is another story. Since I respect the client, I do not want to challenge her by putting her down. On the other hand, I value genuineness and openness. Therefore, I can challenge her by describing her behavior and the impact it has on me and do so without belittling her.

Working values enable the helper to make decisions on how to proceed. Helpers without a set of working values are adrift. Helpers who don't have an explicit set of values have an implicit or "default" set. Human behavior is value driven. By watching helpers interact with their clients—that is, their pattern of behaviors—we could infer their values.

Helping-related values, like other values, cannot be handed to you on a platter. Much less can they be shoved down your throat. Therefore, what is written here is meant to stimulate your thinking about the values that should drive helping. In the final analysis, as you sit with your clients, only those beliefs, values, and norms that you have made your own will make a difference in your helping behavior. Therefore, you need to be proactive in your search for the beliefs, values, and norms that will govern your interactions with your clients. This does not mean that you will invent a set of values different from everyone else's. Tradition is an important part of value formation, and we all learn from the rich tradition of the helping professions.

Values that guide transactions between helpers and their clients can be packaged in a variety of ways. In the following pages, three major values from the tradition of the helping professions — respect, genuineness, and client self-responsibility —are translated into norms to drive behavior in the helping relationship.

## THE VALUE OF RESPECT

Respect is such a fundamental concept that, like most such concepts, it eludes definition. The word comes from a Latin root that includes the idea of "seeing" or "viewing." Indeed, respect is a particular way of viewing oneself and others. Respect, if it is to make a difference, cannot remain just an attitude or a way of viewing others.

## Norms That Translate Respect into Behaviors

Here, then, are some norms, generated by the interaction between a belief in the dignity of the person and the value of respect, that apply to interactions with clients.

**Do no harm.** This is the first rule of the physician and the first rule of the helper. Yet some helpers do harm either because they are unprincipled or because they are incompetent. There is no place for the "caring incompetent" in the helping professions. Helping is not a neutral process — it is for better or for worse. In a world in which such things as child abuse, wife battering, and exploitation of workers is much more common than we care to think, it is important to emphasize a nonmanipulative and nonexploitative approach to clients. Studies show that some instructors do exploit trainees both sexually and in other ways and that some helpers do the same with their clients. Such behavior obviously breaches the code of ethics espoused by all the helping professions.

**Become competent and committed.** Get good at whatever model of helping you use. Get good at the basic problem-management and opportunity-development framework outlined in this book and the skills that make it work because, as noted earlier, they are basic to all helping, even though they need to be adapted to individuals in light of the diversity issues discussed below. The paraprofessionals in the church-based self-help program mentioned in Chapter 2 had mastered a model of helping together with the attitude and skills that made it work. Being people of color themselves and culturally sensitive, they adapted their interventions to the diversity of their clients. In no way did "paraprofessional" for them mean "sloppy." Your clients are your customers and they deserve the best you can give.

**Make it clear that you are "for" the client.** The way you act with clients will tell them a great deal about your attitude toward them. Your manner should indicate that you are "for" the client, that you care for him or her in a down-to-earth, nonsentimental way. It is as if you are saying to the client attitudinally and behaviorally, "Working with you is worth my time and energy." Respect is both gracious and tough-minded. Being for the client is not the same as taking the client's side or acting as the client's advocate. "Being for" means taking clients' points of view seriously even when they need to be challenged. Respect often involves helping clients place demands on themselves. As Fisher and Ury (1981) suggested, "Be soft on the person, hard on the problem." And "tough love" in no way excludes appropriate warmth toward clients.

**Assume the client's goodwill.** Work on the assumption that clients want to work at living more effectively, at least until the assumption is proved false. The reluctance and resistance of some clients, particularly involuntary clients, is not necessarily evidence of ill will. Respect means entering the world of the clients to understand their reluctance and a willingness to help clients work through it. One member of Maria's self-help group said to his fellow group members, "I know I was dragging my feet early on. It's just that I hadn't realized how much work this would be. But you always seemed to expect the best from me. That helped a great deal."

**Do not rush to judgment.** You are not there to judge clients or to shove your values down their throats. You are there to help them — to help them identify, explore, and review and challenge the consequences of the values they have adopted. Let's say that a client during the first session says somewhat arrogantly, "I say whatever I want when I want. If others don't like it, well, that's their problem. My first obligation is to myself, being the person I am." One helper, irked by the client's attitude, might respond judgmentally by saying, "You've just put your finger on the core of your problem! How can you expect to get along with people with this kind of self-centered philosophy?" However, another counselor, taking a different approach, might respond, "So being yourself is one of your top priorities and being totally frank is, for you, part of that picture." The first counselor rushes to judgment; the second neither judges nor condones. At this point she merely tries to understand the client's point of view and let him know that she understands — even if she realizes that this point of view needs to be reviewed and challenged later.

**Keep the client's agenda in focus.** Helpers should pursue their clients' agendas, not their own. Here are three examples of helpers who lost clients because of lack of appreciation of the clients' agendas. One helper recalled, painfully, that he lost a client because he had become too preoccupied with his *theories* of depression rather than the client's painful depressive episodes. Another helper who dismissed as either trivial or irrelevant a client's bereavement over a pet that had died was dumbfounded and crushed when the client made what looked like an attempt on her own life. Later the client related her "gesture," in part, to the loss of the pet. A third helper, a white male who prided himself on his multicultural focus in counseling, went for counseling himself when a Hispanic client quit therapy, saying, perhaps somewhat unfairly, as he was leaving, "I don't think you're interested in me. You're more interested in Anglo-Hispanic politics." This last incident brings us to our next topic.

## Norms Related to Diversity and Multiculturalism

Although dealing knowledgeably and sensitively with diversity and that particular form of diversity called multiculturalism is part of respect, it is given separate attention here because of the emphasis currently being placed on diversity both in the workplace and in the helping professions. The literature on diversity and multiculturalism over the past few years has been vast (Allison, Echemendia, Crawford, & Robinson, 1996; Bernstein, 1994; Bryant, 1994; Chang, 1996; Das, 1995; Fowers & Richardson, 1996; Goodchilds, 1991; Hays, 1996; Helms, 1994; Ivey, Ivey, & Simek-Morgan, 1993; Lee, 1995, 1997; Leigh, Corbett, Gutman, & Morere, 1996; McFadden, 1993, 1996; Patterson, 1996; Pedersen, 1994, 1996; Pedersen & Ivey, 1993; Ponterotto, Casas, Suzuki, & Alexander, 1995; Ridley, Mendoza, Kanitz, Angermeier, & Zenk, 1994; Stone, 1995; Sue, Arrendondo, & McDavis, 1992; Sue & Sue, 1990; Tomes, 1994; Vargas & Willis, 1994; Weinrach & Thomas, 1996; Yutrzenka, 1995; Zane, Sue, Hu, & Kwon, 1991; Zayas, Torres, Malcolm, & DesRosiers, 1996, to name but a handful of those I have found informative, challenging, and at times infuriating).

Many clients come to helpers because they are having difficulties in their relationships with others or because relationship difficulties are part of a larger problem situation. Therefore, understanding clients' different approaches to developing and sustaining relationships is important. Guisinger and Blatt (1994) have made an excellent point:

> Western psychologies have traditionally given greater importance to self-development than to interpersonal relatedness, stressing the development of autonomy, independence, and identity as central factors in the mature personality. In contrast, women, many minority groups, and non-Western societies have generally placed greater emphasis on issues of relatedness. (p. 104)

The authors go on to point out that both interpersonal relatedness and self-definition are essential for maturity. Helping individuals from different cultures achieve the "right balance" depends on understanding what the "right balance" means in any given culture.

As with the "Does helping help?" debate, it is important to come to grips with the debate surrounding diversity and multiculturalism. Many professionals have pointed out that both the differences among and the needs of minority groups — whether race, ethnicity, disability, or some other kind of difference is at issue — together with the contributions such groups make to society have been systematically ignored or misunderstood. This is a social problem that has implications for groups of clients, individual clients within those groups, the helping professions, and individual helpers. This is not the place to go into the details of a debate on the relationship of counseling to social agendas — a debate that has generated as much heat as light. To do so would trivialize an issue of great importance. Rather, we will review some norms derived from the master value of respect that relate to diversity and multiculturalism.

**Understand diversity.** While clients have in common their humanity, they differ from one another in a whole host of ways — accent, age, attractiveness, color, developmental picture, disabilities, economic status, education, ethnicity, gender, group culture, national origin, occupation, personal culture, personality variables, politics, problem type, religion, sexual orientation, social status, to name some of the major categories. This presents several challenges for helpers. For one, it is essential that helpers understand clients and their problem situations contextually. For instance, a life-threatening illness might be one kind of reality for a 20-year-old and quite a different reality for an 80-year-old. We know that homelessness is a complex phenomenon. A homeless client with a history of drug abuse who has dropped out of graduate school is far different from a drifter who hates shelters for the homeless and resists every effort to get him to go to them.

Although it is true that helpers can, over time, come to understand a great deal about the characteristics of the populations with whom they work — for instance, they can and should understand the different development tasks and challenges that take place over the life span, and if they work with the elderly, they should grow in their understanding of the challenges, needs, problems, and opportunities of the aged — still, it is impossible to know everything about every population. This impossibility becomes even more dramatic if the permutations and combinations of characteristics are taken into consideration. How could an African American, middle-class, highly educated, younger, urban, Episcopalian, female psychologist possibly understand a poor, unemployed, homeless, middle-aged, uneducated, lapsed-Catholic male, born of migrant workers? Indeed, how can anybody fully understand anybody else? If the legitimate principles relating to diversity are pushed too far, no one will be able to understand and help anybody else.

**Challenge whatever blind spots you may have.** Since helpers often differ from their clients in many ways, there is often the challenge to avoid diversity-related blind spots that can lead to inept interactions and interventions during the helping process. For instance, a physically attractive and extroverted helper might have blinds spots with regard to the social flexibility and self-esteem of a physically unattractive and introverted client. Much of the literature on diversity and multiculturalism targets such blind spots. Becoming aware of their own cultural values and biases together with taking pains to understand the worldviews of their clients can help counselors dispel such blind spots. Helpers should, as a matter of course, become aware of the key ways in which they differ from their clients and take special care to be sensitive to those differences.

**Tailor your interventions in a diversity-sensitive way.** Both this self-knowledge and this practical understanding of diversity need to be translated into appropriate interventions. The way a Hispanic helper challenges a Hispanic client may be inappropriate if the client is white and vice versa. The way a younger helper shares his own experience with a younger client might be inappropriate for a client who is older and vice versa. Client self-disclosure, especially more intimate disclosure, might be relatively easy for a person from one culture, let us say North American culture, but very difficult for a client from another, let us say Asian or British culture. In this case, interventions that call for intimate self-disclosure may be seen as premature or even extortionistic by the client. Even though a client may be from a culture that is more open to self-disclosure, he or she may be frightened to death by self-disclosure. Therefore, with clients who come from a culture that has a different perception of self-disclosure or with any client who finds self-disclosure difficult, it might make more sense, after an initial discussion of the problem situation in broad terms, to move to what the client wants instead of what he or she currently has (that is, Stage II) rather than to the more intimate details of the problem situation. Once the helping relationship is on firmer ground, the client can move to the work he or she sees as more intimate or demanding.

As noted in Chapter 2, the helping model outlined in this book crosses cultures easily. The model has something to do with the way in which people in all cultures think about constructive change. Although it is used extensively around the world in a wide variety of cultures and subcultures, helpers still need to apply its stages and steps with cultural sensitivity.

**Understand and value the individual.** The diversity principle is clear: The more helpers understand the broad characteristics, needs, and behaviors of the populations with whom they work — African Americans, Caucasian Americans, diabetics, the elderly, you name it — the better positioned they are to adapt these broad parameters and the counseling process itself to the individuals with whom they work. But, whereas diversity focuses on differences both between and within groups — cultures and subcultures, if you will — helpers interact with clients as individuals. Of course, these individuals have *this* set of characteristics, but they do not come as members of a group. One of the principal learnings of social psychology is this: There are as many differences, and sometimes more, *within* groups as *between* groups (see Weinrach & Thomas, 1996, pp. 473–474). A middle-class black male is *this* individual. A poor Asian woman is *this* person. In a very real sense, a conversation between identical twins is a *cross-cultural* event because they are different individuals with differences in personal assumptions, beliefs, values, norms, and patterns of behavior. Genetics and group culture account for commonalities among individuals, but personhood and personal cultures emphasize each person's uniqueness. Focusing excessively on what makes this client different can be just as injurious as ignoring differences.

Finally, valuing diversity is not the same as espousing a splintered, antagonistic society in which one's group membership is more important than one's humanity. On the other hand, valuing individuality is not the same as espousing a "society of one" — radical individualism being the ultimate form of diversity. Moving to a "society of one" makes counseling and other forms of human interaction impossible.

## Guidelines Relating to Diversity and Multiculturalism

Since it is impossible to lay down rules for every possible case in which diversity is an issue, some broad guidelines are called for. The norms previously outlined can serve as guidelines. Ultimately, you have to come to grips with diversity and pull together your own set of guidelines. Here is one set, drawn from an article by Weinrach and Thomas (1996, pp. 475–476) but reworded and reworked a bit.

- Place the needs of the client above all other considerations.
- Identify and focus on whatever frame of reference, self-definition, or belief system is central to any given client, with consideration for, but not limited to, issues of diversity.
- Select counseling interventions on the basis of the client's agenda. Do not impose a social or political agenda on the counseling relationship.
- Make sure that your own values do not adversely affect a client's best interests.
- Avoid cultural stereotyping. Do not overgeneralize. Recognize that within-group differences are often more extensive than between-group differences.
- Do not define diversity narrowly. This client's concern about unattractiveness deserves the helper's engagement just as much as that client's concern about racial intolerance.
- Provide opportunities for practitioners to be trained in the working knowledge and skills associated with diversity-sensitive counseling.

- Subject the assumptions, models, and techniques of diversity-sensitive counseling to the same scrutiny as other aspects of the counseling profession.
- Create an environment that supports professional tolerance.

The fact that not all practitioners would agree with this package highlights the importance of your coming to grips not only with diversity but with the whole range of value questions that permeate the helping.

## The Value of Genuineness: Beyond Professionalism and Phoniness

Like respect, helper genuineness refers to both a set of attitudes and a set of counselor behaviors. Some writers call genuineness "congruence." Genuine people are at home with themselves and therefore can comfortably be themselves in all their interactions. Here are a few principles for translating genuineness into action.

**Do not overemphasize the helping role.** Genuine helpers do not take refuge in the role of counselor. Ideally, relating at deeper levels to others and to the counseling they do is part of their lifestyle, not roles they put on or take off at will. This keeps them far away from being patronizing and condescending. Effective helpers know how to be themselves without becoming "free spirits" who inflict themselves on others. Indeed, "free spirit" helpers can even be dangerous. Being role-free is not license. Freedom from role means that counselors should not use the role or facade of counselor to protect themselves, to substitute for competence, or to fool the client in other ways.

**Avoid defensiveness.** Genuine helpers are nondefensive. They know their own strengths and deficits and are presumably trying to live mature, meaningful lives. When clients express negative attitudes toward them, they examine the behavior that might cause the clients to think negatively, try to understand the clients' points of view, and continue to work with them. In the following example, a client, toward the end of the fourth helping session, says,

**CLIENT:** I don't think I'm really getting anything out of these sessions at all. I still feel drained all the time. Why should I waste my time coming here?

Here are examples of how three different helpers might respond.

**HELPER A:** If you were honest with yourself, you'd see that *you* are the one wasting time. Change is hard and you keep putting it off.
**HELPER B:** Well, that's your decision.

Counselors A and B are both defensive, though in different ways. It is more likely that the client will react to their defensiveness than that she will move forward.

**HELPER C:** So from where you're sitting, there's no payoff for being here. Just a lot of dreary work and nothing to show for it.

Counselor C centers on the experience of the client, with a view to "resetting the system" and helping her explore her responsibility for making the helping process work. Since genuine helpers are at home with themselves, they can allow themselves to examine negative criticism honestly. Counselor C, for instance, would be the most likely of the three to ask herself how she might be contributing to the apparent stalemate.

# THE VALUE OF CLIENT EMPOWERMENT: HELPING CLIENTS DEVELOP SELF-RESPONSIBILITY

The second goal of helping, outlined in Chapter 1, deals with empowerment—that is, helping clients identify, develop, and use resources that will make them more effective agents of change both within the helping sessions and in their everyday lives (Strong, Yoder, & Corcoran, 1995). The opposite of empowerment is dependency (Abramson, Cloud, Kesse, & Keese, 1994), deference (Rennie, 1994), and oppression (McWhirter, 1996). The focus here is on the clients' winning the struggle first with themselves in the context of the helping relationship. This poses a problem. Since helpers are often experienced by clients as relatively powerful people and since even the most egalitarian and client-centered of helpers do influence clients, it is necessary to come to terms with social influence in the helping process itself.

## Helping as a Social-Influence Process

People influence one another every day in every social setting of life. Parents influence each other and their kids. In turn they are influenced by their kids. Teachers influence students and students influence teachers. Bosses influence subordinates and vice versa. Team leaders influence team members, and members influence both one another and the leader. The world is abuzz with social influence. It could not be otherwise. It is also for weal or woe. Parents, teachers, bosses, and helpers all have power and need to be careful how they wield it. Power too often leads to manipulation and oppression.

It is not surprising, then, that helping as a social-influence process has received a fair amount of attention in the helping literature (Dorn, 1986; Heppner & Claiborn, 1989; Heppner & Frazier, 1992; Hoyt, 1996; McNeill & Stolenberg, 1989; Strong, 1968, 1991; Tracey, 1991). Helpers, like all who have power in relationships, can influence clients without robbing them of self-responsibility. Even better, they can exercise their trade in such a way that clients are, to use a bit of current business jargon, "empowered" rather than oppressed in their relationships with themselves and with the world. With empowerment, of course, comes increased self-responsibility.

Imagine a continuum. At one end lies "directing clients' lives" and at the other "leaving clients completely to their own devices." Somewhere along that continuum is "helping clients make their own decisions and act on them." Most forms of helper influence will fall somewhere in between the extremes. Preventing a client from jumping off a bridge moves, understandably, to the controlling end of the continuum, and simply accepting and in no way challenging a client's decision to put off dealing with a troubled relationship because he or she is "not ready" moves toward the other end. As Hare-Mustin and Marecek (1986) noted, there is a tension between the right of clients to determine their own way of managing their lives and the therapist's obligation to help them live more effectively.

## Norms for Empowerment and Self-Responsibility

Helpers don't self-righteously "empower" clients. That is patronizing and condescending. In a classic work, Freire (1970) warned helpers against making helping itself just one more form of oppression for the already oppressed. Effective counselors help clients discover, develop, and use the untapped power within themselves. Here, then, is a range of empowerment-based norms, some adapted from the work of Farrelly and Brandsma (1974).

**Start with the premise that clients can change if they choose.** Clients have more resources for managing problems in living and developing opportunities than they — or sometimes their helpers — assume. The helper's basic attitude should be that clients have the resources both to participate collaboratively in the helping process and to manage their lives more effectively. These resources may be blocked in a variety of ways or simply unused. The counselor's job is to help clients identify, free, and cultivate these resources. They also help clients assess their resources realistically so that aspirations do not outstrip resources.

Even when clients have been victimized by institutions or individuals, don't see them as helpless victims. The cult of victimhood is already growing too fast in society. Even if victimizing circumstances have diminished a client's degree of freedom — the abused spouse's inability to leave a deadly relationship — work with the freedom that is left.

Don't be fooled by appearances. One counselor trainer in a meeting with his colleagues dismissed a reserved, self-deprecating trainee with the words, "She'll never make it. She's more like a client than a trainee." Fortunately, his colleagues did not work from the same assumption. The woman went on to become one of the program's best students. She was accepted as an intern at a prestigious mental-health center and was hired by the center after graduation.

**Share the helping process with clients.** Clients, like helpers, can benefit from maps of the helping process. Helping should not be a "a pig in a poke" or a "black box." Clients have a right to know what they are getting into (Heinssen, 1994; Heinssen, Levendusky, & Hunter, 1995; Hunter, 1995; Somberg, Stone, & Claiborn, 1993; Sullivan, Martin, & Handelsman, 1993). A simple handout with both graphics and text can serve as the starting point. Just what kind of detail will help will differ from client to client. Obviously, clients should not be overwhelmed by distracting detail from the beginning. Nor should highly distressed clients be told to contain their anxiety until helpers teach them the helping model. Rather, the details of the model can be shared over a number of sessions. A simple pamphlet outlining the stages and steps of the helping process can be of great help, provided that it is in language that clients can readily understand. In my opinion, clients should be told as much about the model as they can assimilate.

**Help clients see counseling sessions as work sessions.** Helping is about client-enhancing change — changing self-defeating beliefs, values, and norms and changing the self-defeating patterns of behavior they spawn. Therefore, counseling sessions deal with exploring the need for change, the kind of change needed, creating programs of constructive change, engaging in change "pilot projects," and finding ways of dealing with obstacles to change. This is work pure and simple. It can be arduous, even agonizing, but it can also be deeply satisfying, even exhilarating. Helping clients develop the "work ethic" that makes them partners in the helping process can be one of the helper's most formidable challenges. Some helpers go so far as to cancel counseling sessions until the client is "ready to work." Helping clients discover incentives to work is, of course, less dramatic and hard work in itself.

**Help clients become better problem solvers in their daily lives.** Common sense suggests that problem-solving models, techniques, and skills are important for all of us, since all of us must grapple daily with problems in living of greater or lesser severity. Ask anybody whether problem-management skills are important for day-to-day living, and the answer inevitably is "certainly." Talk about the importance of problem solving is everywhere. Yet if you review the curricula of our primary, secondary, and tertiary schools, you will find that talk outstrips practice. In an ideal world, clients would have learned problem solving when they were young (Elias & Clabby, 1992). But that is the exception rather than the rule. There are those who say that formal courses in problem-solving skills are not found in our schools because such skills are picked up through experience. To a certain extent, that's true. However, if problem-management skills are so important, why does society leave the acquisition of these skills to chance? A problem-solving mentality should become second nature to us. The world may be the laboratory for problem solving, but the skills needed to optimize learning in this lab should be taught; they are too important to be left to chance.

It is no wonder, then, that clients are often poor problem solvers or that whatever problem-solving ability they have tends to disappear in times of crisis. If the second goal of the helping process is to be achieved—that is, if clients are to go away better able to manage their problems in living more effectively on their own—then sharing some form of the problem-management process with them is essential. Helpers add great value to the lives of their clients if they can get them to apply problem-solving techniques to their current problem situations and, at the same time, help them adopt more effective approaches to future problems in living.

**Become a consultant to clients.** Helpers can see themselves as consultants hired by clients to help them face problems in living more effectively. Consultants in the business world adopt a variety of roles. They listen, observe, collect data, report observations, teach, train, coach, provide support, challenge, advise, offer suggestions, and even become advocates for certain positions. But the responsibility for running the business remains with those who hire the consultant. Therefore, even though some of the activities of the consultant can be seen as quite challenging, the decisions are still made by managers. Consulting, then, is a social-influence process, but it is a collaborative one that does not rob managers of the responsibilities that belong to them. In this respect, it is a useful analogy to helping. The best clients, like the best managers, learn how to use their consultants to add value.

Tyler, Pargament, and Gatz (1983) moved a step beyond the consultant role in what they called the "resource collaborator role." Seeing both helper and client as people with defects, they focused on the give-and-take that should characterize the helping process. In their view, either client or helper can approach the other to originate the helping process. The two have equal status in defining the terms of the relationship, in originating actions within it, and in evaluating both outcomes and the relationship itself. In the best case, positive change occurs in both parties.

**Accept helping as a natural, two-way influence process.** Helping is a two-way street. Clients and therapists change one another in the helping process. Even a cursory glance at helping reveals that clients can affect helpers in many ways. Helpers find clients attractive or unattractive and must deal with both positive and negative feelings about them as well as manage their own behaviors. They may have to fight the tendency to be less demanding of attractive clients or not to listen carefully to unattractive clients. On the other hand, some clients trip over their own distorted views of their helpers. A young woman who has had serious problems in her relationship with her mother might begin to have problems relating to a helper who is an older woman. Unskilled helpers can get caught up in both their own and their clients' games. Skilled helpers understand the shadow side of the helping relationship and manage it.

**Focus on learning instead of helping.** Although many see helping as an education process, it is probably better characterized as a *learning* process. Effective counseling helps clients get on a learning track. Both the helping sessions themselves and the time between sessions involve learning, unlearning, and relearning. Howell (1982) gave us a good description of learning when he said that "learning is incorporated into living to the extent that viable options are increased" (p. 14). In the helping process, *learning takes place when options that add value to life are opened up, seized, and acted on.* If the collaboration between helpers and clients is successful, clients learn in very practical ways. They have more "degrees of freedom" in their lives as they open up options and take advantage of them. This is precisely what counseling helped Maria (see Chapter 2) to do. She unlearned, learned, relearned, and acted on her learnings.

**Do not overrate the psychological fragility of clients.** Neither pampering nor brutalizing clients serves their best interests. However, many clients are less fragile than helpers make them out to be. Helpers who constantly see clients as fragile may well be operating in a self-protective model. Driscoll (1984) noted that too many helpers shy away from doing much more than listening early in the helping process. They are fearful of making some kind of irretrievable error. Some caution is appropriate, but one can easily become overly cautious. He suggested that helpers intervene more right from the beginning — for instance, by reasonably challenging the way clients think and act and by getting them to begin to outline what they want and are willing to work for.

# A Working Charter:
# The Client-Helper Contract

Both implicit and explicit contracts govern the transactions that take place between people in a wide variety of situations, including marriage (where some but by no means all of the provisions of the contract are explicit) and friendship (where the provisions are usually implicit). If helping is to be a collaborative venture, then both parties must understand what their responsibilities are. Perhaps the term *working charter* is better than contract. It avoids the legal implications of the latter term and connotes a cooperative venture.

   To achieve these objectives, the working charter should include, generically, the issues that have been covered in Chapters 1 through 3 — that is, (a) the nature and goals of the helping process, (b) an overview of the helping model together with the techniques to be used and a sense of the flexibility built into the process, (c) how this process will help clients achieve their goals, (d) relevant information about yourself and your background, (e) how the relationship is to be structured and the kinds of responsibilities both you and the client will have, (f) the values that will drive the helping process, and (g) procedural issues. "Procedural issues" refers to the nuts and bolts of the helping process, such things as where sessions will be held and how long they will last. Procedural limitations should also be discussed — for instance, how free the client is to contact the helper between sessions. "Ordinarily we won't contact each other between sessions, unless we prearrange it for a particular purpose. However, . . ." Key ground rules should not come as a surprise to clients.

The working charter need not be too detailed, nor should it be rigid. The question is, How much structure will help *this* client at *this* time? Helpers need to provide structure for the relationship and the work to be done without frightening or overwhelming the client. Ideally, the working charter is an instrument that makes clients more informed about the process, more collaborative with their helpers, and more proactive in managing their problems. At its best, a working charter can help client and helper develop realistic mutual expectations, give clients a flavor of the mechanics of the helping process, diminish initial client anxiety and reluctance, provide a sense of direction, and enhance clients' freedom of choice.

## SHADOW-SIDE REALITIES IN THE RELATIONSHIP

There are common flaws in the working alliance that remain in the shadows either because they are not dealt with effectively by the helping professions themselves or because individual helpers are inept at addressing them with clients.

**Ethical flaws.** Little has been said about ethics in the helping process so far, not because it is not important, but because it is so important. There is a vast literature on ethical responsibilities in the helping professions (see Bersoff, 1995; Canter, Bennett, Jones, & Nagy, 1994; Claiborn, Berberoglu, Nerison, & Somberg, 1994; Corey, Corey, & Callanan, 1993; "Ethical Principles," 1992; Gibson & Pope, 1993; Keith-Spiegel, 1994; Meara, Schmidt, & Day, 1996 [plus reactions on pp. 78–104]; Pope & Vasquez, 1991; *Professional Psychology: Research & Practice*, Special Section: The 1992 Ethics Code, 1994). There is also a growing literature on ways in which helpers violate their ethical responsibilities. Since this area is too vast and too important to be given summary treatment here, helpers-to-be are urged to make this part of their professional development program.

**Flaws in the relationship itself.** The helping relationship might be flawed from the beginning. That is, the fit between helper and client is not right. But, for a variety of reasons, it is not easy for a helper to say, "I don't think I'm the one for you." Even if the relationship starts off on the right foot, it can deteriorate. Impasses and ruptures in the relationship are not uncommon, but they are not always dealt with in a straightforward manner. Factors associated with relationship breakdowns include "a client history of interpersonal problems, a lack of agreement between therapists and clients about the tasks and goals of therapy, interference in the therapy by others, transference, possible therapist mistakes, and therapist personal issues" (Hill, Nutt-Williams, Heaton, Thompson, & Rhodes, 1996; Safran & Muran, 1995a, p. 207). If impasses and ruptures are not addressed, the helping process becomes difficult and complicated, often to the point that premature termination takes place. Finally, there is the tendency to blame clients for helping failures: "She wasn't ready," "He didn't want to work," "She was impossible," and so forth.

**Vague and violated values.** Helpers do not always have a clear idea of what their values are. Or the values they say they hold — that is, their *espoused* values — do not always coincide with their actions — their *values-in-use*. Values too often remain "good ideas" and are not translated into specific norms that drive helping behavior. For instance, even though helpers value self-responsibility in their clients, they see them as helpless, make decisions for them, and direct rather than guide. Often they do so out of frustration. Expediency leads them to compromise their values and then compromises are subjected to rationalizing: "I blew up at a client today, but he really deserved it. Probably did more good than my unappreciated patience."

**Failure to share the helping process.** When it comes to sharing the helping process itself, some counselors are reluctant to let the client know what the process is all about. Of course, helpers who "fly by the seat of their pants" can't tell clients what it's all about because they don't know what it's all about themselves. Still others seem to think that knowledge of helping processes is secret or sacred or dangerous and should not be communicated to the client, even though there is no evidence to support such beliefs (Dauser, Hedstrom, & Croteau, 1995; Somberg, Stone, & Claiborn, 1993; Sullivan, Martin, & Handelsman, 1993; Winborn, 1977).

**Flawed contracts.** There is an extensive shadow side to both explicit and implicit contracts. Even when a contract is written, the contracting parties interpret some of its provisions differently. Over time they forget what they contracted to and differences become more pronounced. These differences are seldom discussed. In counseling, the helper-client contract has been, traditionally, implicit, even though the need for more explicit structure has been discussed for years (Proctor & Rosen, 1983). Because of this, the expectations of clients may differ from the expectations of their helpers (Benbenishty & Schul, 1987). Implicit contracts are not enough, but they still abound (Handelsman & Galvin, 1988; Weinrach, 1989; Woody, 1991).

**Warring professionals.** There are not just debates but also conflicts close to internecine wars in the helping professions. For instance, the debate on the "correct" approach to diversity and multiculturalism brings out some of the best and some of the worst in the helping community. Accusations, however subtle or blatant, of cultural imperialism on the one side and "political correctness" on the other fly back and forth. The debate on whether or how the helping professions should take political stands or engage in social engineering generates, as has been noted, more heat than light. The search for the truth gives way at times to the need to be right. It is not always clear how all of this serves the needs of clients. Indeed, clients are often enough left out of the debate. Just as many businesses today are reinventing themselves by starting with their customers and markets, so the helping professions might reinvent themselves by looking at helping through the eyes of clients.

## PART TWO

# Basic Communication Skills for Helping

## The Importance of Communication Skills

Since helping takes place through a dialogue between helper and client, the communication skills of the helper are critical at every stage and step of the helping process. These skills are not the helping process itself, but they are essential tools for developing relationships and interacting with clients. Chapter 4 deals with the first set of these skills, attending and listening. Chapter 5 deals with empathy and Chapter 6 with the art of probing. Chapters 8 through 11 deal with a third set related to challenging, including helping clients identify blind spots and develop new perspectives. All these skills are essential tools for both relationship building and constructive change.

**The skills of everyday life.** These skills are not special skills peculiar to helping. Rather, they are extensions of the kinds of skills all of us need in our everyday interpersonal transactions. Ideally, helpers-to-be would enter training programs with this basic set of interpersonal communication skills in place, and training would simply help them adapt the skills to the helping process. Unfortunately, that is often not the case. Indeed, some of the problems clients have either focus on or are complicated by a lack of interpersonal communication skills.

**The microskills approach to training helpers.** Since communication skills are not ends in themselves but means or instruments to be used in achieving helping outcomes, there has been some concern about the overemphasis on communication skills and techniques. Years ago Carl Rogers (1980) spoke out against an overemphasis on the microskills of helping. Instead of being a fully human endeavor, helping was, in his view, being reduced to its bits and pieces. Some helper training programs focus almost exclusively on these skills. As a result, trainees know how to communicate but not how to help.

This book together with the workbook that accompanies it takes a microskills approach to learning the basics. However, the "overemphasis" noted by Rogers is avoided by continually relating these skills to the needs of clients and by integrating them into the entire helping process. Hills (1984) discussed an "integrative" versus a "technique" approach to training in communication skills. In an integrative approach care is taken to make sure that

- skills and techniques become extensions of the helper's humanity and not just bits of helping technology;
- communication skills and helping techniques serve the goals of the helping process;
- skills and techniques are permeated with and driven by the values discussed in Chapter 3.

Of course, effective helpers weave these communication skills together seamlessly in their interactions with clients. Communication skills need to become "second nature" if they are to serve the helping process and clients' needs. Bob Carkhuff (1987), Allen Ivey (1994, 1997), and Carl Rogers (1951, 1957, 1965) were trailblazers in developing and humanizing communication microskills and integrating them into the helping process. Their influence is seen throughout this book.

# ATTENDING, LISTENING, AND UNDERSTANDING

Two skills, attending and listening at the service of understanding, are the focus of this chapter. These skills are so basic that they should be second nature in all forms of interpersonal communication, including helping. Unfortunately, that is not always the case.

- Attending refers to the ways in which helpers can be with their clients, both physically and psychologically.
- Listening refers to the ability of helpers to capture and understand the messages clients communicate as they tell their stories, whether those messages are transmitted verbally or nonverbally, clearly or vaguely.

These skills, including some of the guiding principles underlying them, are now examined in some detail.

## ATTENDING: BEING VISIBLY TUNED IN TO CLIENTS

At some of the more dramatic moments of life, simply being with another person is extremely important. If a friend of yours is in the hospital, just your being there can make a difference, even if conversation is impossible. Similarly, your being with a bereaved friend can be very comforting to him or her, even if little is said. People appreciate it when others pay attention to them. By the same token, being ignored is often painful: The averted face is too often a sign of the averted heart. Given how sensitive most of us are to others' attention or inattention, it is paradoxical how insensitive we can be at times about attending to others.

Helping and other deep interpersonal transactions demand a certain intensity of presence. Attending, or the way you orient yourself physically and psychologically to clients, contributes to this presence. Effective attending does two things: It tells clients that you are with them, and it puts you in a position to listen carefully to their concerns.

Clients read cues that indicate the quality of your presence to them. Your nonverbal behavior influences clients for better or for worse. Attentive presence can invite or encourage them to trust you, open up, and explore the significant dimensions of their problem situations. Halfhearted presence can promote distrust and lead to clients' reluctance to reveal themselves to you. However, the skills of attending, although they can be learned, will be phony if they are not driven by the attitudes and values discussed in Chapter 3. Your mind-set, what's in your heart, is as important as your visible presence. If you are not "for" your client, if you resent working with a client, this will ooze out into your behavior. I once heard a patient talking with his doctor about his concerns about an invasive diagnostic procedure. The doctor said the right words to reassure the patient, but his physical presence and the way he rushed his words said, "I've heard this dozens of times. I really don't have time for your concerns. Let's get on with this."

# The Microskills of Attending

There are certain microskills helpers can use in attending to clients. These microskills can be summarized in the acronym *SOLER*. Since communication skills are particularly sensitive to cultural differences, care should be taken in adapting what follows to different cultures. It is only a template.

- S: Face the client *Squarely;* that is, adopt a posture that indicates involvement. In North American culture, facing another person squarely is often considered a basic posture of involvement. It usually says, "I'm here with you; I'm available to you." Turning your body away from another person while you talk to him or her can lessen your degree of contact with that person. Even when people are seated in a circle, they usually try in some way to turn toward the individuals to whom they are speaking. The word *squarely* here may be taken literally or metaphorically. What is important is that the bodily orientation you adopt convey the message that you are involved with the client. If, for any reason, facing the person squarely is too threatening, then an angled position may be more helpful. The point is the quality of your attention.

- O: Adopt an *Open* posture. Crossed arms and crossed legs can be signs of lessened involvement with or availability to others. An open posture can be a sign that you're open to the client and to what he or she has to say. In North American culture, an open posture is generally seen as a nondefensive posture. Again, the word *open* can be taken literally or metaphorically. If your legs are crossed, this does not mean that you are not involved with the client. But it is important to ask yourself, "To what degree does my present posture communicate openness and availability to the client?"

- L: Remember that it is possible at times to *Lean* toward the other. Watch two people in a restaurant who are intimately engaged in conversation. Very often they are both leaning forward over the table as a natural sign of their involvement. The main thing is to remember that the upper part of your body is on a hinge. It can move toward a person and back away. In North American culture, a slight inclination toward a person is often seen as saying, "I'm with you, I'm interested in you and in what you have to say." Leaning back (the severest form of which is a slouch) can be a way of saying, "I'm not entirely with you" or "I'm bored." Leaning too far forward, however, or doing so too soon, may frighten a client. It can be seen as a way of placing a demand on the other for some kind of closeness or intimacy. In a wider sense, the word *lean* can refer to a kind of bodily flexibility or responsiveness that enhances your communication with a client.

- E: Maintain good *Eye* contact. In North American culture, fairly steady eye contact is not unnatural for people deep in conversation. It is not the same as staring. Again, watch two people deep in conversation. You may be amazed at the amount of direct eye contact. Maintaining good eye contact with a client is another way of saying, "I'm with you; I'm interested; I want to hear what you have to say." Obviously, this principle is not violated if you occasionally look away. Indeed, you have to if you don't want to stare. But if you catch yourself looking away frequently, your behavior may give you a hint about some kind of reluctance to be with this person or to get involved with him or her. Or it may say something about your own discomfort.

• *R:* Try to be relatively *Relaxed* or natural in these behaviors. Being relaxed means two things. First, it means not fidgeting nervously or engaging in distracting facial expressions. The client may wonder what's making you nervous. Second, it means becoming comfortable with using your body as a vehicle of personal contact and expression. Your being natural in the use of these skills helps put the client at ease.

Given the diversity of clients mentioned in Chapter 3, these guidelines should be read cautiously. People differ both culturally and individually in how they show attentiveness and in how they react to the attentiveness of others.

The point to be stressed is that a respectful, genuine, caring mind-set might well lose its impact if the client does not see these internal attitudes reflected in your external behaviors. In the beginning you may become overly self-conscious about your attending behavior, especially if you are not used to attending carefully to others. Still, the guidelines just presented are just that—guidelines. They should not be taken as absolute rules to be applied rigidly in all cases.

## The Helper's Nonverbal Communication

Much more important than a mechanical application of microskills is an awareness of your body and your voice as a source of communication. Effective helpers are mindful of the cues and messages they are constantly sending through their bodies as they interact with clients. Reading your own bodily reactions is an important first step. For instance, if you feel your muscles tensing as the client talks to you, you can say to yourself, "I'm getting anxious here. What's causing my anxiety? And what cues am I sending the client?" Once you read your own reactions, you can use your body to communicate appropriate messages.

You can also use your body to censor instinctive or impulsive messages that you feel are inappropriate. For instance, if the client says something that instinctively angers you, you can control the external expression of the anger (for instance, a sour look) to give yourself time to reflect. This is not phony because your respect for your client takes precedence over your instinctive reactions. You are not denying your anger. Indeed, your instinctive reactions can provide cues for interactions with clients.

Attending does not mean that you become preoccupied with your body as a source of communication. It means, rather, that you learn to use your body instinctively as a value-driven means of communication. Being aware of and at home with nonverbal communication can reflect an inner peace with yourself, with the helping process, and with this client. The main point

---

BOX 4-1

# Questions on Attending

- What are my attitudes toward this client?
- How would I rate the quality of my presence to this client?
- To what degree does my nonverbal behavior indicate a willingness to work with the client?
- What attitudes am I expressing in my nonverbal behavior?
- What attitudes am I expressing in my verbal behavior?
- To what degree does the client experience me as effectively present and working with him or her?
- To what degree does my nonverbal behavior reinforce my internal attitudes?
- In what ways am I distracted from giving my full attention to this client?
- What am I doing to handle these distractions?
- How might I be more effectively present to this person?

---

is that your nonverbal behavior should enhance rather than stand in the way of your working alliance with your clients. Box 4-1 summarizes, in question form, the main points related to attending. Obviously, helpers are not constantly asking these questions of themselves. Rather, because this skill has become second nature, they are in touch with the quality of their presence to their clients. Both your verbal and nonverbal behavior should indicate a clear-cut willingness to work with the client.

## Active Listening

Effective attending puts helpers in a position to listen carefully to what clients are saying both verbally and nonverbally. Listening carefully to a client's concerns seems to be a concept so simple to grasp and so easy to do that one may wonder why it is given such explicit treatment here. Nonetheless, it is amazing how often people fail to listen to one another. How many times have you heard someone exclaim, "You're not listening to what I'm saying!" When the person accused of not listening answers, almost predictably, "I am listening; I can repeat everything you've said," the accuser is not comforted. What people look for in attending and listening is not the other person's ability to repeat their words. A tape recorder could do that perfectly. People want more than physical presence in human communication; they want the other person to be present psychologically, socially, and emotionally.

Complete listening involves four things: first, listening to and understanding the client's *verbal* messages; second, observing and reading the client's *nonverbal* behavior — posture, facial expressions, movement, tone of voice, and the like; third, listening to the *context* — that is, to the whole person in the context of the social settings of his or her life; fourth, listening to *sour notes* — that is, things the client says that may have to be challenged, at least eventually.

The following case will be used to help you develop a better behavioral feel for both attending and listening.

> Jennie, an African American college senior, was raped by a "friend" on a date. She received some immediate counseling from the university Student Development Center and some ongoing support during the subsequent investigation. But although she knew she had been raped, it turned out that it was impossible for her to prove her case. The entire experience — both the rape and the investigation that followed — left her shaken, unsure of herself, angry, and mistrustful of institutions she had assumed would be on her side (especially the university and the legal system). When Denise, a middle-aged and middle-class African American social worker who was a counselor for a health maintenance organization (HMO), first saw her a few years after the incident, Jennie was plagued by a number of somatic complaints, including headaches and gastric problems. At work, she engaged in angry outbursts whenever she felt that someone was taking advantage of her. Otherwise she had become quite passive and chronically depressed. She saw herself as a woman victimized by society and was slowly giving up on herself.

Even though Denise had excellent attending and listening skills, she took pains to understand both the similarities between herself and Jennie and the differences. For instance, although they were both African American, Jennie had been reared in a conservative, Catholic, semirural environment of poor parents, whereas Denise had grown up in a secular, liberal, urban, and relatively affluent family.

## Listening to and Understanding Verbal Messages: Experiences, Behavior, and Affect

Most immediately, helpers listen to clients' verbal messages — their "stories," if you will — which are mixtures of clients' *experiences, behaviors,* and *affect.* Traditionally, human activity has been divided into three parts — thinking, feeling, and acting. A slightly different approach is taken here.

- Clients talk about their *experiences* — that is, what happens to them. If a client tells you that she was fired from her job, she is talking about her problem situation as an experience. Jennie, of course, talked about being raped.

- Clients talk about their *behavior* — that is, what they do or refrain from doing. If a client tells you that he smokes and drinks a lot or if he says that he spends a great deal of time daydreaming, he is talking about his problem situation as a behavior. Jennie talked about pulling away from her family and friends after the rape investigation.

- Clients talk about their *affect* — that is, the feelings and emotions that arise from or are associated with either experiences or behavior. If a client tells you how depressed she gets after verbal fights with her fiance, she is talking about the affect associated with her problem situation. Jennie talked about her shame and her feelings of betrayal.

All three, of course, are totally interrelated in the day-to-day lives of clients, and in counseling dialogues, clients talk about all three together.

Consider this example. A client says to a counselor in the personnel department of a large company, "I had one of the lousiest days of my life yesterday." At this point the counselor knows that something went wrong and that the client feels bad about it, but she knows relatively little about the specific experiences and behaviors that made the day a horror for the client. However, the client continues, "Toward the end of the day my boss yelled at me in front of some of my colleagues for not getting my work done [an experience]. I lost my temper [emotion] and yelled right back at him [behavior]. He blew up and fired me [an experience for the client]. And now I feel awful [emotion] and am trying to find out if I have really been fired and, if so, if I can get my job back [behavior]." Problem situations are much clearer when they are spelled out as specific experiences, behaviors, and feelings related to specific situations.

**Experiences.** Most clients spend a fair amount of time, sometimes too much time, talking about what happens *to* them.

- "I get headaches a lot."
- "My ulcer acts up when family members argue."
- "My wife doesn't understand me."

Experience talk often focuses on what other people do or fail to do. At times the implication is that others are to blame for one's problems.

- "She doesn't do anything all day. The house is always a mess when I come home from work. No wonder I can't concentrate at work."
- "He tells his little jokes, and I'm always the butt of them. No wonder I feel bad about myself most of the time."

Some clients talk about experiences that are internal and out of their control.

- "These feelings of depression come from nowhere and seem to suffocate me."
- "I just can't stop thinking of him."

One reason that some clients fail to manage the problem situations of their lives is that they see themselves as victims, adversely affected by other people, by the immediate social settings of life such as the family, by society in its larger organizations and institutions such as government or the workplace, by cultural prescriptions, or even by internal forces. They feel that they are no longer in control of their lives or some dimension of life. Therefore, they talk extensively about these experiences.

- "Company policy discriminates against women. It's that simple."
- "The economy is so lousy that there are no jobs. Certainly no jobs I would want."
- "No innovative teacher gets very far around here."

Of course, some clients are treated unfairly; they are victimized by the behaviors of others in the social settings of their lives. Although they can be helped to cope with victimization, full management of their problem situations demands changes in the social settings themselves. One client was helped to cope with a brutal husband, but ultimately the courts had to intervene to keep him at bay.

For other clients, talking constantly about experiences is a way of avoiding responsibility: "It's not my fault. After all, these things are happening *to* me." Sykes (1992) in his book *A Nation of Victims* is troubled by the tendency of the United States to become a "nation of whiners unwilling to take responsibility for our actions."

**Behaviors.** All of us do things that get us into trouble and fail to do things that will help us get out of trouble or develop opportunities. Clients are no different.

- "When he ignores me, I think of ways of getting back at him."
- "I haven't even begun to look for a job. I know there are none in this city. Certainly none I would like."
- "Even though I feel the depression coming on, I don't take the pills the doctor gave me."
- "When I get bored, I find some friends and go get drunk."
- "I have a lot of sexual partners and have unprotected sex whenever my partner will let me."

Some clients talk freely about their experiences, what happens to them, but seem more reluctant to talk about their behaviors.

**Affect.** Affect refers to the feelings and emotions that proceed from, lead to, accompany, underlie, or give color to a client's experiences and behaviors.

- "I finally finished the term paper that I've been putting off for weeks and I feel great!"
- "I've been feeling pretty sorry for myself ever since he left me."
- "I yelled at my mother last night and now I feel very ashamed of myself."
- "I've been anxious for the past few weeks, but I don't know why. I wake up feeling scared and then it goes away but comes back again several times during the day."

Of course, clients often express feelings without talking about them. When a client says, "My boss gave me a raise and I didn't even ask for one!" you can feel the emotion in her voice. A client who is talking listlessly and staring down at the floor may not say, in so many words, "I feel depressed." A dying person may express feelings of anger and depression without talking about them. Other clients feel deeply about things but do their best to hold their feelings back. But often there are cues or hints, whether verbal or nonverbal, of the feelings inside.

The point here is this: When a client tells a full "story," it is a mixture of experiences, behaviors, and affect. Your first job is to listen carefully to what they have to say, the mix of experiences, behaviors, and affect they use to describe their problem situations, what they put in, and what they leave out. In subsequent chapters ways of helping clients "fill out" their stories with essential but missing experiences, behaviors, and feelings will be described and illustrated.

> Denise listens to what Jennie has to say early on about her past and present experiences, actions, and emotions. Jennie told her, "When the investigation began, I had every intention of pushing my case, because I knew that some of the men on campus were getting away with murder. But then it began to dawn on me that people were not taking me seriously because I was an African American woman. First I was angry, but then I just got numb. . . ." Later, Jennie said, "I get headaches a lot now. I don't like taking pills, so I try to tough it out. I have also become very sensitive to any kind of injustice, even in movies or on television. But I've stopped being any kind of crusader. That got me nowhere."

As Denise listens to Jennie speak, questions like this arise in the back of her mind:

- "What are the core messages here?"
- "What themes are coming through?"
- "What is Jennie's point of view?"
- "What is most important to her?"
- "What does she want me to understand?"

Of course, she doesn't distract herself from her client by asking these questions of herself directly, but they symbolize her interest in the world of her client.

# Listening to the Client's Nonverbal
# Messages and Modifiers

Clients send messages through their nonverbal behavior. Helpers need to "learn how to read" these messages without distorting or overinterpreting them.

**Nonverbal behavior as a channel of communication.** Over the years both researchers and practitioners have come to appreciate the importance of nonverbal behavior in counseling (Ekman, 1992; Grace, Kivlighan, Jr., & Kunce, 1995; Highlen & Hill, 1984; Mehrabian, 1972; Russell, 1995; Siegman & Feldstein, 1987). Highlen and Hill suggested that nonverbal behaviors regulate conversations, communicate emotions, modify verbal messages, provide important messages about the helping relationship, give insights into self-perceptions, and provide clues that clients are not saying what they are thinking. Of course, this is not a 20th-century discovery; people of common sense have been reading such messages from time immemorial. This area has taken on even more importance because of the multicultural nature of helping (Sue, 1990).

The face and body are extremely communicative. We know from experience that even when people are together in silence, the atmosphere can be filled with messages. Sometimes the facial expressions, bodily motions, voice quality, and physiological responses of clients communicate more than their words do. An early study illustrates this point. Mehrabian (1971) wanted to know what cues people use to judge whether another person likes them or not. He and his associates discovered that the other person's actual words contributed only 7% to the impression of being liked or disliked, whereas voice cues contributed 38% and facial cues 55%. They also discovered that when facial expressions were inconsistent with spoken words, facial expressions were believed more than the words.

What is significant in Mehrabian's research is not the exact percentages but, rather, the clear importance of nonverbal behavior in the communication process. Effective helpers learn how to listen to and read

- *bodily behavior,* such as posture, body movements, and gestures;
- *facial expressions,* such as smiles, frowns, raised eyebrows, and twisted lips;
- *voice-related behavior,* such as tone of voice, pitch, voice level, intensity, inflection, spacing of words, emphases, pauses, silences, and fluency;
- *observable autonomic physiological responses,* such as quickened breathing, the development of a temporary rash, blushing, paleness, and pupil dilation;
- *physical characteristics,* such as fitness, height, weight, and complexion;
- *general appearance,* such as grooming and dress.

The trick, of course, is to spot the messages in these behaviors without making too little or too much of them.

When Denise said to Jennie, "It's hard talking about yourself, isn't it?" Jennie said, "No, I don't mind at all." But the real answer was probably in her nonverbal behavior, for she spoke hesitatingly while looking away and frowning. Reading such cues helped Denise understand Jennie better. A person's nonverbal behavior has a way of "leaking" messages to others. The very spontaneity of nonverbal behaviors contributes to this "leakage" even in the case of highly defensive clients. It is not easy for clients to fake nonverbal behavior (Wahlsten, 1991). The real messages still tend to leak out.

**Nonverbal behavior as modifiers and "punctuation."**  Besides being a channel of communication in itself, such nonverbal behavior as facial expressions, bodily motions, and voice quality often modify and punctuate verbal messages in much the same way that periods, question marks, exclamation points, and underlining punctuate written language. Nonverbal behavior can punctuate or modify interpersonal communication in the following ways (see Knapp, 1978, pp. 9–12):

- *Confirming or repeating.*  Nonverbal behavior can confirm or repeat what is being said verbally. For instance, once when Denise responded to Jennie with just the right degree of understanding, not only did Jennie say, "That's right!" but also her eyes lit up (facial expression), she leaned forward a bit (bodily motion), and her voice was very animated (voice quality). Her nonverbal behavior confirmed her verbal message.

- *Denying or confusing.*  Nonverbal behavior can deny or confuse what is being said verbally. When challenged by Denise once, Jennie denied that she was upset, but her voice faltered a bit (voice quality) and her upper lip quivered (facial expression). Her nonverbal behavior carried the real message.

- *Strengthening or emphasizing.*  Nonverbal behavior can strengthen or emphasize what is being said. When Denise suggested to Jennie that she discuss the origin of what her boss saw as erratic behavior, Jennie said, "Oh, I don't think I could do that!" while slouching down and putting her face in her hands. Her nonverbal behavior underscored her verbal message. Nonverbal behavior adds emotional color or intensity to verbal messages. Once Jennie told Denise that she didn't like to be confronted without first being understood and then stared at her fixedly and silently with a frown on her face. Jennie's nonverbal behavior told Denise something about the intensity of her feelings.

- *Controlling or regulating.*  Nonverbal cues are often used in conversation to regulate or control what is happening. If, in group counseling, one participant looks at another and gives every indication that she is going to speak to this other person, she may hesitate or change her mind if the person she intends to talk to looks away. Skilled helpers are aware of the ways in which clients send controlling or regulating nonverbal cues.

Of course, helpers can "punctuate" their words with the same kinds of nonverbal messages. Denise, too, without knowing it, sent "silent messages" to Jennie. For instance, she was especially attentive when Jennie talked about actions she could take to do something about her problem situation. This was part of the social-influence dimension of helping.

In reading nonverbal behavior — "reading" is used here instead of "interpreting" — caution is a must. We listen to understand clients rather than to dissect them. There is no simple program available for learning how to read and interpret nonverbal behavior (Russell, 1995). Once you develop a working knowledge of nonverbal behavior and its possible meanings, you must learn through practice and experience to be sensitive to it and read its meaning in any given situation.

Since nonverbal behaviors can often mean a number of things, how can you tell which meaning is the real one? The key is the human context in which they take place. Effective helpers listen to the entire context of the helping interview and do not become overly fixated on details of behavior. They are aware of and use the nonverbal communication system, but they are not seduced or overwhelmed by it. This is the integrative approach. Sometimes novice helpers will fasten selectively on this or that bit of nonverbal behavior. For example, they will make too much of a half-smile or a frown on the face of a client. They will seize upon the smile or the frown and, in over-interpreting it, lose the person.

## Listening to and Understanding Clients in Context

People are more than the sum of their verbal and nonverbal messages. Listening in its deepest sense means listening to clients themselves as influenced by the contexts in which they "live, move, and have their being." Denise tries to understand Jennie's verbal and nonverbal messages, even the core messages, in the context of Jennie's life. As she listens to Jennie's story, Denise says to herself,

> Here is an intelligent African American woman from a conservative
> Catholic background. She was very loyal to the church because it proved
> to be a refuge in the inner city. It was a gathering place for her family and
> friends. It meant a decent primary- and secondary-school education and a
> shot at college. Initially college was a shock. It was her first venture into a
> predominantly white and secular culture. But she chose her friends carefully
> and carved out a niche for herself. Studies were much more demanding, and
> she had to come to grips with the fact that, in this larger environment, she
> was closer to average. The rape and investigation put a great deal of stress on

what proved to be a rather fragile social network. Her life began to unravel. She pulled away from her family, her church, and the small circle of friends she had at college. At a time she needed support the most, she cut it off. After graduation she continued to stay "out of community." She got a job as a secretary in a small company and has remained underemployed.

The helping context is also important. Denise needs to be sensitive about how Jennie might feel about talking to a woman who is quite different from her and also needs to understand that Jennie might well have some misgivings about the helping professions. In other words, Denise tries to pull together the themes she sees emerging in Jennie's story and tries to see those themes in context. She listens to Jennie's discussion of her headaches (experiences), her self-imposed social isolation (behaviors), and her chronic depression (affect) against the background of her social history — the pressures of being religious in a secular society at school, the problems associated with being an upwardly mobile African American woman in a predominantly white male society. Denise sees the rape and investigation as social, not merely personal, events. She listens actively and carefully, because she knows that her ability to help depends, in part, on not distorting what she hears. She does not focus narrowly on Jennie's inner psychology, as if Jennie could be separated from the social context of her life.

## Empathic Listening

Both empathic listening and empathic responding (Eisenberg & Strayer, 1987) are important concepts and skills. However, since there is so much confusion in the psychological literature as to what empathy means (see Duan & Hill, 1996, for an excellent overview), it is essential to point out how it is being used throughout this book. Empathy can be seen as an *intellectual* process that involves understanding correctly another person's emotional state and point of view: "I understand how distressed she feels [emotional state] because her husband won't even talk about having another child [point of view]. Empathy can also refer to *empathic emotions* experienced by the helper: "I feel her agony and I feel for her." In this book the intellectual process is stressed. It is important that helpers understand the feelings and emotions of their clients and their meaning for the clients even though they might not "feel along with" the clients. Furthermore, this intellectual process is presented in these pages as a *communication skill* made up of two parts. In this section empathic listening and understanding is the focus; the ability to communicate this understanding effectively to the client is considered in the next.

Empathic listening centers on the kind of attending, observing, and listening — the kind of "being with" — needed to develop an understanding of clients and their worlds. Although it might be metaphysically impossible to actually get "inside" the world of another person and experience the world as he or she does, it is possible to approximate this. And even an approximation is very useful in helping. Indeed, if people are to care for one another, some form of empathy is essential. Caring for clients and their concerns is part of respect.

Rogers (1980) talked passionately about basic empathic listening — being with and understanding the other — even calling it "an unappreciated way of being" (p. 137). He used the word *unappreciated* because in his view few people in the general population developed this "deep listening" ability and even so-called expert helpers did not give it the attention it deserved. Here is his description of basic empathic listening, or being with:

> It means entering the private perceptual world of the other and becoming thoroughly at home in it. It involves being sensitive, moment by moment, to the changing felt meanings which flow in this other person, to the fear or rage or tenderness or confusion or whatever that he or she is experiencing. It means temporarily living in the other's life, moving about in it delicately without making judgments. (p. 142)

Such empathic listening is selfless because helpers must put aside their own concerns to be fully with their clients. Of course, Rogers pointed out that this deeper understanding of clients remains sterile unless it is somehow communicated to them. Although clients can appreciate how intensely they are attended and listened to, they and their concerns still need to be understood. Empathic listening begets empathic understanding, which begets empathic responding.

Although Rogers's point of view on empathy adds great richness to the idea of listening to and understanding clients, it is not necessary to agree with his contention, mentioned in Chapter 3, that the quality of the helping relationship with respect to the unconditional positive regard, accurate empathy, and genuineness offered by the helper and perceived by the client is both "necessary and sufficient" for therapeutic progress. Empathic participation in the world of another person obviously admits of degrees. As a helper, you must be able to enter clients' worlds deeply enough to understand their struggles with problem situations or their search for opportunities with enough depth to make your participation in problem management and opportunity development valid and substantial. If your help is based on an incorrect or invalid understanding of the client, then your helping may lead him or her astray. If your understanding is valid but superficial, then you might miss the central issues of the client's life.

## Tough-Minded Listening: Hearing the Slant or Spin

Skilled helpers not only observe the experiences, behaviors, and emotions of clients within the helping sessions and listen to their stories but also listen to any *slant* or *spin* that clients might give their stories. Although clients' visions of and feelings about themselves, others, and the world are real and need to be understood, their perceptions of themselves and their worlds are sometimes distorted. For instance, if a client sees herself as ugly when in reality she is beautiful, her experience of herself as ugly is real and needs to be listened to and understood. But her experience of herself does not square with the facts. This, too, must be listened to and understood. If a client sees himself as above average in his ability to communicate with others when, in reality, he is below average, his experience of himself needs to be listened to and understood, but reality cannot be ignored. Tough-minded listening includes detecting the gaps, distortions, and dissonance that are part of the client's experienced reality. This does not mean that helpers challenge clients as soon as they hear any kind of distortion. Rather, they note gaps and distortions and challenge them when it is appropriate to do so (see Chapters 9 and 10).

Denise realizes from the beginning that some of Jennie's understandings of herself and her world are not accurate. For instance, in reflecting on all that has happened, Jennie remarks that she probably got what she deserved. When Denise asks her what she means, she says, "My ambitions were too high. I was getting beyond my place in life." This is the slant or spin Jennie gives to her career aspirations. It is one thing to understand how Jennie might put this interpretation on what has happened; it is another to assume that such an interpretation reflects reality. To be client-centered, helpers must first be reality-centered.

## THE SHADOW SIDE OF LISTENING TO CLIENTS

Active listening is not as easy as it sounds. Obstacles and distractions abound. The following kinds of ineffective listening, as you will see from your own experience, overlap with one another.

**Inadequate listening.** In conversations it is easy for us to be distracted from what other people are saying. We get involved in our own thoughts, or we begin to think about what we are going to say in reply. At such times we may get the "You're not listening to me!" exclamation mentioned earlier. Helpers, too, can become preoccupied with themselves and their own needs in such a way that they are kept from listening fully to their clients. They are attracted to their clients, they are tired or sick, they are preoccupied with their own problems, they are too eager to help, they are distracted because clients have problems similar to their own, or the social and cultural differences between them and their clients make listening and understanding difficult. It is sometimes difficult to remember that each client is new and unique and that each session with the same client is a new challenge.

**Evaluative listening.** Most people, even when they listen attentively, listen evaluatively. That is, as they listen, they are judging what the other person is saying as good-bad, right-wrong, acceptable-unacceptable, likable-unlikable, relevant-irrelevant, and so forth. Helpers are not exempt from this universal tendency. The following interchange took place between Jennie and a friend of hers. Jennie recounted it to Denise as part of her story.

**JENNIE:** Well, the rape and the investigation are not dead, at least not in my mind. They are not as vivid as they used to be, but they are there.

**FRIEND:** That's the problem, isn't it? Why don't you do yourself a favor and forget about it? Get on with life, for God's sake!

That might well be sound advice, but the point here is that Jennie's friend listened and responded evaluatively. Clients should first be understood, then, if necessary, challenged or helped to challenge themselves. Evaluative listening, translated into advice giving, will just put clients off. Of course, understanding the client's point of view is not the same as accepting it. Indeed, a judgment that a client's point of view, once understood, needs to be expanded or transcended or that a pattern of behavior, once listened to and understood, needs to be altered can be quite useful. That is, there are productive forms of evaluative listening. It is practically impossible to suspend judgment completely. Nevertheless, it is possible to set one's judgment aside for the time being in the interest of understanding clients, their worlds, and their points of view.

**Filtered listening.** It is impossible to listen to other people in a completely unbiased way. Through socialization we develop a variety of filters through which we listen to ourselves, others, and the world around us. As Hall (1977) noted: "One of the functions of culture is to provide a highly selective screen between man and the outside world. In its many forms, culture therefore designates what we pay attention to and what we ignore. This screening provides structure for the world" (p. 85). We need filters to provide structure for ourselves as we interact with the world. But personal, familial, sociological, and cultural filters introduce various forms of bias into our listening and do so without our being aware of it.

The stronger the cultural filters, the greater the likelihood of bias. For instance, a white, middle-class helper probably tends to use white, middle-class filters in listening to others. Perhaps this makes little difference if the client is also white and middle class, but if the helper is listening to an Asian client who is well-to-do and has high social status in his community, to an African American mother from an urban ghetto, or to a poor white subsistence farmer, then the helper's cultural filters might introduce bias. Prejudices, whether conscious or not, distort understanding. Like everyone else, helpers are tempted to pigeonhole clients because of gender, race, sexual orientation, nationality, social status, religious persuasion, political preferences, lifestyle, and the like. In Chapter 1 the importance of self-knowledge on the part of the helper was noted. This includes ferreting out the biases and prejudices that distort our listening.

**Labels as filters.** I remember my reaction to hearing a doctor refer to me as the "hernia in 304." Sometimes book learning can act as a distorting filter. The very labels we learn in our training—paranoid, neurotic, sexual disorder, borderline—can militate against empathic understanding. Books on personality theories provide us with pigeonholes. We even pigeonhole ourselves! Diagnostic categories can take precedence over the clients being diagnosed. Helpers forget at times that their labels are interpretations rather than understandings of the client's experience. If what you "hear" is the theory and not the person, you can be "correct" in your diagnosis and once more lose the client. In short, what you learn as you study psychology may help you to organize what you hear, but it may also distort your listening. To use terms borrowed from Gestalt psychology, make sure that your client remains "figure" (in the forefront of your attention) and that models and theories about clients remain "ground" (learnings that remain in the background and are used only in the interest of understanding and helping this unique client).

**Fact-centered rather than person-centered listening.** Some helpers ask many informational questions, as if the client would be cured if enough facts about him or her were known. It's entirely possible to collect facts but miss the person. The antidote is to listen to clients contextually, trying to focus on themes and key messages. Denise, as she listens to Jennie, picks up what is called the "pessimistic explanatory style" theme (Peterson, Seligman, & Vaillant, 1988). This is the tendency to attribute causes to negative events that are stable ("It will never go away"), global ("It affects everything I do"), and internal ("It is my fault"). Denise knows that the research indicates that people who fall victim to this style tend to end up with poorer health than those who do not. There may be a link, she hypothesizes, between Jennie's somatic complaints (headaches, gastric problems) and this explanatory style. This is a theme worth exploring.

**Rehearsing.** When beginning helpers ask themselves, "How am I to respond to what the client is saying?" they stop listening. When experienced helpers begin to mull over the "perfect response" to what their clients are saying, they stop listening. Helping is more than the "technology" found in these pages. It is also an art. Helpers who listen intently to clients and to the themes and core messages embedded in what they are saying, however haltingly or fluently they say it, are never at a loss in responding. They don't need to rehearse. And their responses are much more likely to help clients move forward in the problem-management process. When the client stops speaking, they often pause to reflect on what he or she just said and then speak. But this is not rehearsing.

**Sympathetic listening.** Often clients are people in pain or people who have been victimized by others or by society itself. Such clients can, understandably, arouse feelings of sympathy in helpers. Sometimes these feelings are strong enough to distort the stories that are being told. Consider this case.

> Liz was counseling Ben, a man who had lost his wife and daughter to a tornado. Liz had recently lost her husband to cancer. As Ben talked about his own tragedy during their first meeting, she wanted to hold him. Later that day she took a long walk and realized how her sympathy for Ben had distorted what she heard. She heard the depth of his loss, but, reminded of her own loss, only half heard the implication that his loss now excused him from getting on with his life.

Sympathy has an unmistakable place in human transactions, but its "use," if that does not sound too inhuman, is limited in helping. In a sense, when I sympathize with someone, I become his or her accomplice. If I sympathize with my client as she tells me how awful her husband is, I take sides without knowing what the complete story is. Helpers should not become accomplices in letting client self-pity drive out problem-managing action.

**Interrupting.** I am reluctant to add "interrupting," as some do, to this list of obstacles. Certainly, by interrupting clients, the helper stops listening. And interrupters often deliver messages they have been rehearsing, which means that they have been only partially listening. My reluctance, however, comes from the conviction that helping goes best when it is a dialogue between client and helper. I seldom find monologues, on the part of either client or helper, useful in counseling. Occasionally, monologues that help clients get their stories or significant updates of their stories out are useful. Even then, such a monologue is best followed by a fair amount of dialogue. Therefore, I see benign and malignant forms of interrupting. The helper who cuts the client off in midthought because he has something important to say is using a malignant form. But the case is different when a helper "interrupts" a monologue with some gentle gesture and a comment such as, "You've made several points. I want to make sure that I've understood them." If interrupting promotes the kind of dialogue that serves the problem-management process, then it is useful. Still, care must be taken to factor in cultural differences in storytelling.

## LISTENING TO ONESELF: THE HELPER'S SHADOW CONVERSATION

The conversation helpers have with themselves during helping sessions is the "shadow conversation." To be an effective helper, you need to listen not only to the client but also to yourself. Granted, you don't want to become self-preoccupied, but listening to yourself on a "second channel" can help you identify both what you might do to be of further help to the client and what might be standing in the way of your being with and listening to the client. It is a positive form of self-consciousness.

Some years ago this second channel did not work very well for me. A friend of mine who had been in and out of mental hospitals for a few years and whom I had not seen for over six months showed up unannounced one evening at my apartment. He was in a highly excited state. A torrent of ideas, some outlandish, some brilliant, flowed nonstop from him. I sincerely wanted to be with him as best I could, but I was very uncomfortable. I started by more or less naturally following the guidelines of attending, but I kept catching myself at the other end of the couch on which we were both sitting with my arms and legs crossed. I think that I was defending myself from the torrent of ideas. When I discovered myself almost literally tied up in knots, I would un-twist my arms and legs, only to find them crossed again a few minutes later. It was hard work being with him. In retrospect, I realize I was concerned for my own security. I have since learned to listen to myself on the second chan-nel a little better — to my nonverbal behaviors as well as my internal dia-logues — so that these interactions might serve clients better.

Helpers can use this second channel to listen to their own verbal mes-sages with themselves, their nonverbal behavior, and their feelings and emo-tions. These messages can refer to the helper, the client, or the relationship.

- "I'm letting the client get under my skin. I had better do something to reset the dialogue."
- "My mind has been wandering. I'm preoccupied with what I have to do tomorrow. I had better put that out of my mind.
- "Here's a client who has had a tough time of it, but her self-pity is stand-ing in the way of her doing anything about it. But I had better go slow."
- "It's not clear that this client is interested in changing. It's time to test the waters."

The point is that this shadow conversation goes on all the time. It can be a distraction or it can be another tool for helping. The client, too, is having his or her shadow conversation. One intriguing study (Hill, Thompson, Cogar, and Denman, 1993) suggested that both client and therapist are more or less aware of the other's "covert processes." In this study therapists, though know-ing that clients left things unsaid, were not very good at determining what those things were. At times there are verbal or nonverbal hints as to what the client's internal dialogue might be. Helping clients get key messages from their shadow conversation into the helping dialogue is a key task and will be discussed in Chapters 9, 10, and 11.

## CHAPTER 5

# BASIC EMPATHY

THE THREE DIMENSIONS OF RESPONDING SKILLS:
PERCEPTIVENESS, KNOW-HOW, AND ASSERTIVENESS

BASIC EMPATHY: COMMUNICATING UNDERSTANDING TO
CLIENTS

THE KEY ELEMENTS OF BASIC EMPATHY

The Basic-Empathy Formula

Experiences, Behaviors, and Feelings as Elements of
Empathy

PRINCIPLES TO GUIDE THE USE OF BASIC EMPATHY

Supporting Every Stage and Step

Stimulating Movement

Targeting Core Messages

Responding to Context

Recovering from Inaccuracies

Pretending to Understand

POOR SUBSTITUTES FOR EMPATHY

TACTICS FOR COMMUNICATING EMPATHY

A CAUTION: THE IMPORTANCE
OF EMPATHIC RELATIONSHIPS

The advantage of attending and listening is twofold. First, since attending and listening are ways of translating or implementing the value of respect, they contribute to relationship building. Second, they lay the foundation for the helper's responses to the client. Basic empathy and probing are response skills.

- Basic empathy involves *listening* to clients, *understanding* them and their concerns to the degree that this is possible, and *communicating* this understanding to them so that they might *understand themselves* more fully and *act* on their understanding.
- Probing involves *statements* and *questions* on the part of the helper that enable clients to *explore more fully* any relevant issue at any stage or step of the helping process.

Responding skills such as empathy and probing have three dimensions or requirements as described in the following section.

## THE THREE DIMENSIONS OF RESPONDING SKILLS: PERCEPTIVENESS, KNOW-HOW, AND ASSERTIVENESS

The communication skills involved in responding to clients have three dimensions: perceptiveness, know-how, and assertiveness. These three dimensions are discussed in this chapter as they apply to basic empathy and in Chapters 9 and 10 as they apply to challenging skills.

**Perceptiveness.** Your responding skills are only as good as the accuracy of the perceptions on which they are based. Consider the difference between these two examples.

> Mario, a manager, is counseling Enrique, a relatively new member of his team. During the past week, Enrique has made a significant contribution to a major project, but he has also made one rather costly mistake. Enrique's mind is on his blunder, not his success. Mario, sensing Enrique's discomfort, says, "Your ideas in the meeting last Monday helped us reconceptualize and reset the entire project. It was a great contribution. That kind of 'out of the box' thinking is very valuable here. Your conversation with Acme's purchasing agent on Wednesday made him quite angry. Something tells me that you might be more worried about Wednesday's mistake than delighted with Monday's contribution. I just wanted to let you know that I'm not." Enrique is greatly relieved. They go on to have a useful dialogue about what made Monday so good and what could be learned from Wednesday's blunder.

Mario's perceptiveness and his ability to defuse a tense situation lay the foundation for an excellent dialogue. Note the contrast in the next example.

Beth is counseling Frank in a community mental-health center. Frank is scared to talk about an "ethical blunder" that he made at work. Beth senses his discomfort but thinks that he is angry rather than scared. She says, "Frank, I'm wondering what's making you so angry right now." Since Frank does not feel angry, he says nothing. He's startled by what she says and feels even more insecure. Beth takes his silence as a confirmation of his "anger." She tries to get him to talk about it.

Beth's perception is wrong and disrupts the helping process. She misreads Frank's emotional state and tries to set a course based on her flawed perception.

The kind of perceptiveness needed to be a good helper comes from basic intelligence, social intelligence, effective attending, and careful listening. It does not come automatically with experience.

**Know-how.** Once you are aware of what kind of response is called for, you need to be able to deliver it. For instance, if you are aware that a client is anxious and confused because this is his first visit to a helper, it does little good if your understanding remains locked up inside you. Let's return to Frank and Beth for a moment.

Frank and Beth end up arguing about his "anger." Frank finally gets up and leaves. Beth, of course, takes this as a sign that she was right in the first place. The next day Frank goes to see his minister. The minister sees quite clearly that Frank is scared and confused. His perceptions are right. He says something like this: "Frank, you seem to be very uncomfortable. It may be that whatever is on your mind might be difficult to talk about. But I'd be glad to listen to it, whatever it is. But I don't want to push you into anything." "But it's terrible," Frank blurts out." "I know it's terrible for you, Frank, but my guess is that it's also terribly human." Frank hesitates a bit, then leans back into his chair, takes a deep breath, and launches into his story.

The minister not only is perceptive but also knows how to address Frank's anxiety and hesitation. The minister says to himself in his shadow conversation, "Here's a man who is almost exploding with the need to tell his story, but fear or shame or something like that is paralyzing him. How can I put him at ease, let him know that he won't get hurt here? I need to recognize his anxiety and gently offer an opening."

**Assertiveness.** Accurate perceptions and excellent know-how are meaningless unless they are actually used when called for. If you see that self-doubt is a theme that weaves itself throughout a client's story and search for a better future and if you know how to challenge him to explore this tendency but fail to do so, you do not pass the assertiveness test. Your skills remain locked up inside you. Consider this example.

Nina, a young counselor in the Center for Student Development, is in the middle of the first session with Antonio, a graduate student. During the session, he mentions briefly a very helpful session he had had the previous year with Carl, a middle-aged counselor on the staff. But Carl has accepted an academic position at the university and is no longer involved with the Center. Nina realizes that he is disappointed that he couldn't see Carl and probably has some misgivings about being helped by a relatively new, younger woman. She has faced sensitive issues like this before and would not take it amiss if Antonio were to choose a different counselor. During a lull in the conversation, she says something like this: "Antonio, could we take a 'time-out' here for a momen? I think you were disappointed to find out that Carl was no longer here. Or at least I would be if I were in your shoes. You were just more or less 'assigned' to me and I'm not sure the fit is right. Maybe you can give that a bit of thought. Then, if you think I can be of help, you can schedule another meeting with me. But you're certainly free to review who is on staff and choose whomever you want."

In this case, perceptiveness, know-how, and assertiveness all come together. This is not to suggest that assertiveness is an overriding value in and of itself. To be assertive without perceptiveness and know-how is to court disaster.

## BASIC EMPATHY: COMMUNICATING UNDERSTANDING TO CLIENTS

If attending and listening are the skills that enable helpers to get in touch with the world of the client, then basic empathy is the skill that enables them to communicate their understanding of that world. The term *basic empathy* is used to distinguish it from empathic listening, discussed in Chapter 4, and from another form of empathy, "advanced" empathy, to be discussed in Chapter 10. A secure starting point in helping others is listening to them carefully, struggling to understand their concerns, and sharing that understanding with them. When clients are asked what they find helpful in counseling interviews, being understood gets top ratings.

The confusion surrounding empathy was mentioned in Chapter 4. Some writers wax lyrical about empathy and its place in human commerce: "Empathy, the accepting, confirming, and understanding human echo evoked by the self, is a psychological nutrient without which human life, as we know and cherish it, could not be sustained" (Kohut, 1978, p. 705). Empathy, then, becomes a value, a philosophy, a cause with almost religious overtones. Covey (1989), naming empathic communication one of the "seven habits of highly effective people," said that empathy provides "psychological air"; that is, it helps people breathe more freely in their relationships. Care must be taken, however, not to make a cult out of empathy. It's certainly not a miracle pill. I have a great deal of sympathy with the views expressed by Rogers (see Chapter 4), Kohut, and Covey. Although many people may "feel" empathy for others, the truth is that few know how to put it into words. Empathy as a communication of understanding of the other remains an improbable event in everyday life. Perhaps that's why it is so powerful in helping settings. There is such an unfulfilled need to be understood. The sections that follow ignore the confusion and mysticism surrounding the concept of empathy and continue what was started in the previous chapter—that is, to deal with it as a fully human *communication skill* that is very helpful in promoting both the values and the goals of the helping process.

## THE KEY ELEMENTS OF BASIC EMPATHY

This section is a kind of anatomy lesson. That is, we are going to take basic empathy apart and look at the pieces. Further on, we'll put them back together again.

### The Basic-Empathy Formula

Basic empathic understanding can be expressed in the following stylized formula:

> You *feel* . . . [here name the correct emotion expressed by the client] . . . *because* (or *when*) . . . [here indicate the correct experiences and behaviors that give rise to the feelings]. . . .

The formula is a beginner's tool to get used to the concept of empathy. The formula is used in the following examples. For the moment, ignore how stylized it sounds. Ordinary human language will be substituted later. In the first example, a divorced mother with two young kids is talking to a social worker about her ex-husband.

**CLIENT:** I could kill him! He failed to take the kids again last weekend. This is three times out of the last six weeks.

**HELPER:** You feel furious because he keeps failing to hold up his part of the bargain.

His not taking the kids according to their agreement [an experience for the client] infuriates her [an emotion].

In the following example, a woman who is going to have a hysterectomy is talking with a hospital counselor the night before surgery.

**PATIENT:** God knows what they'll find when they go in tomorrow. I keep asking questions, but they keep giving me vague answers.

**HELPER:** You feel troubled about what they're not telling you.

The accuracy of the helper's response does not solve the woman's problems, but the patient does move a bit. She shares her concerns and perhaps reduces her anxiety by talking about it.

# Experiences, Behaviors, and Feelings as Elements of Empathy

The key elements of an empathic response are the same as the key elements of the client's story — that is, the experiences, behaviors, and feelings that make up that story. This is the next part of our "anatomy" lesson.

**Respond to the client's feelings.** In the formula, "You feel . . ." is to be followed by the correct *family* of emotions and the correct *intensity*.

- The statements "You feel hurt," "You feel relieved," and "You feel enthusiastic" specify different families of emotion.
- The statements "You feel annoyed," "You feel angry," and "You're furious" specify different degrees of intensity in the same family (anger).

The words *sad, mad, bad,* and *glad* refer to four of the main families of emotion, whereas *sad, very sad,* and *extremely sad* refer to different intensities.

Note that the client may actually be talking about emotions felt in the past — that is, at the time of the event being discussed — *or* expressing feelings about the event that arise during the helping session, or both. For instance, consider this interchange between a client involved in a child custody proceeding and a counselor.

CLIENT (calmly): I get furious with him [affect] when he says things, little snide things, that suggest that I don't take good care of the kids [experience].

HELPER: You feel especially angry when he intimates that you're not a good mother.

The client isn't angry right now. Rather, she is talking about her anger. The following example — a woman is talking about one of her colleagues at work — deals with expressed rather than discussed feelings.

CLIENT (enthusiastically): I threw caution to the wind and confronted him about his sarcasm [action] and it actually worked. He not only apologized, but behaved himself the rest of the trip [experiences for the client].

HELPER: You feel great because you took a chance and it paid off.

It goes without saying that clients don't always name their feelings and emotions. However, if they express emotion, it is part of the message and needs to be identified and understood.

Often helpers have to read their clients' emotions — both the family and the intensity — in their nonverbal behavior. In the following example, a North American student comes to you, sits down, looks at the floor, hunches over, and haltingly says,

CLIENT: I don't even know where to start. (He falls silent).

HELPER: You're feeling pretty miserable, though I'm not sure why.

CLIENT (after a pause): Well, let me tell you why. . . .

You see that he is depressed [affect], and his nonverbal behavior indicates that the feelings are quite intense. His nonverbal behavior reveals the broad family ("You feel bad") and the intensity ("You feel very bad"). Of course, you do not yet know the experiences and behaviors that give rise to these emotions.

Naming and discussing feelings threatens some clients. In this case, it might be better to focus on experiences and behaviors and proceed only gradually to a discussion of feelings. The following client, an unmarried man in his mid-30s who has come to talk about "certain dissatisfactions" in his life, has shown some reluctance to express or even to talk about feelings.

CLIENT (in a pleasant, relaxed voice): My mother is always trying to make a little kid out of me. And I'm in my mid-30s! Last week, in front of a group of my friends, she brought out my rubber boots and an umbrella and gave me a little talk on how to dress for bad weather (laughs).

COUNSELOR A: It might be hard to admit it, but I get the feeling that down deep you were furious.

CLIENT: Well, I don't know about that. Anyway, at work. . . .

Counselor A pushes the emotion issue and is met with mild resistance. The client changes the topic.

COUNSELOR B: So she keeps playing the mother role — to the hilt, it would seem.

CLIENT: And the hilt includes not wanting me to grow up. But I am grown up . . . well, pretty grown up.

Counselor B, choosing to respond to the "strong mother" issue rather than the more sensitive "being kept a kid and feeling really lousy about it" issue, gives the client more room to move. This works, for the client himself moves toward the more sensitive issue. Some clients are hesitant to talk about certain emotions. One client might find it relatively easy to talk about his anger but not his hurt. For another client it might be just the opposite. Empathy includes the ability to pick up and deal with these differences.

Finally, keep in mind that in most cases, feelings and emotions arise from the client's experiences and behaviors. Emotions should not be overemphasized or underemphasized. They should be dealt with in an integrated way (see Anderson & Leitner, 1996). Of course, once experienced, emotions go on to drive other behaviors. As Lang (1995) pointed out, they are "action dispositions" (p. 372). They are an important part of the problem situation or the undeveloped opportunity, but they are only a part.

Since clients express feelings in a number of different ways, helpers can communicate an understanding of feelings in a variety of ways.

• **By single words.** You feel good. You're depressed. You feel abandoned. You're delighted. You feel trapped. You're angry.

• **By different kinds of phrases.** You're sitting on top of the world. You feel down in the dumps. You feel left in the lurch. Your back's up against the wall. You're really steaming.

- *By what is implied in behavioral statements.* (What action I feel like taking): You feel like giving up (implied emotion: despair). You feel like hugging him (implied emotion: joy). Now that it's over, you feel like throwing up (implied emotion: disgust).

- *By what is implied in experiences that are revealed.* You feel you're being dumped on (implied feeling: victimized). You feel you're being stereotyped (implied feeling: resentment). You feel you're at the top of her list (implied feeling: elation). You feel you're going to get caught (implied feeling: apprehension). Note that the implication of each could be spelled out: You feel angry because you're being dumped on. You resent the fact that you're being stereotyped. You feel great because it seems that you're at the top of her list.

Since ultimately you must discard formulas and use your own language, words that are *you*, it helps to have a variety of ways of communicating your understanding of clients' feelings and emotions. It keeps you from being wooden in your responses. Consider this example: The client tells you that she has just been given the kind of job she has been looking for for the past two years. Here are some possible responses to her emotion.

- *Single word.* You're really happy.
- *A phrase.* You're on cloud nine.
- *Experiential statement.* You feel you finally got what you deserve.
- *Behavioral statement.* You feel like going out and celebrating.

Obviously, your responses to clients should be you, not canned responses from a textbook. With experience, you can extend your range of expression at the service of your clients.

**Respond to the client's experiences and behaviors.** The "because . . ." in the formula is to be followed by an indication of the experiences and behaviors that underlie the client's feelings. In the following example, the client, a graduate student in law school, is venting his frustration.

CLIENT (heatedly): You know why he got an A? He took my notes and disappeared. I didn't get a chance to study them. And I never even confronted him about it.

HELPER: You feel angry because he stole your course notes and you let him get away with it.

The response specifies both the client's experience [the theft] and his behavior [in this case, a failure to act] that give rise to his distress.

In the next example, the client, who is hearing impaired, has been discussing ways of becoming, in her words, "a full-fledged member of my extended family." The discussion takes place through a combination of lipreading and signing.

CLIENT (enthusiastically): Here's what I intend to do. I'm going to stop fading to the side in family conversation groups. I'll be the best listener there. I'll get my thoughts across even if I have to use props. They're going to find out that I'm actually pretty smart.

HELPER: You feel excited because you've decided to seize your rightful place in family gatherings rather hoping that someone will hand it to you.

The client is setting her agenda (Stage II). The helper's response recognizes that she has made an important decision and expresses her determination to carry it out [decisions as internal actions].

A mugging victim has been talking to a counselor to help cope with his fears of going out. Before the mugging he had given no thought to urban problems. Now he tends to see menace everywhere.

CLIENT: This gradual approach of getting back in the swing seems to be working. Last night I went out without a companion. First time. I have to admit that I was scared. But I think I've learned how to be careful. I'm buying back my freedom bit by bit. Last night was important.

HELPER: You feel satisfied with the approach we've been taking because it paid off last night when you bought back a big chunk of your freedom.

The client is talking about the success of the action phase of the program. The helper's response recognizes the client's satisfaction with the success of the program and how important it is for the client to be both safe and free.

Another client, after a few sessions spread out over six months, says something like this:

CLIENT (talking in an animated way): I really think that things couldn't be going better. I'm doing very well at my new job, and my husband isn't just putting up with it. He thinks it's great. He and I are getting along better than ever, even sexually, and I never expected that. We're both working at our marriage. I guess I'm just waiting for the bubble to burst.

HELPER: You feel great because things have been going better than you ever expected — it's almost too good to be true.

CLIENT: I think there's a difference between being cautious and waiting for disaster to strike. I'll always be cautious, but I'm finding out that I can make things come true instead of sitting around waiting for them to happen as I usually do. I've got to keep making my own luck.

This client, too, talks about her experiences and behaviors and expresses feelings, the flavor of which is captured in the empathic response. The response, capturing as it does both the client's enthusiasm and her lingering fears, is quite useful because the client moves on to her need to make things happen.

The stylized formula "you feel . . . because . . ." has outlived its usefulness and will be dropped in the examples that follow. Experienced trainers use it only when it sounds natural. Otherwise they use ordinary language to express basic empathy. Furthermore, the formula has a flaw. That is, the "because" part comes before the "you feel" part. When ordinary language is used, the right order is easily maintained: "He would always interrupt you at the busiest time of the day and that's what really bugged you."

## Principles to Guide the Use of Basic Empathy

Here are a number of principles that can guide you in your use of empathy. Remember that they are principles, not formulas.

**Use empathy at every stage and step of the helping process.** Basic empathic understanding is a useful response at every stage and every step of the helping process. Here are some examples.

- A teenager in his third year of high school who has just found out that he is moving with his family to a different city: "You feel sad, maybe even a bit betrayed, because moving means leaving all your friends." (Stage I — understanding and clarification)
- A woman who has been discussing the trade-offs between marriage and career: "You feel ambivalent because if you marry Jim, you might not be able to have the kind of career you'd like." (Stage II — options among goals)
- A man who is choosing to try to control his cholesterol level without taking a medicine whose side effects worry him: "You feel relieved because sticking to the diet and exercise might mean that you won't have to take any medicine." (Stage III — choosing action strategies)
- A married couple who have been struggling to put into practice a few strategies to improve their communication with each other: "You feel annoyed with yourselves because you didn't even accomplish the simple active listening goals you set for yourselves." (action phase)

Basic empathy, as a mode of human contact, a relationship builder, a conversational lubricant, a perception-checking intervention, and a mild form of social influence, is always useful. Driscoll (1984), in his common-sense way, referred to empathic statements as "nickel-and-dime interventions which each contribute only a smidgen of therapeutic movement, but without which the course of therapeutic progress would be markedly slower" (p. 90). Since empathy provides a continual trickle of understanding, it is a way of providing support throughout the helping process. It is never wrong to let clients know that you are trying to understand them from their frame of reference. Clients who feel they are being understood participate more effectively and more fully in the helping process. Empathy helps build trust. Basic empathy paves the way for stronger interventions on the part of the helper, such as challenging.

**Respond selectively to core messages.** It is impossible to respond with empathy to everything a client says. Therefore, as you listen to clients, try to identify and respond to what you believe are *core* messages — that is, the *heart* of what the client is saying and expressing, especially if the client speaks at any length. Sometimes this selectivity means paying particular attention to one or two messages even though the client communicates many. For instance, a young woman, in discussing her doubts about marrying her companion, says that she is tired of his sloppy habits, is not really interested in his friends, wonders about his lack of intellectual curiosity, is dismayed at his relatively low level of career aspirations, but vehemently resents the fact that he faults her for being highly ambitious.

COUNSELOR: The whole picture doesn't look very promising, but the mismatch in career expectations is especially troubling.

In this example, the client herself highlights what is core, and the counselor follows her lead. Of course, since clients are not always so obliging, helpers must continually ask themselves, "What is key? What is most important here?" and then find ways of checking it out with the client. This helps clients sort out things that are not clear in their own minds.

At other times selectivity means focusing on experiences *or* actions *or* feelings rather than all three. Consider the following example of a client who is experiencing stress because of his wife's health and because of concerns at work.

CLIENT: This week I tried to get my wife to see the doctor, but she refused, even though she fainted a couple of times. The kids had no school, so they were underfoot almost constantly. I haven't been able to finish a report my boss expects from me next Monday.
HELPER: It's been a lousy, overwhelming week all the way around.
CLIENT: As bad as they come. When things are lousy both at home and at work, there's no place for me to relax. I just want to get the hell out of the house and find some place forget it all. . . . But I can't.

Here the counselor chooses to emphasize the feelings of the client, because she believes that his feelings of frustration and irritation are what is uppermost in his consciousness right now. At another time or with another client, the emphasis might be quite different.

In the next example, a young woman is talking about her problems with her father.

CLIENT: My dad yelled at me all the time last year about how I dress. But just last week I heard him telling someone how nice I looked. He yells at my sister about the same things he ignores when my younger brother does them. Sometimes he's really nice with my mother and other times—too much of the time—he's just awful: demanding, grouchy, sarcastic.
HELPER: The inconsistency is killing you.
CLIENT: Absolutely! It's hard for all of us to know where we stand. I hate coming home when I'm not sure which "dad" will be there. Sometimes I come late to avoid all this. But that makes him even madder.

In this response, the counselor emphasizes the client's experience of her father's inconsistency. It hits the mark and she explores the problem situation further.

Responding to core messages is also the social-influence process. The search for core messages is a selection process. The helper believes that the messages selected for attention are core, not just for himself or herself, but primarily for the client. But the helper also believes, at some level, that certain messages *should* be important for the client. If true dialogue with the client is established, this does not rob clients of their self-responsibility. Everything gets checked out.

**Respond to the context, not just the words.** A good empathic response is not based just on the client's immediate words and nonverbal behavior. It also takes into account the context of what is said, everything that "surrounds" and permeates a client's statement. This client may be in crisis. That client may be doing a more leisurely "taking stock" of where he is in life. You are listening to clients in the context of their lives.

For example, Jeff, a white teenager, is accused of beating a black youth whose car stalled in a white neighborhood. The beaten youth is still in a coma. When Jeff talks to a court-appointed counselor, the counselor listens to what Jeff says in light of Jeff's upbringing and environment. The context includes the racist attitudes of many people in his blue-collar neighborhood, the sporadic violence there, the fact that his father died when Jeff was in primary school, a somewhat indulgent mother with a history of alcoholism, and easy access to soft drugs. The following interchange takes place.

CLIENT: I don't know why I did it. I just did it, me and these other guys. We'd been drinking a bit and smoking up a bit—but not too much. It was just the whole thing.

HELPER: Looking back, it's almost like it's something that happened rather than something you did, and yet you know, somewhat bitterly, that you actually did it.

CLIENT: More than bitter! I've screwed up the rest of my life. It's not like I got up that morning saying that I was going to bash someone that day.

The counselor's response is in no way an attempt to excuse Jeff's behavior, but it does factor in some of the environmental realities. Later on he will challenge Jeff to decide whether his environment—prejudices, gang membership, family history—is to own him or whether, to the degree that this is possible, he is to own his environment.

**Use empathy to stimulate movement in the helping process.** Although empathy is an excellent tool for building the helping relationship, it also needs to serve the goals of the helping process. Therefore, empathy is useful to the degree that it helps the client *move forward*. What does "move forward" mean? That depends on the stage or step in focus. For instance, empathy helps clients move forward in Stage I if it helps them explore a problem situation or an undeveloped opportunity more fully. Empathy helps clients move forward in Stage II to the degree that it helps them identify and explore possibilities for a better future, craft a change agenda, or discuss commitment to that agenda. Moving forward in Stage III means clarifying action strategies, choosing specific things to do, and setting up a plan. In the action phase, moving forward means identifying obstacles to action, overcoming them, and accomplishing goals.

In the following example, a somewhat stressed trainee in a counseling program is talking to his supervisor.

**TRAINEE:** I don't think I'm going to make a good counselor. The other people in the program seem brighter than I am. Others seem to be picking up the knack of empathy faster than I am. I'm still afraid of responding directly to others, even with empathy. I think I should reevaluate my participation in the program.

**TRAINER:** This sense of inadequacy is getting you down, perhaps even enough to make you begin wondering whether you should be here at all.

**TRAINEE:** And yet I know that giving up is part of the problem, part of my style. I'm not the brightest, but I'm certainly not dumb, either. The way I compare myself to others is not very useful. I know that I've been picking up some of these skills. I do attend and listen well. I'm perceptive even though at times I have a hard time sharing these perceptions with others.

When the trainer "hits the mark," the trainee moves forward and explores his tendencies to give up, compare himself unfavorably with others, and underestimate his successes.

In the next example, a young woman visits the student services center at her college to discuss an unwanted pregnancy.

**CLIENT:** And so here I am, two months pregnant. I don't want to be pregnant. I'm not married, and I don't even love the father. To tell the truth, I don't even think I like him. Oh, Lord, this is something that happens to other people, not me! I wake up thinking this whole thing is unreal. Now people are trying to push me toward abortion.

**HELPER:** You're still so amazed that it's almost impossible to accept that it's true. To make things worse, people are telling you what to do.

**CLIENT:** Amazed? I'm stupefied! Mainly, at my own stupidity for getting myself into this. I've never had such an expensive lesson in my life. But I've decided one thing. No one, no one is going to tell me what to do now. I'll make my own decisions.

After the helper's empathy, self-recrimination over lack of self-responsibility leads the client to take a stance on responsible decision making.

Basic empathic statements that hit the mark put pressure on the client to move forward. So basic empathy itself, even though it is a communication of understanding, is also part of the social-influence process discussed in Chapter 3.

**Recover from inaccurate understanding.** Although helpers should strive to be accurate in the understanding they communicate, all helpers can be somewhat inaccurate at times. You may think you understand the client and what he or she has said only to find out, when you share your understanding, that you were off the mark. Therefore, empathy is a perception-checking tool. If the helper's response is accurate, the client often tends to confirm its accuracy in two ways. The first is some kind of verbal or nonverbal indication that the helper is right. That is, the client nods or gives some other nonverbal cue or uses some assenting word or phrase such as "that's right" or "exactly." This happens in the following example, in which a client who has been arrested for selling drugs is talking to his probation officer.

**HELPER:** So your neighborhood makes it easy to do things that can get you into trouble.

**CLIENT:** You bet it does! For instance, everyone's selling drugs. You not only end up using them, but you begin to think about pushing them. It's just too easy.

The second and more substantive way in which clients acknowledge the accuracy of the helper's response is by moving forward in the helping process — for instance, by clarifying the problem situation or preferred-scenario possibilities more fully. In the preceding example, the client not only acknowledges the accuracy of the helper's empathy verbally — "you bet it does" — but, more important, also outlines the problem situation in greater detail. If the helper again responds with empathy, this leads to the next cycle. The problem situation becomes increasingly clear in light of specific experiences, behaviors, and feelings.

On the other hand, when a response is inaccurate, the client often lets the counselor know in different ways: He or she may stop dead, fumble around, go off on a new tangent, tell the counselor "That's not exactly what I meant," or even try to provide empathy for the counselor to get him or her back on track. A helper who is alert to these cues can get back on track. Ben, a man who lost his wife and daughter in a train crash, has been talking about the changes that have taken place since the accident.

**HELPER:** So you don't want to do a lot of the things you used to do before the accident. For instance, you don't want to socialize much anymore.

**BEN** (pausing a long time): Well, I'm not sure that it's a question of wanting to or not. I mean that it takes much more energy to do a lot of things. It takes so much energy for me just to phone others to get together. It takes so much energy sometimes being with others that I just don't try. It's as if there's a weight on my soul a lot of the time.

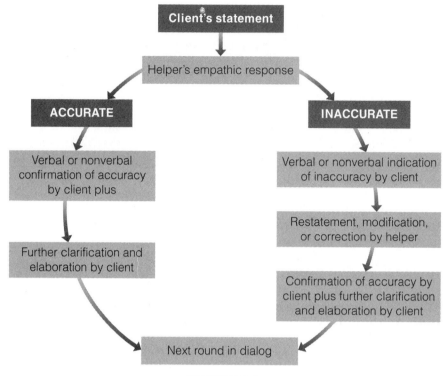

**FIGURE 5-1**
**The Movement Caused by Accurate and Inaccurate Empathy**

**HELPER:** It's like a movie of a man in slow motion — it's so hard to do almost anything.

**BEN:** I'm in low gear, grinding away. And I don't know how to get out of it.

Ben says that it is not a question of motivation but of energy. The difference is important to him. By picking up on it, the helper gets the interview back on track. If you are intent on understanding your clients, they will not be put off by occasional inaccuracies on your part. Figure 5-1 indicates two different paths — one made when helpers hit the mark in their communication of empathy, the other when they are inaccurate and then recover.

**Use empathy as a way of bridging diversity gaps.** This principle is a corollary of the preceding two. Empathy based on effective attending and listening is one of the most important tools you have in interacting with clients who differ from you in significant ways. In this context Scott and Borodovsky (1990) referred to empathic listening as "cultural role taking." They could have said "diversity role taking." In the following example, a younger white male counselor is talking with an elderly African American woman who has recently lost her husband. She is in the hospital with a broken leg.

CLIENT: I hear they try to get you out of these places as quick as possible. But I seem to be lying around here doing nothing. Jimmy [her late husband] wouldn't even recognize me.

HELPER: It's pretty depressing to have this happen so close to losing your husband.

CLIENT: Oh, I'm not depressed. I just want to get out of here and get back to doing things at home. Jimmy's gone, but there's plenty of people around there to help me take care of myself.

HELPER: Getting back into the swing of things is the best medicine for you.

CLIENT: Now you got it right. What I need right now is to know when I can go home and what I need to do for my leg once I get there. I've got to mobilize things.

The helper makes assumptions that might be true for him and his culture, but they miss the mark with the client. She's taking her problems in stride and counting on her social system and a return to everyday household life to keep her going. The helper's second response hits the mark and she, in Stage II fashion, outlines some of her needs.

**Don't pretend to understand.** Clients are sometimes confused, distracted, and in a highly emotional state. All these conditions affect the clarity of what they are saying about themselves. Helpers may fail to pick up what the client is saying because of the client's confusion or because they themselves have become distracted in one way or another. In any case, it's a mistake to feign understanding. Genuine helpers admit that they are lost and then work to get back on track again. A statement like "I think I've lost you. Could we go over that once more?" indicates that you think it important to stay with the client. It is a sign of respect. Admitting that you're lost is infinitely preferable to such clichés as "uh-huh," "ummmm," and "I understand."

## POOR SUBSTITUTES FOR EMPATHY

Many responses that novice or inept helpers make are really poor substitutes for accurate empathy. An example will be used to illustrate a range of poor responses. Robin is a young woman who has just started a career in law. This is her second visit to a counselor in private practice. In the first session she said she wanted to "talk through" some issues relating to the "transition" from school to business life. She appeared quite self-confident. In this session, after talking about a number of transition issues, she begins speaking in a rather strained voice and avoids eye contact with the counselor.

ROBIN: Something else is bothering me a bit. . . . Maybe it shouldn't. After all, I've got the kind of career that a lot of women would die for. Well—I'm glad that none of my feminist colleagues is around—I don't like the way I look. I'm neither fat nor thin, but I don't really like the shape of my body. And I'm uncomfortable with some of my facial features. Maybe this is a strange time of life to start thinking about this. In two years I'll be thirty. . . . I bet I seem like an affluent, self-centered yuppie . . . .

Robin pauses and looks at a piece of art on the wall. What would *you* do or say? Here are some possibilities that are better avoided.

- **No response.** It can be a mistake to say nothing, though cultures differ widely in how they deal with silence (Sue, 1990). In North American culture, generally speaking, if the client says something significant, respond to it, however briefly. Otherwise the client may think that what he or she has just said doesn't merit a response. Don't leave Robin sitting there stewing in her own juices. A skilled helper would realize that a woman's nonacceptance of her body could generalize to other aspects of her life (Dworkin & Kerr, 1987; Worsley, 1981) and therefore should not be treated as just a "vanity" problem.

- **Distracting questions.** A counselor might ask something like, "Is this something new now that you've started working?" This response ignores what Robin has said and the feelings she has expressed and focuses rather on the helper's agenda to get more information.

- **Clichés.** A counselor might say, "The workplace is competitive. It's not uncommon for issues like this to come up." This is cliché-talk. It turns the helper into an instructor and may sound dismissive to the client. Clichés are hollow. The helper is saying, in effect, "You don't really have a problem at all, at least not a serious one."

- **Interpretations.** A counselor might say something like this: "Robin, my bet is that your body-image concerns are probably just a symptom. I've got a hunch that you're not really accepting yourself. That's the real problem." The counselor fails to respond to the client's feelings and also distorts the content of the client's communication. The response implies that what is really important is hidden from the client.

- **Advice.** Another counselor might say, "Hey, don't let this worry you. You'll be so involved with work issues that these concerns will disappear." Advice giving at this stage is out of order and, to make things worse, the advice given has a cliché flavor to it. The values of self-responsibility suggest that advice giving be kept to a minimum.

- **Parroting.** Empathy does not consist of merely repeating what the client has said. Such parroting is a parody of empathy.

COUNSELOR: So, Robin, even though you have a great job, one that many people would envy, it's your feelings about your body that bother you. The feminist in you recoils a bit from this news. But there are things you don't like—your body shape, some facial features. You're wondering why this is hitting you now. You also seem to be ashamed of these thoughts. "Maybe I'm just self-centered," is what you're saying to yourself.

This may be verbally accurate, but it sounds awful. Mere repetition carries no sense of real understanding of, no sense of being with, the client. Since real understanding is in some way "processed" by you, since it passes through you, it should convey some part of yourself. Empathy always adds *something*. To avoid parroting, come at what the client has said from a slightly different angle, use different words, change the order, refer to an expressed but unnamed emotion — in a word, do whatever you can to let the client know that you are *working* at understanding.

   • *Sympathy and agreement.* Being empathic is not the same as agreeing with the client or being sympathetic. An expression of sympathy has much more in common with pity, compassion, commiseration, and condolence than with empathic understanding. Although these are fully human traits, they are not particularly useful in counseling. Sympathy denotes agreement, whereas empathy denotes understanding and acceptance of the person of the client. At its worst, sympathy is a form of collusion with the client. Note the difference between Counselor A's response to Robin and Counselor B's response.

**COUNSELOR A:** This is not an easy thing to struggle with. It's even harder to talk about. It's even worse for someone who is as self-confident as you usually are.

**ROBIN:** I guess so.

Note that Robin responds with halfhearted collusion-talk. The helping process does not move forward.

**COUNSELOR B:** You've got some misgivings about your bodily characteristics, yet you wonder whether you're even justified talking about it.

**ROBIN:** I know. It's like I'm ashamed of my being ashamed. What's worse, I get so preoccupied with my body that I stop thinking of myself as a person. It blinds me to the fact that I more or less like the person I am.

Counselor B's response gives Robin the opportunity to deal with her immediate anxiety and then to explore her problem situation more fully.

## TACTICS FOR COMMUNICATING EMPATHY

The principles outlined previously provide strategies for the use of basic empathy. Here are a few hints — tactics, if you will — to help you improve the quality of your empathic responses.

   • *Give yourself time to think.* Beginners sometimes jump in too quickly with an empathic response when the client pauses. "Too quickly" means that they do not give themselves enough time to reflect on what the client has just said in order to identify the core message being communicated. Even the experts pause and allow themselves to assimilate what the client is saying.

   • *Use short responses.* I find that the helping process goes best when I engage the client in a dialogue rather than give speeches or allow the client to ramble. In a dialogue the helper's responses can be relatively frequent, but lean and trim. In trying to be accurate, the beginner is often long-winded, especially if he or she waits too long to respond. Again, the question "What is the core of what this person is saying to me?" can help you make your responses short, concrete, and accurate.

• *Gear your response to the client, but remain yourself.* If a client speaks animatedly, telling the helper of his elation over various successes in his life, and she replies accurately but in a flat, dull voice, her response is not fully empathic. This does not mean that helpers should mimic their clients. It means that part of being with the client is sharing in a reasonable way in his or her emotional tone. Consider this example:

**TWELVE-YEAR-OLD CLIENT:** My teacher started picking on me from the first day of class. I don't fool around more than anyone else in class, but she gets me anytime I do. I think she's picking on me because she doesn't like me. She doesn't yell at Bill Smith, and he acts funnier than I do.

**COUNSELOR A:** This is a bit perplexing. You wonder why she singles you out for so much discipline.

Counselor A's language is stilted, not in tune with the way a 12-year-old speaks.

**COUNSELOR B:** You're mad because the way she picks on you seems unfair.

On the other hand, helpers should not adopt a language that is not their own just to be on the client's wavelength. A white counselor speaking African American slang or vice versa sounds ludicrous.

# A Caution: The Importance of Empathic Relationships

In day-to-day interpersonal communication, empathy is a tool of civility. Making an effort to get in touch with another's frame of reference sends a message of respect. Therefore, empathy plays an important part in building relationships. It gives concrete expression to the value of respect. However, the communication skills as they are practiced in helping settings don't automatically transfer to the ordinary social settings of everyday life. In everyday life, empathy does not necessarily have to be put into words. Given enough time, people can establish empathic relationships with one another in which understanding is communicated in a variety of rich and subtle ways without necessarily being put into words. A simple glance across a room as one spouse sees the other trapped in a conversation with a person he or she does not want to be with can communicate worlds of understanding. The glance says, "I know you feel caught. I know you don't want to hurt the other person's feelings. I can feel the struggles going on inside you. I also know that you'd like me to rescue you, if I can do so tactfully."

People with empathic relationships often express empathy in actions. An arm around the shoulders of someone who has just suffered a defeat can

BOX 5-1
# Suggestions for the Use of Empathy

1. Remember that empathy is, ideally, a way of being and not just a professional role or communication skill.

2. Attend carefully, both physically and psychologically, and listen to the client's point of view.

3. Try to set your judgments and biases aside for the moment and walk in the shoes of the client.

4. As the client speaks, listen especially for core messages.

5. Listen to both verbal and nonverbal messages and their context.

6. Respond fairly frequently, but briefly, to the client's core messages.

7. Be flexible and tentative enough that the client does not feel pinned down.

8. Use empathy to keep the client focused on important issues.

9. Move gradually toward the exploration of sensitive topics and feelings.

10. After responding with empathy, attend carefully to cues that either confirm or deny the accuracy of your response.

11. Determine whether your empathic responses are helping the client remain focused while developing and clarifying important issues.

12. Note signs of client stress or resistance; try to judge whether these arise because you are inaccurate or because you are too accurate.

13. Keep in mind that the communication skill of empathy, however important, is a tool to help clients see themselves and their problem situations more clearly with a view to managing them more effectively.

be filled with both support and empathy. I was in the home of a poor family when the father came bursting through the front door shouting, "I got the job!" His wife, without saying a word, went to the refrigerator, got a bottle of beer with a makeshift label on which "Champagne" had been written and offered it to her husband. Beer never tasted so good. Some people do enter caringly into the world of another and are "with" him or her but are unable to communicate understanding through words. Of course, the more frequent use of verbal empathy in everyday life is highly desirable. Verbal empathy can play an important role in developing empathic relationships. Box 5-1 summarizes the main points about the use of empathy as a communication skill.

# THE ART OF PROBING AND SUMMARIZING

In most of the examples used in the discussion of empathy, clients have demonstrated a willingness to explore themselves and their behavior relatively freely. Obviously, this is not always the case. Although it is essential that helpers respond with empathy to their clients when they do reveal themselves, it is also necessary at times to encourage, prompt, and help clients to explore their concerns when they fail to do so spontaneously. Therefore, the ability to use prompts and probes well is another important communication skill. Prompts and probes are verbal and sometimes nonverbal tactics for helping clients talk about themselves and define their concerns more concretely through specific experiences, behaviors, and feelings and the themes that emerge from an exploration of these. Since the purpose of probes is to help clients explore issues more fully, they are useful in every stage and step of the helping process. They can be used to help clients explore different goals just as they can be used to help them explore initial concerns. Probes, judiciously used, provide focus and direction for the entire helping process. They make it more efficient, and that benefits everyone.

**The different forms of probes.** Some authors make a distinction between prompts and probes. Here the simple term *probe* will suffice, since prompting and probing have the same objective—to help clients name, take notice of, explore, clarify, or further define some issue in the interest of moving forward in the process of constructive change. Probes can take different forms.

- *Statements:* "It's not clear to me which of these two options you would choose."
- *Requests:* "Tell me what you mean when you say that 'three's a crowd' at home."
- *Questions:* "What obstacles might interfere with your implementing this plan?"
- *Single words or phrases* that are, reductively, requests or questions.
- *Nonverbal prompts:* For example, the helper says nothing but, rather, simply leans forward attentively in response to what the client has said.

Often empathy that really hits the mark also has the impact of a probe. It says, in effect, "I've understood your point of view; tell me more; move forward." Although these forms are illustrated in the examples that follow, they are all, directly or indirectly, questions or requests.

**Some cautions about questions.** Before we review the principles that guide the use of probes, a word about questions, one form of probe, is in order. Helpers often ask too many questions. When in doubt about what to say or do, novice or inept helpers tend to ask questions that add no value. It is as if gathering information were the goal of the helping interview. Social intelligence calls for restraint. It is not that questions, judiciously used, cannot be an important part of your interactions with clients. Here are two guidelines.

• *Do not ask too many questions.* When clients are asked too many questions, they feel grilled, and that does little for the helping relationship. Furthermore, many clients instinctively know when questions are just filler, used because the helper does not have anything better to say. I have caught myself asking questions the answers to which I didn't even want to know! Let's assume that the helper working with Robin, the young woman exploring her concerns about looks and body image, asks her a whole series of questions:

- ○ "When did you first feel like this?"
- ○ "Have you discussed this with anyone?"
- ○ "What do you do to improve your looks?"
- ○ "What is it about your looks that you think others don't like?"

Robin would have every right to say, "Goodbye, no thanks" in response to these intrusions. Helping sessions were never meant to be question-and-answer sessions that go nowhere.

• *Ask open-ended questions.* As a general rule, ask open-ended questions—that is, questions that require more than a simple yes or no or similar one-word answer. Not "Now that you've decided to take early retirement, do you have any plans?" but "Now that you've decided to take early retirement, what are your plans?" Counselors who ask closed questions find themselves asking more and more questions. One closed question begets another. Of course, if a specific piece of information is needed, then a closed question may be used. A career counselor might ask, "How many jobs have you had in the past two years?" The information is relevant to helping the client draw up a résumé and a job-search strategy. Open-ended questions help clients fill in things that are being left out of their story, whether experiences, behaviors, or feelings. Of course, occasionally, a sharp closed question can have the right impact: "Is that what you *really* want?"

## PRINCIPLES IN THE USE OF PROBES

Here, then, are some principles that can guide you in the use of probes. They apply to all probes, including questions.

### Use Probes to Help Clients Achieve Concreteness and Clarity

Probes can help clients turn what is abstract and vague into something concrete and clear—something you can get your hands on and work with.

CLIENT: He treats me badly, and I don't like it!

HELPER: What does he actually do?

CLIENT: He talks about me behind my back—I know he does. Others tell me what he says. He also cancels dates when something more interesting comes up.

In this example, the helper's probe leads to a clearer statement of the client's experience. In the next example, a simple probe leads to a significant revelation.

**CLIENT:** I do funny things that make me feel good.

**HELPER:** What kinds of things?

**CLIENT:** Well, I daydream about being a hero, a kind of tragic hero. In my daydreams I save the lives of people whom I like but who don't seem to know I exist. And then they come running to me but I turn my back on them. I choose to be alone! I come up with all sorts of variations of this theme.

The helper's probe leads to a clearer statement of the client's internal behaviors. The client's fantasy life as outlined could be an important part of the problem situation.

The next client has become the breadwinner since her husband suffered a stroke.

**CLIENT:** Since my husband had his stroke, coming home at night is rather difficult for me.

**HELPER:** It gets you down. . . . What's it like?

**CLIENT:** When I see him sitting immobile in the chair, I'm filled with pity for him and the next thing I know it's pity for myself and it's mixed with anger or even rage, but I don't know what or whom to be angry at. I don't know how to focus my anger. Good God, he's only forty-two and I'm only forty!

In this case, the helper's probe leads to a fuller description of the client's feelings and emotions. In each of these cases, the client's story gets more specific. Of course, the goal is not to get more and more detail. Rather, it is to get the kind of detail that makes the problem or unused opportunity clear enough to see what can be done about it.

## Use Probes to Help Clients Fill in Missing Pieces of the Picture

Probes help clients identify missing pieces of the puzzle—experiences, behaviors, and feelings that would help both clients and helpers get a better fix on the problem situation. The client in the following example is at odds with his wife over his mother-in-law's upcoming visit.

**HELPER:** I realize now that you often get angry when your mother-in-law stays for more than a day. But I'm still not sure what she does that makes you angry.

**CLIENT:** First of all, she throws our household schedule out and puts in her own. Then she provides a steady stream of advice on how to raise the kids. My wife sees this as an "inconvenience." For me it's a total family disruption. When she leaves, there's a residue.

This client's experience of his mother-in-law's behavior has been missing. Once this has been spelled out in some detail, it is easier to understand why he gets so angry. It is still not clear what he does when he sees his mother-in-law taking over.

In the next example, a divorced woman is talking about the turmoil that takes place when her ex-husband visits the children.

**HELPER:** The Sundays your husband exercises his visiting rights with the children end in his taking verbal potshots at you, and you get these headaches. I've got a fairly clear picture of what *he* does when he comes over, but it might help if *you* could describe what you do.

**CLIENT:** Well, I do nothing.

**HELPER:** So last Sunday he just began letting you have it for no particular reason. Or just to make you feel bad.

**CLIENT:** Well . . . not exactly. I asked him about increasing the amount of the child-support payments. And I asked him why he's dragging his feet about getting a better job. He's so stupid. He can't even take a bit of sound advice.

Through probes the counselor helps the client fill in a missing part of the picture—her own behavior. She keeps describing herself as total victim and her ex-husband as total aggressor, and that may not be the full story.

Next we have a mother of four young kids talking to a friend who does a lot of volunteer work at their church.

**MOTHER** (in a matter-of-fact voice): At the end of the day, what with the kids and dinner and cleaning up, I'm not at my best.

**FRIEND:** Meaning that . . .

**MOTHER:** Meaning that I'm not just tired . . . (she becomes agitated) . . . but also angry, even hurt. . . . My husband and the kids do practically nothing to help me! . . . I shouldn't have to ask.

A phrase helps her reveal unnamed, unexpressed, and, most likely, unmanaged emotions—anger, hurt, and resentment.

## Use Probes to Help Clients Get a Balanced View of Problem Situations

This is part of filling out the picture. Clients, in their eagerness to discuss an issue or make a point, often describe one side of a picture or one viewpoint. Probes can be used to help them fill out the picture. Consider the following exchange between Robin and her counselor.

**ROBIN:** I'm finding some dilemmas in all of this. I know I'm ambitious, but I never saw myself as really materialistic. . . . Getting preoccupied with looks doesn't seem to fit.

**COUNSELOR:** So your principles say don't fall into the materialism trap. . . . What do your principles say on the positive side?

**ROBIN** (pausing): Well, I don't know. . . . Well, that I *should* care about how I look. As a professional. . . . Just as a human being.

Robin has given a negative tone to the whole discussion. The counselor asks her to look at the other side of the coin.

## Use Probes to Help Clients Move into Beneficial Stages and Steps of the Helping Process

Clients do not naturally move into whatever stage or step of the helping process might be most useful for them. Probes can help them do so. Even such responses as "uh-huh," "mmm," "yes," "I see," "ah," and "oh," as well as nods and silence, can serve as prompts, provided they are used intentionally and are not a sign that the helper's attention is flagging or that he or she does not know what else to do.

**CLIENT** (hesitatingly): I don't know if I can tell you this. I haven't told it to anyone.

**HELPER** (The helper says nothing but maintains good eye contact and leans slightly forward.)

**CLIENT** (pauses): Well, I had an abortion a few years ago. I was living in another state. But no one here knows. The men I've become friendly with don't know. I feel that a man would think less of me if he were to find out that I had been keeping this from him. . . . But I also think that he would think less of me if I *did* tell him. I'm caught.

The helper's simple gesture conveys an invitation to discuss a sensitive issue, to move more fully into Stage I. It is a low-key way of saying, "Tell me; you can trust me." But the responsibility is left to the client. Probes, though a form of influence, are not ways of extorting from clients things they don't want to give. Statements that have the flavor of "Oh, come on, tell me! It's really not going to hurt" are forms of extortion. Use probes to help increase rather than decrease the client's initiative.

In the next example, the counselor has helped a middle-aged couple, Sean and Fiona, move beyond complaining about each other to finding ways they might "reinvent" their marriage (Stages II and III).

**COUNSELOR:** What attitudes or points of view or pastimes do you have in common that might provide a basis for doing things together? Does that make any sense?

**FIONA:** I can think of something, though it might sound stupid to him. We both care about other people. Before we were married we talked about spending some time in the Peace Corps together, though it never happened.

**SEAN:** Come on. It doesn't sound stupid. We've never talked about it, but it's something we both wanted to do. . . . But those days are past.

COUNSELOR: Are they? The Peace Corps may not be an option, but are there other possibilities? (Neither Fiona nor Sean says anything.) Here are a couple of pieces of paper. Jot down three ways of helping others. Do your own list. Forget what the other one might think.

The counselor uses probes to get Sean and Fiona to brainstorm possibilities for some kind of service to others. This moves them away from what was proving to be tortuous problem exploration and toward opportunity development. Developing unused opportunities is often a way of *transcending* instead of solving problems.

The next client has been talking endlessly about his problems with a colleague at work. He has been going around in circles.

COUNSELOR: If you had just the kind of relationship with him you wanted, what would it look like?

CLIENT: Well, let me think. . . . He'd be less self-centered. He'd be more interested in my concerns. He'd show more willingness to help me when I get overloaded. After all, I help him. . . . He'd suggest things like getting a beer after work. I make most of the invitations now.

The helper's question not only puts the ball in the client's court but also asks the client to move from merely telling the story to creating the preferred scenario. Note, however, that he puts the entire burden on the other person. Further probes can get at a more balanced view of the preferred relationship.

In the following example, Jill, the helper, and Justin, the client, have been discussing how Justin is letting his impairment—he has lost a leg in a car accident—stand in the way of his picking up his life again. The session has bogged down a bit.

JILL: If you had to ask yourself one question right now about all of this, what would it be?

JUSTIN (pausing a long time): Why are you taking the coward's way out? Why are you on the verge of giving up?" (His eyes tear up.)

Jill's question puts the ball in Justin's court. It's her way of asking Justin to "move forward" and take responsibility for his part of the session. Justin uses a question to challenge himself in a way he might not have done otherwise. There is no magic to this. Justin might well have said, "I don't know what I'd ask myself. That's your job."

## Use Probes to Help Clients Move Forward within Some Step of the Process

Probes can be used not only to help clients move to a different stage or step but also to move within a step. Contrast the two following approaches to probing. The client, a woman in the middle of an acrimonious divorce, has recently learned that her breast cancer has reappeared. She is seeing the counselor after a long interlude.

**CLIENT** (toward the end of the session): Well, now we're up to date. You know the full miserable story.

**COUNSELOR A:** I haven't seen you for a while. When did you find out about the reappearance of the cancer?

**CLIENT:** Let's see. . . . Oh, who knows and who cares! . . . Well, I have to go.

The probe is a useless one. Mere filler. Its does nothing but annoy the client. Let's replay the scene with another counselor.

**COUNSELOR B:** I haven't seen you for a while. I've been wondering whether the reappearance of the cancer has altered your thinking about the divorce in any way.

**CLIENT:** I don't want to die married to him. I just don't. It would be dishonest. Our relationship, as you know, ended a long time ago.

**COUNSELOR B:** So since the principle of the thing hasn't changed, you're sticking to your guns. . . . I think you know why I ask. The proceedings up to now have been pretty bitter for you. . . . I don't know whether there are some trade-offs that might benefit you. Or is the word "trade-offs" off limits?

**CLIENT:** Hmm. . . . "Caving in" is off limits. Let's talk about "trade-offs" the next time we meet. I have to think about it.

The helper probes to see whether her decision to pursue the divorce (a Stage II activity) is irrevocable. There might be other options that won't compromise her principles and will save her much needless pain.

## Use Probes to Have Clients Ask Themselves, "What's Going On?"

Probes can be excellent tools for helping clients draw conclusions for themselves. Let's look in on Robin once more. In this session, Robin is talking about her relationship with one of her male friends at work. She is still talking about her body-image problem.

**ROBIN:** He has no idea that I'm bothered about all this. Just no idea.

**COUNSELOR:** Is he the kind of person who is oblivious to things that are important for you?

**ROBIN:** Just the opposite! Actually he's very sensitive. More than the average man, I bet. . . .

**COUNSELOR:** Then how come he's not bothered?

**ROBIN:** I'd have to say that he's not bothered because he doesn't even think that way about me. He doesn't seem to have any problem with my looks. Part of my preoccupation with this is to think that it's on everyone else's mind. . . . But I have to say that there's no real indication of that.

A simple probe or two helps Robin see that her assumption—her preoccupation—is not shared by the whole world. This is an important learning. She can now ask herself, "Why am I so bothered when even important others are not?" Answering that question could help her move forward.

In the following example, the client, a widow, learned that she has inoperable cancer. This sent her into a flurry of activity.

**HELPER:** When the diagnosis of cancer came in two weeks ago, you said that you were both relieved, because now you knew, and depressed, because of what you were going to have to face. You've mentioned that your behavior has been a bit chaotic since then. Tell me what you've been doing.

**CLIENT:** I've been to the lawyer to get a will drawn up. I'm trying to get all my business affairs in order. I've been writing letters to relatives and friends. I've been on the phone a lot. I guess I've been filling my days with lots of activity to forget about it all.

**HELPER:** So if you're just busy enough . . . How is all of that working out?

**CLIENT:** Not too well at all. It's like business as usual. I'm ignoring the fact that I'm going to die. There's something else I've got to do, but I'm not entirely sure what it is.

The helper's probe, "How is all of that working out?" has the impact of a "What's going on?" probe. The client says, in effect, "I'm running away."

As is clear from many of the foregoing examples, the use of probes, like the use of empathy, is part of the social-influence process. The helper chooses a theme and invites the client to explore it. Indeed, many probes are not just requests for relevant information. They also edge toward challenges of one kind or another. Probes then also serve as a bridge between empathy and challenge. Given the power of probes, helpers have to keep reminding themselves that clients' agendas are central, not their own.

## THE ART OF SUMMARIZING: PROVIDING FOCUS AND DIRECTION

The communication skills of attending, listening, empathy, and probing need to be orchestrated in such a way that they help clients focus their attention on issues that make a difference. The ability to summarize and to help clients summarize the main points of a helping interchange or session is a skill that can be used to provide both focus and challenge.

Brammer (1973) listed a number of goals that can be achieved by judicious use of summarizing: "warming up" the client, focusing scattered thoughts and feelings, bringing the discussion of a particular theme to a close, and prompting the client to explore a theme more thoroughly. There are certain times when summaries prove particularly useful: at the beginning of a new session, when the session seems to be going nowhere, and when the client gets stuck.

• *At the beginning of a new session.* When summaries are used at the beginning of a new session, especially when clients seem uncertain about how to begin, they prevent clients from merely repeating what has already been said before. They put clients under pressure to move on. Consider this example: Liz, a social worker, began a session with a rather overly talkative man with a summary of the main points of the previous session. This served several purposes. First, it showed the client that she had listened carefully to what he had said and that she had reflected on it after the session. Second, the summary gave the client a jumping-off point for the new session. It gave him an opportunity to add to or modify what was said. Finally, it placed the responsibility for moving on with the client. Clients might well need help to move on at times, but summaries give them the opportunity to exercise initiative.

• *In the course of a session that is going nowhere.* A summary can be used to give focus to a session that seems to be going nowhere. One of the main reasons sessions go nowhere is that helpers allow clients to keep "going 'round the mulberry bush"—that is, saying the same things over and over again—instead of helping them either go more deeply into their stories or spell out the implications of what they have said, especially the implication for action.

• *When a client gets stuck.* Summaries can be used when clients seem to have exhausted everything they have to say about a particular issue and seem to be stuck. However, the helper does not always have to provide the summary. Often it is better to ask the client to pull together the major points. This helps the client own the process, pull together the salient points, and move on. Since this is not meant to be a way of testing clients, the counselor should provide clients whatever help they need to stitch the summary together.

Often when scattered elements are brought together, the client sees the "bigger picture" more clearly. In the following example, a man who has been reluctant to go to a counselor with his wife has agreed to a couple of sessions "to please her." At this point he has described a great deal of his behavior at home. Much of it is caring.

COUNSELOR: I'd like to pull a few things together. You've encouraged your wife in her career, especially when things are difficult for her at work. You also encourage her to spend time with her friends as a way of enjoying herself and letting off steam. You also make sure that you spend time with the kids. In fact, time with them is important for you.

IGNATIUS: Yeah. That's right.

COUNSELOR: Also, if I have heard you correctly, you currently take care of the household finances. You are usually the one who accepts or rejects social invitations, because your schedule is tighter than Julia's. And now you're about to ask her to move because you can get a better job in Boston.

IGNATIUS: When you put it all together like that, it sounds as if I'm running her life.

COUNSELOR: What would your reaction be if the picture were reversed?

IGNATIUS: Hmmm . . . Well, I'd . . . hmm . . . (laughs). I'm giving myself away! I guess I wouldn't mind any one part of the picture, but I think I would resent the whole package.

Helping Ignatius explore the "I am making all the big decisions for her" theme in the context of his contributions to family life is a step forward. When he puts himself in her shoes, he doesn't like it.

Like probes, summaries can have the impact of a challenge, since they prevent a client from "just talking" and apply pressure for more focus. They help clients see the bigger picture. They are invitations to get at more substantive issues and to move forward to more advanced stages and steps of the helping process.

## INTEGRATING COMMUNICATION SKILLS: THE SEAMLESS USE OF ATTENDING, LISTENING, UNDERSTANDING, EMPATHY, PROBING, AND SUMMARIZING

The trouble with dealing with skills one at a time is that each skill is taken out of context. In the give-and-take of any given helping session, however, the skills must be intermingled in a natural way. In actual sessions, skilled helpers continually attend and listen and use a mix of probes and empathy to help clients clarify and come to grips with their concerns. There is no formula for the right mix. That depends on the client, client needs, the problem situation, possible opportunities, the stage, and the step.

A word about the relationship between empathy and probes. After using a probe to which a client responds, use basic empathy rather than another probe as a way of encouraging further exploration. The logic of this is straightforward. First, if a probe is effective, it will yield information that needs to be listened to and understood. Empathy is called for. Second, empathy, if accurate, tends to place a demand on clients to explore further. It puts the ball back in the client's court. In a workshop, Bob Carkhuff suggested with his usual edge that if helpers find themselves asking two questions in a row, without any intervening empathy, they may just have asked two stupid questions.

In the following example, the client is a young Chinese American woman whose father died in China and whose mother is now dying in the United States. She has been talking about the traditional obedience of Chinese women and her fears of slipping into a form of passivity in her American life. She talks about her sister, who gives everything to her husband without looking for anything in return.

COUNSELOR: Is this self-effacing role rooted in your genes?

CLIENT: Well, certainly in my cultural genes. And yet I look around and see many of my North American counterparts adopt a different style. A style that frankly appeals to me. But last year, when I took a trip back to China with my mother to meet my half sisters, the moment I landed I wasn't American. I was totally Chinese again.

COUNSELOR A: What did you learn there?

CLIENT: That I was Chinese!

The client says something significant about herself, but instead of responding with understanding, the helper uses a probe. This elicits only a repetition of what she had just said. Now a different approach.

COUNSELOR B: You learned just how deep your roots go.

CLIENT: And if these roots are so deep, what does that mean for me here? I love my Chinese culture. I want to be Chinese and American at the same time. How to do that, well, I haven't figured that out yet. I thought I had, but I haven't.

In this case, an empathic statement works much more effectively than another probe. Counselor B helps the client move forward.

Both empathy and probing are used throughout the helping process. Both help "keep things moving," helping clients explore issues more fully and in a more focused way, take responsibility, say what they want, discover how to get what they want, and move to action.

In the next example, a single middle-aged man working for a company that has "downsized" finds himself under tremendous pressure at work.

CLIENT: Well, I suppose that I should be grateful for even having a job. I didn't get downsized. But the additional work means additional hours both at the office and at home. My life is no longer mine!

HELPER: So the extra pressure and stress makes you wonder just how "grateful" you should feel.

CLIENT: Precisely.

HELPER: How could you go about reclaiming your life?

CLIENT: Hmmm. . . . Well, I know one way. We all keep complaining to one another at work. And this seems to make things even worse. I can get out of that loop. It's a simple way of getting back a bit of my freedom.

HELPER: So one way is to stop contributing to your own misery by staying away from the complaining chorus. Maybe there are other ways.

CLIENT: Well, there's no use sitting around hoping that the downsizing is going to be reversed. Not in the short term, anyway. So it makes sense to ask myself how I could get a better life by doing things that are realistic and within my reach.

HELPER: So it's a question of repositioning yourself at work.

CLIENT: Repositioning. Hmm, I like that word. It makes a lot of pictures dance through my mind. . . . Yes, I need to reposition myself. For instance . . .

BOX 6-1

# Suggestions for the Use of Probes

1. Keep in mind the goals of probing:

    a. To help nonassertive or reluctant clients tell their stories and engage in other behaviors related to the helping process.

    b. To help clients remain focused on relevant and important issues.

    c. To help clients identify experiences, behaviors, and feelings that give a fuller picture of the issue at hand.

    d. To help clients understand themselves and their problem situations more fully.

2. Use a mix of probing statements, open-ended questions, and interjections.

3. Do not engage clients in question-and-answer sessions.

4. If a probe helps a client reveal relevant information, follow it up with basic empathy rather than another probe.

5. Use whatever judicious mixture of empathy and probing is needed to help clients clarify problems, identify blind spots, develop new scenarios, search for action strategies, formulate plans, and review outcomes of action.

6. Remember that probing is a communication tool that is effective to the degree that it serves the stages and steps of the helping process.

There is a payoff in this combination of empathy and probing. Instead of focusing on the misery of the present situation, the client names a broad goal, "getting my life back," a Stage II activity, and then begins to explore ways of achieving that goal, a Stage III activity. Box 6-1 summarizes the main points about the use of probing as a communication skill.

Helpers should be careful not to become "empathy machines," grinding out one empathic response after another. All responses to clients, including probes and challenges as we shall see in Chapters 9 and 10, can be *implicitly* empathic if they are based on a solid understanding of the client's core messages and the client's point of view. For instance, a question such as "How did that happen?" can, in context, indicate that you have been listening intently and have captured the client's point of view. Responses that build on and add to the client's remarks are implicitly empathic. Since these responses are empathic in effect, they cut down on the need for a steady stream of directly empathic responses.

# BECOMING PROFICIENT
# AT COMMUNICATION SKILLS

Understanding communication skills and how they fit into the helping process is one thing. Becoming proficient in their use is another. Some trainees think that these "soft" skills should be learned easily and fail to put in the kind of hard work and practice that makes them "fluent" in them (Binder, 1990; Georges, 1988). Doing the exercises on communication skills in the manual that accompanies this book and practicing these skills in training groups can help, but that isn't enough to make the skills second nature. Attending, listening, empathy, and probing that are trotted out, as it were, for helping encounters are likely to have a hollow ring. These skills must become part of your everyday communication style.

After providing some initial training in communication skills, I tell students, "Now, go out into your real lives and get good at these skills. I can't do that for you. But you cannot be certified in this program unless and until you demonstrate competency in these skills." In the beginning it may be difficult to practice all these skills in everyday life, not because they are so difficult, but because they are "improbable events" in human communication. Take empathy. Listen to the conversations around you. Using an unobtrusive counter, press the plunger every time you hear empathy used. You may go days without pressing the plunger. In the end, however, you can make empathy a reality in *your* everyday life. And those who interact with you will notice the difference. They probably will not call it empathy. Rather, they will says such things as "She really listens to me" or "He takes me seriously." Empathy, provided that it is not a gimmick, goes far in fostering caring human relationships. On the other hand, you will hear many probes in everyday conversations. People are much more comfortable asking questions than providing understanding. However, many if not most of these probes will be aimless. Learning how to integrate purposeful probes with empathy demands practice in everyday life.

I once ran a training program on these skills for a CPA firm. Although the director of training believed in their value in the business world, many of the account executives did not. They resisted the whole process. I got a call one day from one of them. He had been one of the most virulent resisters. He said, "I owe you this call." "Really?" I replied with an edge of doubt in my voice. "Really," he said. He went on to tell me how he had called on a potential customer, a large company that was thinking of changing firms. He said to himself, "Since we don't have the slightest chance of getting this account, why don't I amuse myself by trying these skills?" In his phone call to me, he went on to say, "This morning I got a call from that customer. He gave us the account, but he said, "You're not getting the account because you were the low bidder. You were not. You're getting the account because we thought that you were the only one that really understood our needs.""

# THE SHADOW SIDE
## OF COMMUNICATION SKILLS

Some helpers tend to overidentify the helping process with the communication skills — that is, with the tools — that serve it. This is true not only of attending, listening, empathy, and probing but also of the skills of challenging to be treated in Chapters 9 and 10. Being good at communication skills is not the same as being good at helping. Moreover, an overemphasis on communication skills can make helping a great deal of talk with very little action. Communication skills are essential, of course, but they still must serve both the process and the outcomes of helping. These skills certainly help you establish a good relationship with clients. And a good relationship is the basis for the kind of social-emotional reeducation that has been outlined earlier. But you can be good at communication, good at relationship building, even good at social-emotional reeducation and still shortchange your clients, because they need more than that. Some who overestimate the value of communication skills tend to see a skill such as empathy as some kind of "magic bullet." Others overestimate the value of information gathering. This is not a broad indictment of the profession. Rather, it is a caution for beginners.

On the other hand, some practitioners underestimate the need for solid communication skills. There is a subtle assumption that the "technology" of their approach to helping suffices. They listen and respond through their theories and constructs rather than through their humanity. They are like some medical doctors who become more and more proficient at the use of medical technology and less and less in touch with the humanity of their patients. I recently spent ten days in a hospital. The staff were magnificent in addressing my medical needs. But the psychological needs that sprang from my anxiety about my illness were not addressed at all. Unfortunately, my anxieties were often expressed through physical symptoms. Then those symptoms were treated medically. I asked, "When you have conferences during which patients are discussed, do you say, 'Well, we've thoroughly reviewed his medical status and needs. Now let's turn our attention to what he's going through. What can we do to help him through this experience?'" One resident said, "No, we don't have time." Don't get me wrong. These were dedicated, generous people who had my interests at heart. We have a long way to go.

# STAGE I OF THE HELPING MODEL AND ADVANCED COMMUNICATION SKILLS

The communication skills reviewed in Part Two are critical tools; with them, you can help clients engage in the stages and steps of the helping model. But those communication skills are not the helping process itself. Part Three is a detailed exposition and illustration of Stage I of the helping model, together with its three steps. The first of these steps is discussed in Chapter 7. Advanced communication skills — those related to helping clients challenge themselves — are discussed and illustrated in Chapters 9, 10, and 11. In Chapter 12, the concept of leverage — helping clients choose the right issues to work on — is explained and illustrated.

# Stage I: Helping Clients Identify and Clarify Problem Situations and Unused Opportunities

Clients come to helpers because they need help in managing their lives more effectively. Stage I illustrates three ways in which counselors can help clients understand themselves, their problem situations, and their unused opportunities with a view to managing them more effectively. There are three principles.

**Step I-A:** Help clients tell their stories in terms of problem situations and unused opportunities.

**Step I-B:** Help clients move beyond blind spots and develop new perspectives on their problem situations and opportunities.

**Step I-C:** Help clients discover and work on issues that will make a difference in their lives.

These principles are not restricted to Stage I. Rather, they encompass skills and processes to be used at all stages and steps. Figure I highlights these three steps to the helping process. Another way of conceptualizing the stages and steps of the helping process is as different ways of "being with" clients in the interest of constructive change.

## Stage I and Assessment

Stage I in the helping process can be seen as the assessment stage—finding out, or rather learning, what's wrong, what resources are not being used, what opportunities lie fallow. Recall that learning takes place when options are identified, increased, and acted on. Assessment makes sense to the degree that it contributes to learning, to increasing the client's options. Client-centered assessment is the ability to understand clients, to spot "what's going on" with them, to see what they do not see and need to see, to make sense out of their chaotic behavior and help them make sense out of it—all in the interest of helping them manage their lives and develop their resources more effectively. Assessment, then, is not something helpers do to clients: "Now that I have my secret information about you, I can fix you." Rather, it is a kind of learning in which, ideally, both client and helper participate.

Assessment is a way of asking, What's really going on here? It takes place, in part, through the kind of reality-testing listening discussed in Chapter 4. As helpers listen to clients, they hear more than the client's point of view. They listen to the client's deep conviction that he is "not an alcoholic"—that is his "slant"—and at the same time they see the trembling hand, smell the alcohol-laden breath, and hear the desperate tone of voice. The purpose of reality-oriented listening is not to place the client in the

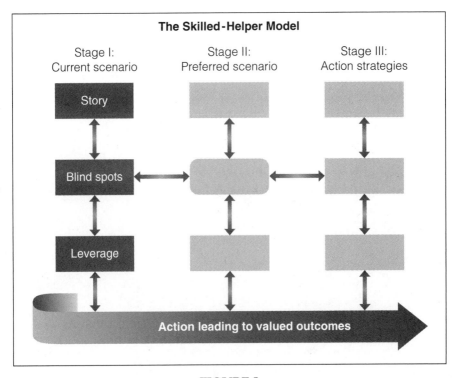

**FIGURE I**
The Helping Model — Stage I

proper diagnostic category. The purpose is to be open to any kind of information or understanding that will enable helpers to help the client.

In medicine, assessment is often a separate phase. In helping, there is an interplay between assessment and intervention. Helpers who continually listen to clients in context are engaging in a kind of ongoing assessment that is in keeping with the clients' interests. Assessment is thus part and parcel of all stages and steps of the helping model. As in medicine, though, some initial assessment of the seriousness of the client's concerns is called for. Members of the helping profession use many different procedures to assess clients; however, the "clinical interview," the dialogue between client and helper, is the most common assessment procedure. Other forms of assessment such as psychological tests are outside the scope of this book. Courses in abnormal psychology provide templates for identifying a wide range of social-emotional problems, but that, too, is beyond the scope of this book.

## CHAPTER 7

# STEP I-A: HELPING CLIENTS TELL THEIR STORIES

THE GOALS OF STEP I-A

HELPING CLIENTS EXPLORE PROBLEM SITUATIONS
AND UNEXPLOITED OPPORTUNITIES
 Learn to Work with All Styles of Storytelling
 Help Clients Clarify Key Issues
 Assess the Severity of the Client's Problems
 Help Clients Talk Productively about the Past
 As Clients Tell Their Stories, Search for Resources
 Help Clients Spot and Explore Unused Opportunities

STEP I-A AND ACTION

THE SHADOW SIDE OF STEP I-A
 Clients as Storytellers
 The Nature of Discretionary Change

EVALUATION QUESTIONS FOR STEP I-A

# THE GOALS OF STEP I-A

In specifying the goals of Step I-A, it must be remembered once more that "telling the story" may be first in a logical sense, but it may not be first in time. As mentioned in the overview of the helping model, clients do not necessarily start at the beginning nor do they tell their stories all at once.

The importance of helping clients tell their stories well should not be underestimated. As Pennebaker (1995b) has noted, "An important . . . feature of therapy is that it allows individuals to translate their experiences into words. The disclosure process itself, then, may be as important as any feedback the client receives from the therapist" (p. 3). From the point of view of problem managment and opportunity development, client self-disclosure has two benefits. First, it can and often does have a cathartic effect. "Getting things out on the table" can help reduce the number of negative feelings and emotions. This is part of the social-emotional reeducation process alluded to earlier. Second, self-disclosure provides the grist for the mill of problem solving. This book avoids the predictable debate as to whether or not self-disclosure in and of itself "cures" (see Pennebaker, 1995a). That said, we can list some broad principles related to storytellling.

• **Clarity.** Help clients spell out their problem situations and unexploited opportunities with the kind of concrete detail—specific experiences, behaviors, and emotions—that enables them to do something about them. Clients who are helped to tell their stories well learn about themselves. This learning opens the door to more creative options in living. On the other hand, vague stories lead to vague options and actions.

• **Relationship building.** Help clients tell their stories in such a way that the helping relationship develops and strengthens. This means translating the values of respect, genuineness, and empowerment into attending, listening, understanding, empathy, and probing. These are the basic tools for both clarity and relationship building.

• **Action.** Right from the beginning help clients begin to act on what they learn. Clients do not need "grand plans" before they can act on their own behalf. Later in this chapter, more will be said about the "bias toward action" needed in the helping process.

Clients differ radically in their ability to talk about themselves and their problem situations. Reluctance to disclose oneself within counseling sessions is often a window into the client's inability to share himself or herself and to be reasonably assertive in the social settings of everyday life. Then counseling may require helping clients develop the skills, confidence, and courage they need to share themselves appropriately (see Lynch & Gussel, 1996).

## HELPING CLIENTS EXPLORE PROBLEM SITUATIONS AND UNEXPLOITED OPPORTUNITIES

As with values and communication skills, there are a number of principles that can guide you as you help clients tell their stories.

## Learn to Work with All Styles of Storytelling

There are both individual and cultural differences (Wellenkamp, 1995) in clients' willingness to talk about themselves. Both affect storytelling. Some clients are highly verbal and quite willing to tell almost everything at the first sitting. Take the case of Martina.

> Martina, 27, asks a counselor in private practice for an appointment to discuss "a number of issues." Martina is both verbal and willing to talk, and her story comes tumbling out in rich detail. Although the helper uses the skills of attending, listening, empathy, and probing, she does so sparingly. Martina is too eager to tell her story.
>
> Although trained as a counselor, Martina is currently working in her uncle's business because of an offer she "could not turn down." She is doing very well financially, but she feels guilty because service to others has always been a value for her. And although she likes her current job, she also feels hemmed in by it. A year of study in Europe has whetted her appetite for "adventure." She feels that she is nowhere near fulfilling the great expectations she has for herself.
>
> She also talks about her problems with her family. Her father is dead. She has never gotten along well with her mother, and now that she has moved out of the house, she feels that she has abandoned her younger brother, whom she loves very much. She is afraid that her mother will "smother" her brother with too much maternal care.
>
> Martina is also concerned about her relationship with a man who is two years younger than she. They have been involved with each other for about three years. He wants to get married, but she feels that she is not ready. She still has "too many things to do" and would like to keep the arrangement they have.
>
> This whole complex story comes tumbling out in a torrent of words. Martina feels free to skip from one topic to another. The way Martina tells her story is part of her impulsive style. At one point she stops, smiles, and says, "My, that's quite a bit, isn't it!" Although all these issues bother her, she does not feel overwhelmed by them.

As the helper listens to Martina, she learns a number of things about her: that she is young, bright, and verbal and has many resources; that she is eager and impatient; that some of her problems are probably of her own making; that she has some blind spots that could stand in the way of her grappling more creatively with her problems; that she has many unexplored options, many unexploited opportunities. However, it is not clear why she came for counseling or how serious she is. (Is she there out of curiosity?) She also surmises that Martina would make her way in life, however erratically, with no counseling at all.

Contrast that example with the following one of a man who comes to a local mental-health center because he feels he can no longer handle his 9-year-old boy.

> Nick is referred to the clinic by a minister because of the trouble he is having with his son. He has been divorced for about two years and is living in a housing project on public assistance. After introductions and preliminary formalities have been taken care of, he just sits there and says nothing; he does not even look up. Since Nick offers almost nothing spontaneously, the counselor uses a relatively large number of probes to help him tell his story. Even when the counselor responds with empathy, Nick volunteers very little. Every once in a while, he tears up a bit. When asked about the divorce, he says he does not want to talk about it. "Good riddance" is all he can say about his former wife. Gradually, almost torturously, the story gets pieced together. Nick talks mostly about the "trouble" his son is getting into, how uncontrollable he seems to be getting, and how helpless he feels.

Martina's story is full of possibilities, whereas Nick's is mainly about limitations.

Each client is different and will approach the telling of the story in a different way. Some clients will come voluntarily; others will be sent. Some of the stories you will help clients tell will be long and detailed, others short and stark. Some will be filled with emotion; others will be told coldly, even though they are stories of terror. Some stories will be, at least at first blush, single-issue stories—"I want to get rid of these headaches"—whereas others, like Martina's, will be multiple-issue stories. Some will be told readily, tumbling out with practically no help, and others will be told only haltingly or grudgingly. Some clients will tell the core story immediately; others will tell a secondary story first to test the helper's reactions. Some stories will deal with the inner world of the client almost exclusively—"I hate myself," "I hear voices"—whereas others will deal with the outer world, problems in the social settings of life; still others will deal with a combination of inner and outer concerns. Some clients will make it clear that they trust you just because you are a helper; you will read mistrust in the eyes of others, sometimes just because you are a helper. In all these cases, your job is to establish a working relationship with your clients and help them tell their stories as a prelude to helping them manage the problems and take advantage of the opportunities buried in those stories. A story that is brought out into the open is the starting point for possible constructive change. Often the very airing of the story is a solid first step toward a better life.

When clients like Martina pour out their stories, you may let them go on or you may insist on some kind of dialogue. If the client tells the "whole" story in a more or less nonstop fashion, it will be impossible for you to respond with empathy to everything the client has said. But you can then help

BOX 7-1
# Problem Finding

Here are some questions counselors can help clients ask them-selves to "find," discover, and specify problem situations.

- What's going on?
- What are my problems?
- What issues do I need to face?
- What are my concerns?
- What's troubling me?
- What do those who know me tell me?
- What's keeping me back?
- What problem situation needs to be resolved?

the client review the most salient issues in some orderly way. Expressions like the following can be used to help the client review the core parts of the story.

> You've said quite a bit. Let's see if I've understood some of the main points you've made. First of all . . .

At this point your empathy will let the client know that you have been listening intently and that you are concerned about him or her. With clients like Nick, however, it's a different story. Clients who lack the skills needed to tell their stories well or who are reluctant to do so constitute a different kind of challenge. Empathy and probing can be tough work. Box 7-1 provides questions clients can ask themselves to identify problem situations.

## Help Clients Clarify Key Issues

To clarify means to discuss problem situations and unused opportunities in terms of specific experiences, specific behaviors, and specific feelings. Vagueness and ambiguity lead nowhere. It is up to the counselor to help the client move to clarity.

Janice's husband has been suffering from severe depression for over a year. All the attention is on him. One day, after a fainting spell, she too talks with a counselor. At first, feeling guilty about her husband, she is hesitant to discuss her own problems. In the beginning she says only that her social life is "a bit restricted by my husband's illness." With the help of empathy and probing on the part of the helper, her story emerges. "A bit restricted" turns, bit by bit, into the following, much fuller story. This is a summary. Janice did not say this all at once.

John has some sort of "general fatigue" illness that no one has been able to figure out. It's like nothing I've ever experienced. I move from guilt to anger to
indifference to hope to despair. I have no social life. Friends avoid us because
it is so difficult being with John. I feel I shouldn't leave him, so I stay at home.
He's always tired, so we have little interaction. I feel like a prisoner in my own
home. Then I think of the burden he's carrying and the roller-coaster emotions
start all over again. Sometimes I can't sleep, then I'm as tired as he. He is always saying how hopeless things are and even though I don't feel like he does,
some kind of hopelessness creeps into my bones. I feel that a stronger woman,
a more selfless woman, a smarter woman would find ways to deal with all of
this. But I end up feeling that I'm not dealing with it at all. From day to day I
think I cover most of this up, so he doesn't see it nor do the few people who
come around see it. I'm as alone as he is.

Here, then, is a whole series of negative experiences and emotions. The actions Janice takes — staying at home, covering her feelings up — are part of
the problem, not the solution. But now that the story is out in the open, there
is some possibility of doing something about it.

In another case, a bulimic woman, now under psychiatric care, says that
she acted "a little erratically at times" with some of her classmates in law
school. The counselor, using empathy and probes, helps her tell her story in
much greater detail. Like Janice, she does not say all of this at once, but this
is the fuller picture of "a little erratic."

I usually think about myself as plain even though when I take care of myself
some say that I don't look that bad. Ever since I was a teenager, I've preferred
to go it alone, because it was safer. No fuss, no muss, and especially no rejec-
tion. In law school, right from the beginning I entertained romantic fantasies
about some of my classmates who I didn't think would give me a second look.
I pretended to have meals with those who attracted me and then I'd have fantasies of having sex with them. Then I'd purge, getting rid of the fat I got from
eating and getting rid of the guilt. In my mind I picked them off one at a time.
I would go out of my way to meet my latest imagined partner. And then I'd be
rude to him to get back at him for what he did to me. That was my way of getting rid of him. He wouldn't know what hit him.

She was not really delusional, but gradually her external behaviors with a
kind of twisted logic began to reinforce her internal fantasizing behavior.
However, once her story became "public" — that is, reviewed by both herself
and her helper — she began to take back control of her life.

## Assess the Severity of the Client's Problems

Clients come to helpers with problems of every degree of severity. Objec-
tively, problems run from the inconsequential to the life-threatening.
Subjectively, however, even a relatively inconsequential problem can be experienced as severe by a client. If a client thinks that a problem is critical,

even though by objective standards the problem does not seem to be that bad, then for him or her it is critical. In such a case, the client's tendency to "catastrophize," to judge a problem situation to be more severe than it actually is, itself becomes an important part of the problem situation. One of the tasks of the counselor in this case will be to help the client put the problem in perspective or to teach him or her how to distinguish between degrees of problem severity. Howard (1991, p. 194) put it well.

> In the course of telling the story of his or her problem, the client provides the therapist with a rough idea of his or her orientation toward life, his or her plans, goals, ambitions, and some idea of the events and pressures surrounding the particular presenting problem. Over time, the therapist must decide whether this problem represents a minor deviation from an otherwise healthy life story. Is this a normal, developmentally appropriate adjustment issue? Or does the therapist detect signs of more thorough-going problems in the client's life story? Will therapy play a minor, supportive role to an individual experiencing a low point in his or her life course? If so, the orientation and major themes of the life will be largely unchanged in the therapy experience. But if the trajectory of the life story is problematic in some fundamental way, then more serious, long-term story repair might be indicated. So, from this perspective, part of the work between client and therapist can be seen as life-story elaboration, adjustment, or repair.

Savvy therapists not only gain an understanding of the severity of a client's problem or the extent of the client's unused resources, but also understand the limits of helping. Life-story adjustment or repair is not the same as attempting to redo the client's personality.

Mehrabian and Reed (1969) suggested the following formula as a way of determining the severity of any given problem situation:

$$\text{Severity} = \text{Distress} \times \text{Uncontrollability} \times \text{Frequency}$$

The multiplication signs in the formula indicate that these factors are not just additive. Even low-level anxiety, if it is uncontrollable or persistent, can constitute a severe problem; that is, it can severely interfere with the quality of a client's life.

One way of viewing helping is to see it as a process in which clients are helped to control the severity of their problems in living. The severity of any given problem situation will be reduced if the stress can be reduced, if the frequency of the problem situation can be lessened, or if the client's control over the problem situation can be increased. Consider the following example.

> Indira is greatly distressed because she experiences migrainelike headaches, sometimes two or three times a week, and seems to be able to do little about them. No painkillers seem to work. She has even been tempted to try strong narcotics to control the pain, but she fears she might become an addict. She

feels trapped. For her the problem is quite severe because stress, uncontrollability, and frequency are all high.

Eventually, Indira is referred to a doctor who specializes in treating headaches. He is an expert in both medicine and human behavior. He first helps her to see that the headaches are getting worse because of her tendency to "catastrophize" whenever she experiences one. That is, the self-talk she engages in ("How intolerable this is!") and the way she fights the headache actually add to its severity. He helps her control her stress-inducing self-talk and teaches her relaxation techniques. Neither of these gets rid of the headaches, but they help reduce the degree of stress she feels.

Second, he helps her identify the situations in which the headaches seem to develop. They tend to come at times when she is not managing other forms of stress well, as when she lets herself become overloaded and gets behind at work. He helps her see that, although her headaches constitute a very real problem, they are part of a larger problem situation. Once she begins to control and reduce other forms of stress in her life, the frequency of the headaches begins to diminish.

Third, the doctor also helps Indira spot cues that a headache is beginning to develop. She finds that, though she can do little to control the headache once it is in full swing, there are things she can do during the beginning stage. For instance, she learns that medicine that has no effect when the headache is in full force does help if it is taken as soon as the symptoms appear. Relaxation techniques are also helpful if they are used soon enough. Indira's headaches do not disappear completely, but they are no longer the central reality of her life.

The doctor helps the client manage a severe problem situation much better than she has been doing by helping her reduce stress and frequency while increasing control. In other words, he helps the client learn. Of course, if you believe that a client's problem is too severe for you to handle, you need to refer the client to another helper.

## Help Clients Talk Productively about the Past

Some schools of psychology suggest that problem situations are not clear and fully understood until they are understood in the context of their historic roots. Therefore, helpers in these schools spend a great deal of time helping clients uncover the past. Others disagree with that point of view. For instance, Fish (1995) suggested that attempts to discover the hidden root causes of current problem behavior may be unnecessary, misguided, or even counterproductive. Constructive change does not depend on causal connections in the past. There is evidence to support Fish's contention. Long ago Deutsch (1954) noted that it is often almost impossible, even in carefully controlled laboratory situations, to determine whether event B, which follows event A in time, is actually caused by event A. Therefore, trying to connect present complicated patterns of behavior with past complicated events is an exercise in frustration. Asking clients to come up with causal connections between current unproductive behavior and past events could be an exercise in futility because such causal connections remain hypothetical, because understanding past causes does little to change present behavior, because talking about the past often focuses mostly on what happened to clients (their experiences) rather than on what they did about what happened to them (their behaviors) and therefore interferes with the "bias toward action" clients need to manage current problems.

This is not to say that a person's past does not influence current behavior. Nor does it imply that a client's past should not be discussed. But the fact that past experiences influence current behavior does not mean that they necessarily *determine* present behavior. Kagan (1996) has challenged what may be called the "scarred for life" assumption: "If orphans who spent their first years in a Nazi concentration camp can become productive adults and if young children made homeless by war can learn adaptive strategies after being adopted by nurturing families" (p. 901), that means that there is hope for us all. As you can imagine, this is one of those issues that members of the helping professions argue about endlessly. Therefore, this is not a debate that is to be settled with a few words here. Instead, there are a couple of principles based on hope that might help.

**Help clients talk about the past to make sense of the present.** Many clients come expecting to talk about the past or wanting to talk about the past. There are ways of talking about the past that help clients make sense of the present. But making sense of the present needs to remain center stage. Thus, how the past is discussed is more important than whether it is discussed. The following man has been discussing how his interpersonal style gets him into trouble.

**HELPER:** So your father's unproductive interpersonal style is, in some ways, alive and well in you.

**CLIENT:** Until we began talking I had no idea about how alive and well it is. For instance, even though I hated his cruelty, it lives on in me in much smaller ways. He beat my brother. I just cut him down to size verbally. He told my mother what she could do and couldn't do. I try to get my mother to adopt my set of living arrangements for her own good. There's a whole pattern that I haven't noticed. I've inherited more than his genes.

**HELPER:** That's quite an inheritance. . . . But now what?

**CLIENT:** Well, now that I see what's happened, I'd like to change things. A lot of this is ingrained in me, but it's not genetic.

It really does not make any difference whether the client's behavior has been "caused" by his father or not. In fact, if, by hooking the present into the past, he feels in some way that his current nasty style is not his fault, what difference does it make? Helping is about the future. The point is that now that the problem has been named and is out in the open, it is possible to do something about it. In sum, if the past can add clarity to current experiences, behaviors, and emotions, then let it be discussed. If it can provide hints as to how self-defeating thinking and behaving can be changed now, let it be discussed. The past, however, should never become the principal focus of the client's self-exploration. When it does, helping tends to slow down needlessly.

**Help clients talk about the past to be liberated from it.** A potentially dangerous logic can underlie discussions of the past. It goes something like this: "I am what I am today because of my past; I cannot change my past. So how can I be expected to change today?"

CLIENT: I was all right until I was about 13. I began to dislike myself as a teenager. I hated all the changes — the awkwardness, the different emotions. I
was so impressionable. I began to think that life actually must get worse and
worse instead of better and better. I just got locked into that way of thinking.

That is not liberation talk. The past is still casting its spell. Helpers need to understand that clients may see themselves as prisoners of their past, but then, in the spirit of Kagan's earlier comments, they must help them move beyond such a self-defeating mind-set.

The following case is quite different. It is that of the father of a boy who has been sexually abused by a minister of their church. Dealing with the in-cident helps him review his own abuse by his father. In a tearful session he tells the whole story. In a second interview he has this to say:

CLIENT: Someone said that good things can come from evil things. What
happened to my son was evil. But we'll give him all the support he needs.
Though I had the same thing happen to me, I kept it all in until now. It
was all locked up inside. I was so ashamed and my shame became part of
me. When I let it all out last week, it was like throwing off a dirty cloak that
I'd been wearing for years. Getting it out was so painful, but now I feel so
different, so good. I wonder why I had to hold it in for so long.

That is liberation talk. When counselors help or encourage clients to talk about the past, they should have a clear idea of what their objective is. Is it to learn from the past? Is it to be liberated from it? To assume that there is some "silver bullet" in the past that will solve today's problem is to search for magic.

## As Clients Tell Their Stories, Search for Resources

Incompetent helpers concentrate on clients' deficits. Skilled helpers, as they listen to and observe clients, do not blind themselves to deficits, but they are quick to spot clients' resources, whether used, unused, or even abused. These resources can become the building blocks for the future. Consider this example: Terry, a young woman in her late teens who has been arrested several times for street prostitution, is an involuntary client. She ran away from home when she was 16 and is now living in a large city on the East Coast. Like many other runaways, she was seduced into prostitution "by a halfway decent pimp." Now she is very street-smart. She is also cynical about herself, her occupation, and the world. She is forced to see a counselor as part of the probation process. As might be expected, Terry is quite hostile during the interview. She has seen other counselors and sees the interview as a game. The counselor already knows a great deal of the story because of the court record. The dialogue is not easy. Some of it goes like this:

**TERRY:** If you think I'm going to talk with you, you've got another think coming. What I do is my business.

**COUNSELOR:** You don't have to talk about what you do outside. We could talk about what we're doing here in this meeting.

**TERRY:** I'm here because I have to be. You're here either because you're dumb and like to do stupid things or because you couldn't get a better job. You people are no better than the people on the street. You're just more "respectable." That's a laugh.

**COUNSELOR:** So nobody could really be interested in you.

**TERRY:** I'm not even interested in me!

**COUNSELOR:** And if you're not interested in yourself, then no one else could be, including me.

Terry has obvious deficits. She is engaged in a dangerous and self-defeating lifestyle. But as the counselor listened to Terry, he spotted many different resources. Terry is a tough, street-smart woman. The very virulence of her cynicism and self-hate, the very strength of her resistance to help, and her almost unchallengeable determination to go it alone are all signs of resources. Many of her resources are currently being used in self-defeating ways, but they are still resources.

Helpers need a resource-oriented mind-set in all their interactions with clients. Contrast two different approaches.

**CLIENT:** I practically never stand up for my rights. If I disagree with what anyone is saying—especially in a group—I keep my mouth shut.

**HELPER:** So clamming up is the best policy. . . . What happens when you do speak up?

**CLIENT** (pausing): I suppose that on the rare occasions when I do speak up, the world doesn't fall in on me. Sometimes others do actually listen to me. But I still don't seem to have much impact on anyone.

**COUNSELOR A:** So speaking up, even when it's safe, doesn't get you much.

**CLIENT:** No, it doesn't.

Counselor A misses the resource mentioned by the client. Although it is true that the client habitually fails to speak up, he has some impact when he does speak — others do listen, at least sometimes, and this is a resource. Counselor A emphasizes the deficit. Let's try the other counselor.

**COUNSELOR B:** So when you do speak up, you don't get blasted, you even get a hearing. Tell me what makes you think you don't exercise much influence when you speak?

**CLIENT** (pauses): Well, maybe influence isn't the issue. Usually I don't want to get involved. Speaking up gets you involved.

Note that both counselors use empathy, but they focus on different parts of the client's message. Counselor A emphasizes the deficit; Counselor B notes the asset and follows up with a probe. This produces a significant clarification of the client's problem situation.

The search for resources is especially important when the story being told is bleak. I once listened to a man's story that included a number of bone-jarring life defeats — a bitter divorce, false accusations that led to his dismissal from a job, months of unemployment, serious health concerns, months of homelessness, and more. The only emotion the man exhibited during his story was depression. Toward the end of the session, we had this interchange:

**HELPER:** Just one blow after another, grinding you down.

**CLIENT:** Grinding me down, and almost doing me in.

**HELPER:** Tell me a little more about the "almost" part of it.

**CLIENT:** Well, I'm still alive, still sitting here talking to you.

**HELPER:** Despite all these blows, you haven't fallen apart. That seems to say something about the fiber in you.

At the word *fiber*, the client looked up, and there seemed to be a glimmer of something besides depression in his face. I put a line down the center of a newsprint pad. On the left side I listed the life blows the man had experienced. On the right I put "Fiber." Then I said, "Let's see if we can add to the list on the right side." We came up with a list of the man's resources — his "fiber," his musical talent, his honesty, his concern for and ability to talk to others, and so forth. After about a half hour, he smiled — weakly, but he did smile. He said that it was the first time he could remember smiling in months.

## Help Clients Spot and Develop Unused Opportunities

Early in the history of modern psychology, William James remarked that few people bring to bear more than about 10% of their human potential on the problems and challenges of living. Others since James, though changing the percentages somewhat, have said substantially the same thing, and few have challenged their statements (Maslow, 1968). It is probably not an exaggeration to say that unused human potential constitutes a more serious social problem than emotional disorders, since it is more widespread. Maslow (1968) suggests that what is usually called "normal" in psychology "is really a psychopathology of the average, so undramatic and so widely spread that we don't even notice it ordinarily" (p. 16). Many clients you will see, besides having more or less serious problems in living, will also probably be chronic victims of self-inflicted psychopathology of the average.

Clients are much more likely to talk about problem situations than about unused opportunities. However, the flip side of just about every problem is an opportunity. William Miller (1986) talked about one of the worst days of his life. Everything was going wrong at work: Projects were not working out, people were not responding, the work overload was bad and getting worse — nothing but failure all around. Later that day, over a cup of coffee, he took some paper, put the title "Lessons Learned and Relearned" at the top, and wrote down as many entries as he could. Some hours and seven pages later, he had listed 27 lessons. The day turned out to be one of the best of his life. Both his attitude and his ability to come up with solutions to his problems helped so much that he began writing a "Lessons Learned" journal every day. He learned not to get caught up in self-blame and defeatism. Subsequently, on days when things were not working out, he would say to himself, "Ah, this will be a day filled with learnings!"

Although there is no justification for romanticizing pain, the flip side of human problems is human opportunities. Here are a few examples.

- Joe used his diagnosis of AIDS as a starting point for reintegrating himself into his extended family and challenging the members of his family to come to grips with some of their own problems, problems they had been denying for a long time.

- Beatrice used her divorce as an opportunity to develop a new approach to men based on mutuality. She also became an entrepreneur, starting an arts and crafts company.

- Jerome used a ten-day stay in the hospital as an opportunity to reset some assumptions about life itself and to review and revise his career goals.

- Sheila used her incarceration as an opportunity to finish her high school degree and get a head start on college.

- An actor suffering from a traumatic disability found new life by becoming a public advocate for those suffering a similar fate.

- A couple mourning the death of their only child started a day-care center in conjunction with other members of their church.

Box 7-2 provides some questions that clients can be helped to ask themselves to identify unused resources and opportunities.

**BOX 7-2**
## Opportunity Finding
Here are some questions counselors can help clients ask themselves to identify unused opportunities.

- What are my unused skills/resources?
- Which of these am I failing to use?
- What opportunities do I let go by?
- What ambitions remain unfulfilled?
- What could I be doing that I'm not doing?
- What am I failing to accomplish?
- What could I become if I tried?
- Which opportunities should I be developing?
- Which role models could I be emulating?

# STEP I-A AND ACTION

Shakespeare, in the person of Hamlet, talks about important enterprises, what he calls "enterprises of great pith and moment," losing "the name of action." Helping, an enterprise "of great pith and moment," can lose the name of action. One of the principal reasons clients do not manage the problem situations of their lives effectively is their failure to *act* intelligently, forcefully, and prudently in their own best interests. Covey (1989), in his immensely popular *The Seven Habits of Highly Effective People*, named "proactivity" as the first habit: "It means more than merely taking initiative. It means that as human beings, we are responsible for our own lives. Our behavior is a function of our decisions, not our conditions. We can subordinate feelings to values. We have the . . . responsibility to make things happen" (p. 71). Inactivity can be bad for body, mind, and spirit. Consider the following workplace example.

> A counselor at a large manufacturing concern realized that inactivity did not benefit injured workers. If they stayed at home, they tended to sit around, gain weight, lose muscle tone, and suffer from a range of psychological symptoms such as psychosomatic complaints unrelated to their injuries. Taking a people-in-systems approach, she worked with management, the unions, and doctors to design temporary, physically light jobs for injured workers. In some cases, nurses or physical therapists visited these workers on the job. Counseling sessions helped to get the right worker into the right job. The workers, active again, felt better about themselves and the company benefited.

Counselors add value by helping their clients become proactive. Helping too often becomes a process of too much talking and too little action.

Every stage and every step of the helping process should be conducted in such a way as to become a stimulus for problem-managing and opportunity-developing action. There are a number of ways of looking at the relationship between this first step of the model and action. Some clients seem to need only this first step of the helping model. That is, they spend a relatively limited amount of time with a helper, they tell their story in greater or lesser detail, and then they go off and manage quite well on their own. Here are two ways in which clients may need only the first step of the helping process.

**A declaration of intent and the mobilization of resources.** For some clients, the very fact that they approach someone for help may be sufficient to help them begin to pull together the resources needed to manage their problem situations more effectively. For these clients, going to a helper is a declaration not of helplessness but of intent: "I'm going to do something about this problem situation."

> Declan was a young man from Ireland living illegally in New York. Even though he had a community of fellow countrymen to provide him support, he felt, as he put it, "hunted." He lived with the fear that something terrible was going to happen. He talked to a priest about his concerns. The priest listened carefully but had no advice to give him. But the talk provided the stimulus Declan needed. He acted. He returned to Ireland, became skilled in computers, moved to Britain, and got a job with the English arm of a German computer firm. Just as important, he felt "whole" again.

Merely seeking help can trigger a resource-mobilization process in some clients. Once they begin to mobilize their resources, they begin to manage their lives quite well on their own.

**Coming out from under self-defeating emotions.** Some clients come to helpers because they are incapacitated, to a greater or lesser degree, by feelings and emotions. Often when helpers show such clients respect, listen carefully to them, and understand them in a nonjudgmental way, those self-defeating feelings and emotions subside. That is, the clients benefit from the counseling relationship as a process of social-emotional reeducation and repair. Once that happens, they are able to call on their own inner and environmental resources and begin to manage the problem situation that precipitated the incapacitating feelings and emotions. In short, they move to action. These clients, too, seem to be "cured" merely by telling their story.

> One woman, let's call her Katrina, was very depressed after undergoing a hysterectomy. She had only two relatively brief sessions with a counselor. Katrina, helped to sort through the emotions she was feeling, discovered that the predominant emotion was *shame*. She felt wounded and exposed *to herself* and guilty about her incompleteness. Once she saw what was going on, she was able to pick up her life once more.

Such clients may even say something like "I feel relieved even though I haven't solved my problem." Often they feel more than relieved; they feel empowered by their interactions, however brief, with a caring helper. Of course, not all clients fall into these two categories. Many need the kind of help provided by one or more of the other stages and steps of the helping model.

# THE SHADOW SIDE OF STEP I-A

There is a deep shadow side to Step I-A. One important part refers to the ways in which clients tell their stories. Another is the fact that they are telling their stories in the hope of changing in some way.

## Clients as Storytellers

When clients present themselves to helpers, there is no way of reading what is in their hearts. This, as we shall see, unfolds over the course of helping sessions, but it is not visible in the beginning. There is no single mold of storytelling. There are several continuums involved. One deals with the quality of the story told. At one end of the spectrum are clients who try to tell their stories up front, clearly, and in detail; at the other end are clients who tell stories that are general, partial, and ambiguous. A second continuum deals with concern about self-presentation. At one end are clients who are not especially concerned about what their helpers think of them; they have no particular need to be seen in a favorable light. At the other end are clients who are extremely concerned about what their helpers think of them and will skew their stories to present themselves in the best light. A third continuum deals with the veracity of the story. At one end are clients who tell their stories as honestly as possible; at the other end are clients who, for whatever reason, lie. The latter might not give a hoot about what the helper thinks of them, but they still lie. In all three continuums there is every shade in between. Throw all of these together with their various combinations and permutations, add every other possible motivation, and you have an "infinite" number of storytelling styles. Rather, each client has his or her own style. Add the enormous diversity found in clients and in story content, and it is clear to savvy helpers that each client represents an N = 1 (sole subject) research project. The "research" deals with this individual and constructive change. The competent and savvy helper tackles each project without naivete and without a hint of cynicism. Each client, of whatever stripe, engages her or his humanity. The picture is further complicated by the nature of the change in most helping situations. It is essential, therefore, to understand the nature of *discretionary* change.

## The Nature of Discretionary Change

The distinction between discretionary and nondiscretionary change is critical for helpers. Nondiscretionary change is mandated. If the courts say to a divorced man negotiating visiting rights with his children, "You can't have visiting rights unless you stop drinking," then the change is nondiscretionary. There will be no visiting rights without the change. In contrast, a man and wife having difficulties with their marriage are not under the gun to change the current pattern. Change here is discretionary. The shadow-side principle here is quite challenging.

> *In both individual and organizational affairs, the track record for discretionary change is quite poor.*

This fact is central to the psychopathology of the average. If we don't have to change, very often we don't. We need merely review the track record of our New Year's resolutions. Unfortunately, in helping situations, clients probably see most change as discretionary. They may talk about it as if it were nondiscretionary, but deep down "I don't really have to change" pervades the helping process. The sad state of discretionary change is not meant to discourage you but to make you more realistic about the challenges you face — or the challenges you help your clients face.

A pragmatic bias toward client action on the part of helpers — rather than merely talking about action — is a cardinal value. Effective helpers tend to be active with clients and see no particular value in mere listening and nodding. They engage clients in a dialogue. During that dialogue, they constantly ask themselves, "What can I do to raise the probability that this client will act on her own behalf intelligently and prudently?" I know a man who years ago went "into therapy" (as into "another world") because, among other things, he was indecisive. Over the years he became engaged several times and each time broke it off. So much for decisiveness. Again, savvy helpers know that in some ways the deck is stacked against them from the start. But if we can believe the outcome studies reviewed in Chapter 1, effective helpers consistently win against the odds.

**?**

# Evaluation Questions for Step I-A

How effectively am I doing the following?

## Establishing a Working Alliance

- Developing a collaborative working relationship with the client
- Using the relationship as a vehicle for social-emotional reeducation
- Not doing for clients what they can do for themselves

## Helping Clients Tell Their Stories

- Using a mix of attending, listening, empathy, and probing to help clients tell their stories in terms of specific experiences, behaviors, and feelings
- Using probes when clients get stuck, wander about, or lack clarity
- Understanding blocks to client self-disclosure and providing support for clients having difficulty talking about themselves
- Helping clients talk productively about the past

## Building Ongoing Client Assessment into the Helping Process

- Getting an initial feel for the severity of a client's problems and his or her ability to handle them
- Noting client resources, especially unused resources
- Understanding clients' problems and opportunities in the larger context of their lives

## Helping Clients Move to Action

- Helping clients develop an action orientation
- Helping clients spot early opportunities for changing self-defeating behavior

## Integrating Evaluation into the Helping Process

- Keeping an evaluative eye on the entire process with the goal of adding value through each interaction and making each session better
- Finding ways of getting clients to participate in and own the evaluation process

# RELUCTANT AND RESISTANT CLIENTS

It is impossible to be in the business of helping people for long without encountering both reluctance and resistance (Clark, 1991; Ellis, 1985; Fremont & Anderson, 1986; Friedlander & Schwartz, 1985; Kottler, 1992; Otani, 1989). Mahalik (1994) developed a Client Resistance Scale to measure clients' opposition to dealing with painful emotions, disclosing material relevant to the problem situation, collaborating with the helper, dealing with blind spots and developing new perspectives, and embracing constructive change.

- *Reluctance* refers to clients' *hesitancy* to engage in the work demanded by the stages and steps of the helping process. Problem management and opportunity development involve a great deal of work. Therefore, there are sources of reluctance in all clients—indeed, in all human beings. For instance, part of problem management is trying new behaviors. Many clients are reluctant to do so.
- *Resistance* refers to the *push-back* by clients when they feel they are being *coerced*. Clients who believe that their cultural beliefs, values, and norms—whether group or personal—are being violated by the helper can be expected to resist. For instance, since individual and cultural norms regarding self-disclosure differ widely (see Wellenkamp, 1995), clients who believe that self-disclosure is being extorted from them might well resist.

Although the behaviors through which reluctance and resistance are expressed might either seem or be the same, the distinction is still a useful one. The seeds of reluctance are in the client, whereas the stimulus for resistance is in the helper (Bischoff & Tracey, 1995) or the social setting surrounding the helping process. It is easier for helpers to see push-back on the part of clients as reluctance. It takes them off the hook. In practice a mixture of reluctance and resistance may be found in the same client.

## Reluctance — Misgivings about Change

Reluctance refers to the ambiguity clients feel when they know that managing their lives better is going to exact a price. Clients are not sure whether they want to pay that price. The incentives for not changing drive out the incentives for changing. This accounts for the sad record for discretionary change mentioned in the last chapter. Since reluctance and resistance can be encountered at any stage or step of the helping process, it is useful to consider them up front.

**Reluctant clients.** Clients exercise reluctance in many, often covert, ways. They talk about only safe or low-priority issues, seem unsure of what they want, benignly sabotage the helping process by being overly cooperative, set unrealistic goals and then use them as an excuse for not working, don't work very hard at changing their behavior, and are slow to take responsibility for themselves. They tend to blame others or the social settings and systems of their lives for their troubles and play games with helpers. Reluctance admits of degrees; clients come "armored" against change to a greater or lesser degree.

**Some of the roots of reluctance.** The reasons for reluctance are many. They are built into our humanity, plagued as it is by the "psychopathology of the average," described by Maslow (1968, p. 16) as "so undramatic and so widely spread that we don't even notice it ordinarily." Here is a sampling.

• *Fear of intensity.* If the counselor uses high levels of attending, listening, empathy, and probing and if the client cooperates by exploring the feelings, experiences, and behaviors related to his or her problems in living, the helping process can be an intense one. This intensity can cause both helper and client to back off. Skilled helpers know that counseling is potentially intense. They are prepared for it and know how to support a client who is not used to such intensity.

• *Lack of trust.* Some clients find it very difficult to trust anyone, even a most trustworthy helper. They have irrational fears of being betrayed. Even when confidentiality is an explicit part of the client-helper contract, some clients are very slow to reveal themselves.

• *Fear of disorganization.* Some people fear self-disclosure because they feel that they cannot face what they might find out about themselves. The client feels that the facade he or she has constructed, no matter how much energy must be expended to keep it propped up, is still less burdensome than exploring the unknown. Such clients often begin well but retreat once they begin to be overwhelmed by the data produced in the problem-exploration process. Digging into one's inadequacies always leads to a certain amount of disequilibrium, disorganization, and crisis. But growth takes place at crisis points. A high degree of disorganization immobilizes the client, whereas very low disorganization is often indicative of a failure to get at the central issues of the problem situation.

• *Shame.* Shame is a much overlooked variable in human living (Egan, 1970; Kaufman, 1989; Lynd, 1958). It is an important part of disorganization and crisis. The root meaning of the verb *to shame* is "to uncover, to expose, to wound," a meaning that suggests the process of painful self-exploration. Shame is not just being painfully exposed to another; it is primarily an exposure of self to oneself. In shame experiences, particularly sensitive and vulnerable aspects of the self are exposed, especially to one's own eyes. Shame is often sudden—in a flash, one sees heretofore unrecognized inadequacies without being ready for such a revelation. Shame is sometimes touched off by external incidents, such as a casual remark someone makes, but it could not be touched off by such insignificant incidents unless, deep down, one was already ashamed. A shame experience might be defined as an acute emotional awareness of a failure to *be* in some way.

• *Fear of change.* Some people are afraid of taking stock of themselves because they know, however subconsciously, that if they do, they will have to change — that is, surrender comfortable but unproductive patterns of living, work more diligently, suffer the pain of loss, acquire skills needed to live more effectively, and so on. For instance, a husband and wife may realize, at some level of their being, that if they see a counselor, they will have to reveal themselves and that once the cards are on the table, they will have to go through the agony of changing their style of relating to each other.

In a counseling group, I once dealt with a man in his 60s who complained about constant anxiety that was making his life quite painful. He told the story of how he had been treated brutally by his father until he finally ran away from home. We saw his logic in the previous chapter: "No one who grows up with scars like mine can be expected to take charge of his life and live responsibly." He had been using his mistreatment as a youth as an excuse to act irresponsibly at work (he had an extremely poor job record), in his relationship with himself (he drank excessively), and in his marriage (he had been uncooperative and unfaithful and yet expected his wife to support him). The idea that he could change, that he could take responsibility for himself even at his age, frightened him, and he wanted to run away from the group. But since his anxiety was so painful, he stayed. And he changed.

Each one of us needs only to look at his or her own struggles with growth, development, and maturity to add to this list of the roots of reluctance.

## Resistance — Reacting to Coercion

Reluctance is often passive. Resistance can be active. Resistance refers to the reaction of clients who in some way feel coerced. It is their way of fighting back (see Dimond and his associates, 1978, and Driscoll, 1984). For instance, spouses who feel forced to come to marriage counseling sessions are often resistant. Resistance is the client's reaction to a power play. It may well arise from a misperception — the client's seeing coercion where it does not exist. But since people act on their perceptions, the result is still some form of fighting back.

Resistant clients are likely to present themselves as not needing help, feel abused, show little willingness to establish a relationship with the helper, and con helpers whenever possible. They may be resentful, make active attempts to sabotage the helping process, terminate the process at the earliest possible moment, and be either testy or actually abusive and belligerent. In marriage counseling, one spouse may be there willingly whereas the other is there because he or she feels pressured by the helper, the spouse, or both. Resistance to helping is, of course, a matter of degree, and not all these behaviors in their most virulent forms are seen in all resistant clients.

Involuntary clients — "mandated" clients — are often resisters. Helpers can expect to find a large proportion of such clients in settings where clients are forced to see a counselor (a high school student gets into trouble with a teacher and sees being sent to a counselor as a form of punishment) or in settings where a reward can be achieved only on the condition of being involved in some kind of counseling process (going to a counselor is the price that must be paid to get a certain job). Clients like these are found in schools, especially schools below college level, in correctional settings, in marriage counseling, especially if it is court-mandated, in employment agencies, in welfare agencies, in court-related settings, and in other social agencies. But any client who feels that he or she is being coerced can become a resister, perhaps not to the entire helping process, but at least with respect to the perceived coercion.

There are all sorts of reasons for resisting, which is to say that clients can experience coercion in a wide variety of ways. The following kinds of clients are likely to be resistant.

- Clients who see no reason for going to the helper in the first place.
- Clients who resent third-party referrers (parents, teachers, agencies) and whose resentment carries over to the helper.
- Clients in medical settings who are asked to participate in counseling.
- Clients who feel awkward in participating, who do not know how to be "good" clients.
- Clients who have a history of being rebels.
- Clients who see the goals of the helper or the helping system as different from their own. For instance, the goal of counseling in a welfare setting may be to help the clients become financially independent, whereas some clients may be satisfied with financial dependency.
- Clients who have developed negative attitudes about helping and helping agencies and who harbor suspicions about helping and helpers. Sometimes these clients refer to helpers in derogatory and inexact terms ("shrinks").
- Clients who believe that going to a helper is the same as admitting weakness, failure, and inadequacy. They feel that they will lose face by going. By resisting the process, they preserve their self-esteem.
- Clients who feel that counseling is something that is being done to them. They feel that their rights are not being respected.
- Clients who feel that they have not been invited by helpers to be participants in the decisions that are to affect their lives. This includes expectations for change and decisions about procedures to be used in the helping process.
- Clients who feel a need for personal power and find it through resisting a powerful figure or agency: "I may be relatively powerless, but I still have the power to resist."
- Clients who dislike their helpers but do not discuss their dislike with them.
- Clients who differ from their helpers about the degree of change needed.

Clients like these have in common a feeling of being coerced to engage in the process, even if they are not sure precisely of the source of the coercion. Resistance, of course, can be a healthy sign. Clients are fighting back.

Many sociocultural variables—gender, prejudice, race, religion, social class, upbringing, cultural and subcultural blueprints, and the like—can play a part in resistance. For instance, a man might instinctively resist being helped by a woman and vice versa. An African American might instinctively resist being helped by a white and vice versa. A person with no religious affiliation might instinctively think that help coming from a minister will be "pious" or will automatically include some form of proselytizing.

## PRINCIPLES FOR MANAGING RELUCTANCE AND RESISTANCE

Because both reluctance and resistance are such pervasive phenomena, helping clients manage them is part and parcel of all our interactions with clients (Kottler, 1992). Here are some principles that can act as guidelines.

### Avoid Unhelpful Responses to Reluctance and Resistance

Helpers, especially beginning helpers who are unaware of the pervasiveness of reluctance and resistance, are often disconcerted when they encounter them. They may feel confused, panicked, irritated, hostile, guilty, hurt, rejected, meek, or depressed. Distracted by these unexpected feelings, they react in any of several unhelpful ways.

- They accept their guilt and try to placate the client.
- They become impatient and hostile and manifest these feelings either verbally or nonverbally.
- They do nothing in the hope that the reluctance or the resistance will disappear.
- They lower their expectations of themselves and proceed with the helping process, but in a halfhearted way.
- They try to become warmer and more accepting, hoping to win the client over by love.
- They blame the client and end up in a power struggle with him or her.
- They allow themselves to be abused by clients, playing the role of a scapegoat.
- They lower their expectations of what can be achieved by counseling.
- They hand the direction of the helping process over to the client.
- They give up and terminate counseling.

In short, when helpers engage "difficult" clients, they experience stress, and some give in to self-defeating "fight or flight" approaches to handling it.

The source of this stress is not just the behavior of clients; it also comes from the helper's own self-defeating attitudes and assumptions about the helping process. Here are some of these attitudes and assumptions.

- All clients should be self-referred and adequately committed to change before appearing at my door.
- Every client must like me and trust me.
- I am a consultant and not a social influencer; it should not be necessary to place demands on clients or even help them place demands on themselves.
- Every unwilling client can be helped.
- No unwilling client can be helped.
- I alone am responsible for what happens to this client.
- I have to succeed completely with every client.

Effective helpers neither court reluctance and resistance nor are surprised by them.

## Develop Productive Approaches to Dealing with Reluctance and Resistance

In a book like this, it is impossible to identify every possible form of reluctance and resistance, much less provide a set of strategies for managing each. Here are some principles and a general approach to managing reluctance and resistance in whatever forms they take.

- *See some reluctance and resistance as normal.* Help clients see that they are not odd because they are reluctant. Beyond that, help them see the positive side of resistance. It may well be a sign of their affirmation of self.

- *Accept and work with the client's reluctance and resistance.* This is a central principle. Start with the frame of reference of the client. Accept both the client and his or her reluctance or resistance. Do not ignore it or be intimidated by it. Let clients know how you experience it and then explore it with them. Model openness to challenge. Be willing to explore your own negative feelings. The skill of direct, mutual talk (called immediacy, to be discussed later) is extremely important here. Help clients work through the emotions associated with reluctance and resistance. Avoid moralizing. *Befriend* the reluctance or the resistance instead of reacting to it with hostility or defensiveness.

- *See reluctance as avoidance.* Reluctance is a form of avoidance that is not necessarily tied to client ill will. Therefore, you need to understand the principles and mechanisms underlying avoidance behavior, which is often discussed in texts dealing with the principles of behavior (see Watson & Tharp,

1993). Some clients avoid counseling or give themselves only halfheartedly to it because they see it as lacking in suitable rewards or even as punishing. If that is the case, then counselors have to help them in their search for suitable rewards. Constructive change is more rewarding than the status quo, but that might not be the perception of any given client.

• *Explore your own reluctance and resistance.* Examine reluctance and resistance in your own life. How do you react when you feel coerced? How do you run away from personal growth and development? If you are in touch with the various forms of reluctance and resistance in yourself and are finding ways of overcoming them, you are more likely to help clients deal with theirs.

• *Examine the quality of your interventions.* Without setting off on a guilt trip, examine your helping behavior. Ask yourself whether you are doing anything to coerce clients, thereby running the risk of provoking resistance. For example, you may have become too directive without realizing it. Furthermore, take stock of the emotions that are welling up in you because clients lash back or drag their feet and the ways in which these emotions might be "leaking out." Do not deny these feelings. Rather, own them and find ways of coming to terms with them. Do not overpersonalize what the client says and does. If you are allowing a hostile client to get under your skin, you are probably reducing your effectiveness.

• *Be realistic and flexible.* Remember that there are limits to what a helper can do. Know your own personal and professional limits. If your expectations for growth, development, and change exceed the client's, you can end up in an adversarial relationship. Rigid expectations of the client and yourself become self-defeating.

• *Establish a "just society" with your client.* Deal with the client's feelings of coercion. Provide what Smaby and Tamminen (1979) called a "two-person just society" (p. 509). A just society is based on mutual respect and shared planning. Therefore, establish as much mutuality as is consonant with helping goals. Invite participation. Help clients participate in every step of the helping process and in all the decision making. Share expectations. Discuss and get reactions to helping procedures. Explore the helping contract with your clients and get them to contribute to it.

• *Help clients search for incentives for moving beyond resistance.* Help clients find incentives for participating in the helping process. Use client self-interest as a way of identifying these. Use brainstorming as a way of discovering possible incentives. For instance, the realization that he or she is going to remain in charge of his or her own life may be an important incentive for a client.

• *Tap significant others as resources.* Do not see yourself as the only helper in the lives of your clients. Engage significant others, such as peers and family members, in helping clients face their reluctance and resistance. For instance, lawyers who belong to Alcoholics Anonymous may be able to deal with a lawyer's reluctance to join a treatment program more effectively than you can.

• *Employ clients as helpers.* If possible, find ways to get reluctant and resistant clients into situations where they are helping others. The change of perspective can help clients come to terms with their own unwillingness to work. One tactic is to take the role of the client in the interview and manifest the same kind of reluctance or resistance he or she does. Have the client take the counselor role and help you overcome your unwillingness to work or cooperate. Group counseling, too, is a forum in which clients become helpers. One person who did a great deal of work for Alcoholics Anonymous had a resistant alcoholic go with him on his city rounds, which included visiting hospitals, nursing homes for alcoholics, jails, flophouses, and down-and-out people on the streets. The alcoholic saw through all the lame excuses other alcoholics offered for their plight. After a week he joined AA himself.

In summary, do not avoid dealing with reluctance and resistance, but do not reinforce these processes in clients. Work with your client's unwillingness and become inventive in finding ways of dealing with it.

## CHAPTER 9

# STEP I-B:
# I. THE NATURE OF CHALLENGING — HELPING CLIENTS CHALLENGE THEMSELVES

# INTRODUCTION TO CHALLENGING: HELPING CLIENTS DEAL WITH THEIR BLIND SPOTS

We have noted that effective helpers not only are committed to understanding clients, including the ways in which they experience themselves and the world, but also are reality testers. Contrary to one version of the client-centered credo, empathic understanding is quite often not enough. As we have already seen, probing can help clients move forward in a number of ways. Now we move to a new principle and a new set of skills. The principle is this:

> *Invite clients to challenge themselves to change ways of thinking and acting that keep them mired in problem situations and prevent them from identifying and developing opportunities. If they do not accept the invitation, then challenge them directly to change.*

Indeed, writers who have emphasized helping as a social-influence process have always seen some form of challenge as central to helping (Bernard, 1991; Dorn, 1984, 1986; Strong & Claiborn, 1982). Kiesler (1988) has an interesting way of distinguishing Step I-A from Step I-B in Stage I. In Step I-A client's launch into a rather subjective and sometimes rehearsed version of their stories. Kiesler calls this the "hooked" stage because clients are experts in their stories. They live them every day. I might add that since bits and pieces of the client's story can emerge throughout the helping process, the "hooked" stage can crop up at any time. Helpers listen to these stories and very often watch dysfunctional patterns of behavior unfold during the presentation. Helpers become "unhooked" in Stage I when they use probing and challenging to get at the story behind the story.

Martin (1994) put it well when he suggested that helping conversations may add the most value when they are perceived by clients as relevant, helpful, interested, *and* "somehow inconsistent (discordant) with their current theories of themselves and their circumstances" (pp. 53–54). The same point is made by Trevino (1996, p. 203) in the context of cross-cultural counseling:

> Certain patterns of congruency and discrepancy . . . between client and counselor facilitate change. There is a significant body of research suggesting that congruency between counselor and client enhances the therapeutic relationship, whereas discrepancy between the two facilitates change. In a review of the literature on this topic, Claiborn (1982, p. 446) concluded that the presentation of discrepant points of view contributes to positive outcomes by changing "the way the client construes problems and considers solutions."

If, like most people, clients use only a fraction of their potential in managing their problem situations and developing unused resources, then challenge has a place in helping. Weinrach (1995, 1996), in touting the virtues of Rational Emotive Behavior Therapy (REBT), which incorporates challenge into its core, suggested that some members of the helping profession object to that form of therapy because it is a "tough-minded therapy for a tender-minded profession" (1995, p. 296).

Note that the word *challenge* is used here rather than the more hard-hitting term *confrontation*. Many see both confronting and being confronted as unpleasant experiences. The history of psychology has seen periods when irresponsible confrontation was justified as "honesty" (Egan, 1970). On the other hand, there can be responsible forms of confrontation. Challenging clients to change attitudes and behaviors after they have rejected invitations to do so themselves is a form of confrontation.

Here is an example of challenging. Alicia, a woman who experiences herself as unattractive, is a member of a counseling group.

GROUP COUNSELOR: You say that you're unattractive, and yet you have talked about how you get asked out a lot. I don't find you unattractive myself. And, if I'm not mistaken, I see people here react to you in ways that say they like you. Help me pull all this together.

ALICIA: What you say is probably true, and it helps me clarify what I mean. First of all, I'm no raving beauty, and when others find me attractive, I think that means that they find me intellectually interesting, a caring person, and things like that. At times I wish I were more physically attractive, though I feel ashamed when I say things like that. The real issue is that much of the time I *feel* unattractive. And sometimes I feel most unattractive at the very moment people are telling me directly or indirectly that they find me attractive.

GROUP MEMBER: So you've gotten into the habit of telling yourself in various ways that you're unattractive. . . . It sounds like a bad habit.

ALICIA: It is a lousy habit. If I look at my early home life and my experiences in grammar school and high school, I could probably give you the long, sad story of how it happened. But that was then and this is now.

Since the counselor's experience of Alicia is so different from her experience of herself, he invites her to explore this discrepancy in the group. Her self-exploration clarifies the issue greatly. A couple of questions remain to be answered. First, what costs or, in a backhanded way, benefits are there in Alicia's keeping her "lousy habit"? Second, what does she need to do to get rid of this way of thinking? At least now the issue is on the table.

I-B is a "step" in the logical movement from attending to listening to understanding to empathy to probing to challenging. Note that the term *blind spots* is used. *This refers broadly to ways of thinking and acting that clients fail to see or don't want to see.* However, although challenge is discussed here as a separate step, in reality it involves a set of communication skills that are to be used in all stages and steps of the helping process. Clients can be helped to challenge themselves to

- clarify problem situations by describing specific experiences, behaviors, and feelings when they are being vague or evasive;
- talk about issues — problems, opportunities, goals, commitment, strategies, plans, actions — when they are reluctant to do so;

---

BOX 9-1
**Questions to Uncover Blind Spots**
These are the kinds of questions you can help clients ask themselves to develop new perspectives, change internal behavior, and change external behavior.

- What problems am I avoiding?
- What opportunities am I ignoring?
- What's really going on?
- What am I overlooking?
- What do I refuse to see?
- What don't I want to do?
- What unverified assumptions am I making?
- What am I failing to factor in?
- How am I being dishonest with myself?
- What's underneath the rocks?
- If others were honest with me, what would others tell me?

---

- develop new perspectives on themselves, others, and the world when they prefer to cling to distortions;
- review alternative scenarios, critique them, develop goals, and commit themselves to reasonable agendas when they continue to wallow in their problems;
- search for ways of getting what they want, their preferred scenarios, even in the face of obstacles;
- spell out specific plans instead of taking a scattered, hit-or-miss approach to change;
- persevere in the implementation of these plans when they are tempted to give up;
- review what is and what is not working in their pursuit of change "out there."

In sum, counselors can help clients challenge themselves to engage more effectively in all the stages and steps of problem management during the sessions themselves and in the changes they are pursuing in everyday life. This chapter discusses the nature of challenging. Chapter 10 focuses on the communication skills and strategies involved in challenging together with a range of principles to guide challenges. Box 9-1 reviews the kinds of questions counselors can help their clients ask themselves to uncover blind spots.

# THE GOALS OF CHALLENGING

There are three general areas of human behavior that often need to be challenged — dysfunctional mind-sets and perspectives, self-limiting internal actions, and problematic external actions. These are *blind spots*—that is, self-limiting ways of thinking and acting that clients fail to see or don't want to see.

• *Dysfunctional mind-sets and perspectives.* Invite clients to transform outmoded, self-limiting mind-sets and perspectives into self-liberating and self-enhancing *new perspectives*. For instance, Candace is having a great deal of trouble with a colleague at work who happens to be Jewish. As she grew up, she picked up the idea that "Jews are treacherous businessmen." The counselor invites her to rethink this prejudicial stereotype. Her colleague may or may not be treacherous, but it's not because he is Jewish.

• *Self-limiting internal actions.* Invite clients to challenge and change self-limiting and self-defeating *internal behaviors*. For instance, John daydreams a lot. Thinking about success has taken the place of working for success. When given a project, Minerva immediately thinks of why it won't work. She first has to see that her internal behavior is self-limiting and then do something to change it.

• *Problematic external actions.* Invite clients to challenge and change self-defeating *external behaviors*. Ryan is having trouble relating to his college classmates. He is aggressive, hogs conversations, tries to get his own way when events are being planned, and criticizes others freely. He first needs to understand that his external behavior is self-defeating and then find ways of doing something to change it.

In practice these three categories are often mixed together. Minerva believes that the world is filled with dishonest people (an idea needing a new perspective). Whenever she meets someone new, she reviews that person's behavior through this lens and thinks that he or she is guilty until proved innocent (internal behavior in need of reform). Therefore, when she meets someone new, she is defensive and often questions that person's actions (external actions needing change). Saul, a counseling-psychology trainee, is helped to see how his hesitancy to participate actively in the training group makes him an observer of rather than a contributor to (one new perspective) the group. In fact, some members experience him as a detractor (another new perspective) because his reluctance to participate is discussed frequently and soaks up the time and energy of the group. Saul had justified his stance to himself by telling himself that he was "learning a lot" (internal behavior). Once he realizes this, he develops an action program that helps him to become a contributor (new external behavior) instead of an observer (old lack of external behavior). Use the following principles to guide your use of challenging.

# Help Clients Develop New Perspectives

Over the past few years a number of psychological theoreticians have turned to philosophy to understand human beings more fully, to discover the intellectual underpinnings of their own psychological theories, and to use philosophical theories as a way of pushing back the frontiers of psychology. Since human beings are thinking animals, much of this theorizing has focused on epistemology, the philosophy of knowing. Under the rubric "constructivism," theorists have rediscovered that both cultures and individuals, to a greater or lesser extent, *construct* the realities to which they respond (see Borgen, 1992, pp. 120–121; Mahoney, 1991; Mahoney & Patterson, 1992, pp. 671–674; Neimeyer, 1993; Neimeyer & Mahoney, 1995). This view does not deny that there is a "reality out there," nor does it deny that this reality affects us. But cultures over time create "worldviews" and act on them. Within these cultures individuals, over time, create subsets of these worldviews—that is, their own personal worldviews. These personal worldviews drive behavior. Of course, if each culture and each human being were totally constructivist, this would lead to chaos and eliminate the ability to communicate—"We're all mad here." Still, the commonsense view that perceptions are important because people act on them has merit.

The tendency to construct reality or perceive the world in an individualist way produces a great deal of diversity among individuals, which, as outlined in Chapter 3, is a challenge to helpers. The construction of reality can contribute both to cultural and individual richness and to the creation of social and individual problems. The focus here is the individual. Some mental constructions are self-limiting. Challenging means, in part, helping clients explore their constructions and the actions and impact that follow from them. Then, using clients' ability to "construct" reality, counselors can help them *reconstruct* their views of themselves and their worlds in more self-enhancing ways.

There are many upbeat names for this process: seeing things more clearly, getting the picture, getting insights, developing new perspectives, spelling out implications, transforming perceptions, developing new frames of reference, looking for meaning, shifting perceptions, seeing the bigger picture, developing different angles, seeing things in context, context breaking, rethinking, getting a more objective view, interpreting, overcoming blind spots, second-level learning, double-loop learning (Argyris, 1982), thinking creatively, reconceptualizing, discovery, having an "aha" experience, developing a new outlook, questioning assumptions, getting rid of distortions, relabeling, and making connections. Add such terms as framebreaking, framebending, and reframing. All of these imply some kind of cognitive restructuring, developing understanding, or awareness that is needed or useful for engaging in problem-managing action. Research has shown that the development of new perspectives is highly prized by clients (Elliott, 1985). In the following example, the client, Leslie, an 83-year-old woman, is a resident of a nursing home. She is talking to one of the nurses.

**LESLIE:** I've become so lazy and self-centered. I can sit around for hours and just reminisce . . . letting myself think of all the good things of the past — you know, the old country and all that. Sometimes a whole morning can go by.

**NURSE:** I'm not sure what's so self-indulgent about that.

**LESLIE:** Well, it's in the past and all about myself. . . . I don't know if it's right.

**NURSE:** When you talk to me about it, the reminiscing sounds almost like a kind of meditation for you.

**LESLIE:** You mean like a prayer?

**NURSE:** Well, yes . . . like a prayer.

The client sees reminiscing as laziness. The nurse helps her develop a new, more positive perspective. Effective helpers assume that clients have the resources to see the world in a less distorted way and to act on what they see. Another way of putting it is that skilled counselors help clients move from what the Alcoholics Anonymous movement calls "stinkin' thinkin'" to "healthy thinking" (Kendall, 1992). For instance, it is time to challenge the outmoded medical view that many women have unwittingly adopted of menopause as a "deficiency disease" and as a sign of "getting old." Rather, menopause is a natural developmental stage of life. Although it indicates the ending of one phase, it also opens up new life-stage possibilities.

## Help Clients Link New Perspectives to Constructive Behavioral Change

Although new perspectives are important, they are not magic. Overstressing insight and self-understanding — that is, new perspectives — can actually stand in the way of action instead of paving the way for it. The search for insight can too easily become a goal in itself. Unfortunately, in psychology, especially the pop version, much more attention has been devoted to developing insight than to linking insight to action. The former is often presented as "sexy," whereas the latter is work. I do not mean to imply that achieving useful insights into oneself and one's world is not hard work and often painful. But if the pain is to be turned into gain, constructive behavioral change is required, whether the behavior is internal, external, or both.

Ned, a manager who thought that he was a leader, comes to realize that despite a company "empowerment" program, he still keeps making all the decisions. This slows down the work and makes the department less efficient than it might be. Once he gets over the shock of reading the review from his team members, he uses their suggestions to help him translate his discovery into action — that is, real delegation, less "telling," and more coaching and counseling. He gets reports regularly from his team on his progress.

# Help Clients Challenge and Change Self-Limiting Internal Behavior

Internally we daydream, pray, ruminate on things, believe, make decisions, formulate plans, make judgments, question motives, approve of self and others, disapprove of self and others, wonder, value, imagine, ponder, think through, create standards, fashion norms, mull over, worry, panic, ignore, forgive, rehearse—we do all sorts of things. These are behaviors, not experiences. These are things we do; they are not things that happen to us.

**Self-talk.** Included here is the self-talk in which all of us, including clients, engage. The way we talk to ourselves internally has a distinct impact on the way we act. As mentioned earlier, self-talk is not the same as the shadow conversation we have with ourselves when talking to others. Self-talk refers to the thematic messages we send ourselves. Some people keep saying to themselves, "You're no good." Others send a completely different message: "You're special; you're exempt from the ordinary rules." Worriers engage in a great deal of dysfunctional self-talk. The theme is that things are going to go wrong. Of course, self-talk can also be upbeat. One client learned to say to himself, "When others poke fun at you in a good-natured way, join in, enjoy it, and defuse it. That works much better than getting angry."

**Self-limiting beliefs.** Ellis (1987a, 1987b, 1991; see also Bernard & DiGiuseppe, 1989; Ellis & Dryden, 1987; Huber & Baruth, 1989), in his rational-emotional-behavioral approach to helping, claims that one of the most useful interventions helpers can make is to challenge clients' irrational beliefs. Since clients in some way "talk themselves" into these dysfunctional beliefs, that is a form of self-talk. Some of the common beliefs that Ellis believes get in the way of effective living are these:

- *Being liked and loved.* I must always be loved and approved by the significant people in my life.
- *Being competent.* I must always, in all situations, demonstrate competence, and I must be both talented and competent in some important area of life.
- *Having one's own way.* I must have my way, and my plans must always work out.
- *Being hurt.* People who do anything wrong, especially those who harm me, are evil and should be blamed and punished.
- *Being in danger.* If anything or any situation is dangerous in any way, I must be anxious and upset about it.
- *Being problemless.* Things should not go wrong in life, and if by chance they do, there should be quick and easy solutions.

- *Being a victim.* Other people and outside forces are responsible for any misery I experience. No one should ever take advantage of me.
- *Avoiding.* It is easier to avoid facing life's difficulties than to develop self-discipline; making demands of myself should not be necessary.
- *Tyranny of the past.* What I did in the past, and especially what happened to me in the past, determines how I act and feel today.
- *Passivity.* I can be happy by avoiding, by being passive, by being uncommitted, and by just enjoying myself.

Ellis suggests that if any of these beliefs are violated in a person's life, he or she tends to see the experience as terrible, awful, even catastrophic. But "catastrophizing" gets clients nowhere. In this approach there is, of course, the danger that the helper will try, overtly or covertly, to substitute his or her values for the client's.

**Internal actions as life-limiting.** Sometimes our internal actions add value; at other times they are a sign of or lead to trouble. In a manufacturing company a supervisor was fired when she kept making snap decisions based on inaccurate information. Her thinking about herself and her style had become "automated." She was in a rut. She thought she was being a "hotshot." Sometimes our internal actions—for instance, our judgments of people—flow from a good understanding of ourselves, others, and the world around us. At other times, our internal actions stem from prejudice, ignorance, and misunderstanding. Usually, new perspectives are needed before clients see the need for changes in internal behavior. For instance, what one client saw as "carefully thinking things through" was really fear of making decisions. This led to a pervasive procrastination that others saw as laziness. He was engaged in a great deal of activity, but it was all inside his head.

The following example deals with Kris, a student in a counseling-psychology M.A. program. He is having his biweekly review session with his instructor-trainer. Training in counseling skills takes place in small groups. The issue is the status of his proficiency in communication skills.

TRAINEE: I'm surprised to hear that you don't think that my communication skills are improving. I thought they were.

TRAINER: Kris, you know that's not what the feedback forms from your fellow group members are saying.

TRAINEE: Well, they're not all doing that well themselves.

TRAINER: I have a hunch that I'd like to share with you. You're very bright. That's no secret. And you're an attractive guy. There could be a person inside you saying something like this: "This stuff is really easy. You already have most of these skills. And what you lack now can be made up by personality and smarts." Comment?

TRAINEE: Ouch!

TRAINER: Come on.

**TRAINEE:** I said "ouch" because there's something to what you said. I know I can be cocky, but I don't like being called on it.

**TRAINER:** How much time do you spend practicing these skills in your everyday life?

**TRAINEE:** I think you know the answer to that. I'm also hearing that I'm kidding myself if I think that I'm getting away with it. . . . Okay. I've got the message.

The heart of this exchange is the trainer's hypothesis about Kris's self-talk, and he hits the mark. Later he challenges Kris on the part of the training contract dealing with making communication skills second nature through practice in day-to-day interactions. To this point Kris thought that such a program was beneath him. Note that there is little explicit empathy in the exchange. There are two reasons for this. First, by now the relationship is itself fairly empathic. Second, there is an implicit empathic spirit to the trainer's comments.

Changing internal or external behavior includes the following generic actions:

- *Starting* activities related to managing problems and developing opportunities
- *Continuing* and *increasing* activities that contribute to problem management and opportunity development
- *Stopping* activities that either cause problems and limit opportunities or stand in the way of problem management and opportunity development.

Take an example. Roberto is unhappy in his marriage. Before he and Maria got married, they talked a great deal about the cultural difficulties they might face. He still veered toward the Hispanic culture, whereas she had become quite "Anglo" in her attitudes and behavior. She was especially worried about norms relating to the role of the women. Roberto said he would enjoy being married to someone with a "pioneer" spirit. They thought that they had things worked out. That was then. Now Maria has broken through a number of cultural taboos. She has put herself through college, gotten a job, developed it into a career, and assumed the role of mother and co-breadwinner. She makes more money than Roberto. His woes include losing face in the community, feeling belittled by his wife's success, and being forced into an overly "democratic" marriage.

If Roberto is going to manage the conflict between himself and his wife better, he needs to challenge himself to change a number of internal behaviors. Here are some possibilities.

- *Start* thinking of his wife as an equal in the relationship, *start* understanding her point of view, and *start* imagining what an improved relationship with her might look like
- *Continue* to take stock of the ways he contributes to their difficulties and *increase* the number of times he tells himself to let her live her life as fully as he wants to live his own

- *Stop* telling himself that she is the one with the problem, *stop* seeing her as the offending party when conflicts arise, and *stop* telling himself there is no hope for the relationship

That is, he has to put his internal life in order and begin to mobilize internal resources. Of course, given that any relationship is a two-way street, Maria probably has to develop some new perspectives and change some of her internal and external behavior.

## Help Clients Challenge and Change Self-Limiting External Behavior

Clients may develop new perspectives and change their internal behavior and still get nowhere. Usually the ultimate payoff lies in changing external behaviors. The following example deals with challenging external behavior in a counselor-training program. Trainees A and B are feedback partners; that is, outside group sessions they meet and comment on the quality of each other's participation in the group with a view to helping the other improve his or her performance. In this conversation they focus on each other's behavior in what is called the counseling program's lifestyle group. A lifestyle group is one in which trainees talk about issues in their own lives that might stand in the way of becoming effective helpers.

**TRAINEE A:** You are very insightful. When you provide empathy in practice counseling sessions, it almost always hits the mark. You could add a lot of value if you used this skill during the lifestyle-group discussions.

**TRAINEE B:** Somehow I see the practice sessions as real and the lifestyle-group discussions as, well, not phony, but not essential.

**TRAINEE A:** You participate a lot, but mainly to challenge what others say. Sometimes there's a bit of an edge in your voice when you speak. Also, you don't say much about yourself.

**TRAINEE B:** I don't like talking about myself in the group. . . . It's phony.

**TRAINEE A:** I'm not sure what's phony about discussing issues that can stand in the way of our effectiveness as counselors. Maybe it's personally phony for you.

**TRAINEE B:** There are a couple of things that might affect my effectiveness, but I don't think I'm ready to talk about them in an open group. Maybe I mean that talking when you're not ready is phony.

**TRAINEE A:** So the group might not be the right place for discussing some personal issues. . . . What forum do you think would be the best for you? I'm not sure whether you've found one yet.

**TRAINEE B:** To tell you the truth, I don't talk about some things with anyone. I don't trust any of the trainers enough to talk to them one-on-one, at least not yet. . . . I talk to myself (laughs).

**TRAINEE A:** So talking can be tough. . . . In the practice counseling sessions, when students use real issues, you give the impression that that's what you expect. They trust you, and you don't betray that trust. You talk to everyone in a caring way. But you are no pushover.

**TRAINEE B** (pauses): Why don't you just say it? I don't do what I expect others to do.

**TRAINEE A:** Well . . .

**TRAINEE B** (interrupting): I apologize. I'm doing it right here. I'm getting angry at you because you're speaking the truth.

In this exchange, Trainer A challenges Trainer B by describing B's behavior and by pointing out some skills and resources B has but does not use; that is, he challenges some of B's actions and some of his unused resources.

One way of helping clients challenge both internal and external actions is to help them explore the consequences of their actions. Let's return to Roberto and his mild attempts at sabotaging his wife's career. He refers to his actions as "delaying tactics."

**HELPER:** It might be helpful to see where all of this is leading.

**ROBERTO:** What do you mean?

**HELPER:** I mean let's review the impact your delaying tactics have had on Maria and your marriage. And then let's review where these tactics are most likely to lead.

**ROBERTO:** Well, I can tell you one thing. She's become even more stubborn.

Through their discussion Roberto discovers not only that his sabotage is not working for him but also that it is actually working against him. His campaign is headed in the wrong direction. Compromise is becoming less and less likely.

Roberto needs to start, continue, increase, and stop certain external behaviors if he is going to do his part in developing a better relationship with his wife. Here are some possibilities.

- *Start* activities that will help him develop his own career, *start* sharing his feelings with her instead of just expressing them in negative ways, *start* engaging in mutual decision making, and *start* taking more initiative in household chores and child care

- *Continue* visiting her parents with her and *increase* the number of times he goes to business-related functions with her

- *Stop* criticizing her in front of others, *stop* creating crises at home and assigning the blame for them to her, and *stop* making fun of her business friends

With the counselor's help, Roberto might challenge himself to develop and implement a set of possibilities that would help him do his part to reset the marriage.

## Help Clients Find Strengths amid Their Weaknesses

In a backhanded way, helping clients identify self-limiting blind spots by exploring cognitive perspectives, dysfunctional behavior, and life-limiting external behavior can also be a search for resources. Driscoll (1984), as we have seen, has pointed out that helpers can show clients that their "irrationalities" actually make sense. Instead of forcing clients to see how stupidly they are thinking and acting, helpers can challenge them to find the logic embedded even in seemingly dysfunctional ideas and behaviors. Then clients can use that logic as a resource to manage problem situations instead of perpetuating them. A psychiatrist friend of mine helped a client see the "beauty," as it were, of a very carefully constructed self-defense system. The client, through a series of mental gymnastics and external behaviors, was cocooning himself from real life. My friend helped the client see how inventive he had been and how powerful the system that he had created was. He went on to help him redirect that power into more life-enhancing channels.

## THE CONTENT OF CHALLENGE: COMMON DYSFUNCTIONAL MIND-SETS AND BEHAVIOR FOUND IN HELPING SETTINGS

Self-limiting perspectives, dysfunctional internal behavior, and problematic external behavior, whether related to the helping sessions themselves or to the client's everyday life, provide grist for challenging. Although challenge has to be adapted to the needs and style of each client, there are some common attitudes and behaviors that often need to be challenged. Here are some common areas. Invite clients to challenge

- failure to own problems;
- failure to define problems in solvable terms;
- faulty interpretations;
- everyday dishonesties: evasions, distortions, and game playing.

This list is not exhaustive, but it's a start. We human beings are very inventive when it comes to self-deception and avoidance.

### Invite Clients to Own Their Problems and Unused Opportunities

It is all too common for clients to refuse to take responsibility for their problems and unused opportunities. Instead, there is a whole list of outside forces and other people who are to blame. Therefore, clients need to challenge themselves to own the problem situation. Here is the experience of one counselor who had responsibility for about 150 young men in a youth prison within the confines of a larger central prison.

> I believe I interviewed each of the inmates. And I did learn from them. What I learned had little resemblance to what I had found when I read their files containing personal histories, including the description of their crimes. What I learned when I talked with them was that they didn't belong there. With almost universal consistency they reported a "reason" for their incarceration that had little to do with their own behavior. An inmate explained with perfect sincerity that he was there because the judge who sentenced him had also previously sentenced his brother, who looked very similar; the moment this inmate walked into the courtroom he knew he would be sentenced. Another explained with equal sincerity that he was in prison because his court-appointed lawyer wasn't really interested in helping him. (Miller, 1984, pp. 67–68)

Some people are truly victimized by the systems of society. They can be helped by people willing to fight for the rights of others and by structural changes in society. Other people are victimized by themselves. We all do this from time to time. Some helpers have seen their private practices wither because of the rise of managed care in the United States. It is one thing to criticize managed-care approaches to mental-health service to the degree that they shortchange clients. It is another thing to blame managed care for one's professional woes. Upheavals in all industries, including the helping professions, are part of the North American picture.

Not only problems but also opportunities need to be owned by clients. As Wheeler and Janis (1980) noted, "Opportunities usually do not knock very loudly, and missing a golden opportunity can be just as unfortunate as missing a red-alert warning." (p. 18)

Carkhuff (1987) talked about owning problem situations and unused opportunities in terms of "personalizing." Suppose a client feels that her business partner has been pulling a fast one on her. Consider how the following helper statements differ.

**STATEMENT A:** You feel angry because he unilaterally made the decision to close the deal on his terms.

**STATEMENT B:** You're angry because your legitimate interests were ignored.

**STATEMENT C:** You're furious because you were ignored, your interests were not taken into consideration, maybe you were even financially victimized, and you let him get away with it.

These three statements become progressively more personal. Personalizing means helping clients understand that in some situations, they may have some responsibility for creating or at least perpetuating their problem situations. Statement C does precisely that. It is about ownership.

# Invite Clients to State Their Problems as Solvable

Jay Haley (1976, p. 9) said that if "therapy is to end properly, it must begin properly — by negotiating a solvable problem." It is not uncommon for clients to state problems so that they seem unsolvable. This justifies a "poor-me" attitude and a failure to act.

> **UNSOLVABLE PROBLEM:** In sum, my life is miserable now because of my past. My parents were indifferent to me and at times even unjustly hostile. If only they had been more loving, I wouldn't be in this mess. I am the failed product of an unhappy environment.

Of course, clients will not use this rather stilted language, but the message is common enough. The point is that the past cannot be changed. As we have seen, clients can change their attitudes about the past and deal with the present consequences of the past. Therefore, when a client defines the problem exclusively as a result of the past, the problem cannot be solved. "You certainly had it rough in the past and are still suffering from the consequences now" might be the kind of response that such a statement is designed to elicit. The client needs to move beyond such a point of view.

This brings us to a key principle that has its origin in the value of empowerment and self-responsibility.

> *Help clients state problems or unused opportunities as results of their own behavior, what they do or fail to do.*

A solvable or manageable problem is one that clients can do something about. Consider a different version of the foregoing unsolvable problem.

> **SOLVABLE PROBLEM:** Over the years I've been blaming my parents for my misery. I still spend a great deal of time feeling sorry for myself. As a result, I sit around and do nothing. I don't make friends, I don't involve myself in the community, I don't take any constructive steps to get a decent job.

This message is quite different from that of the previous client. The problem is now open to being managed because it is stated almost entirely as something the client does or fails to do. The client can stop wasting her time blaming her parents, since she cannot change them; she can increase her self-esteem through constructive action and therefore stop feeling sorry for herself; and she can develop the interpersonal skills and courage she needs to enter more creatively into relationships with others.

This does not mean that all problems are solvable by the direct action of the client. A teenager may be miserable because his self-centered parents are constantly squabbling and seem indifferent to him. He certainly can't solve the problem by making them less self-centered, stopping them from fighting, or getting them to care for him more. But he can be helped to find ways to cope with his home situation more effectively by developing fuller social opportunities outside the home. This could mean helping him develop new perspectives on himself and family life and challenging him to act both internally and externally in his own behalf.

## Invite Clients to Move beyond Flawed Interpretations

Often enough clients fail to manage problem situations and develop opportunities because of the ways in which they interpret, understand, or label their experiences, behaviors, and feelings. For instance, Lila, a competent person, may (1) be unaware of her competence, (2) see herself as incompetent, or (3) label her competence as "pushiness" when it is, in fact, assertiveness. Or Dudley, an aggressive and sometimes even violent person, may (1) be unaware that he is aggressive, (2) see himself as merely asserting his rights, or (3) label the assertive actions of others as forms of aggression. Both need to move beyond their faulty interpretations. Here are a number of examples of helpers inviting clients to challenge their interpretations.

Dwayne is a single, middle-aged man who is involved in an affair with a married woman. The helper is getting the feeling that the client is being used but does not realize it.

**HELPER:** How often does she call you?

**DWAYNE:** She doesn't. I always contact her. That's the way we've arranged it.

**HELPER:** I'm curious why you've arranged it that way.

**DWAYNE:** Well, she wanted it that way.

**HELPER:** In mutual relationships, people are eager to get in touch with each other. I don't sense that in your relationship, but I could be wrong.

**DWAYNE:** Well, it's true that in some ways I'm more eager—maybe more dependent.

The helper is inviting the client to explore the relationship more fully without telling the client that there is something wrong with it.

## Invite Clients to Challenge the Predictable Dishonesties of Life

"Predictable dishonesties of life" refers to the discrepancies, distortions, evasions, games, tricks, excuse making, and smoke screens that keep clients mired in their problem situations. All of us have ways of defending ourselves from ourselves, from others, and from the world. We all have our little dishonesties. But defenses are two-edged swords. Daydreaming may help me cope with the dreariness of my everyday life, but it also keeps me from doing something to better myself. Blaming others for my misfortunes helps me save face, but it disrupts interpersonal relationships and prevents me from developing a healthy sense of self-responsibility. The purpose of helping clients challenge themselves is not to strip clients of their defenses, which in some cases could be dangerous, but to help them cope with their inner and outer worlds more effectively.

**Challenge discrepancies.** It can be helpful to zero in on discrepancies

- between what clients think or feel and what they say,
- between what they say and what they do,
- between their views of themselves and the views that others have of them,
- between what they are and what they wish to be,
- between their stated goals and their behavior
- between their expressed values and their actual behavior.

For instance, a helper might challenge the following discrepancies that take place outside the counseling sessions.

- Tom sees himself as witty; his friends see him as biting.
- Minerva says that physical fitness is important, but she overeats and underexercises.
- George says he loves his wife and family, but he is seeing another woman and stays away from home a great deal.
- Clarissa, unemployed for several months, wants a job, but she doesn't want to participate in a retraining program.

Let's use the example of Clarissa to illustrate how this kind of discrepancy can be challenged.

COUNSELOR: I thought that the retraining program would be just the kind of thing you've been looking for.

CLARISSA: Well . . . I don't know if it's the kind of thing I'd like to do. . . . The work would be so different from my last job. . . . And it's a long program.

COUNSELOR: So you feel the fit isn't good.

CLARISSA: Yeah.

COUNSELOR: Clarissa, you say that you don't like being unemployed and this program is designed to help you get what you want. What's going on?

CLARISSA (pauses): You know, I've gotten a bit lazy. . . . I don't like being out of work, but I've gotten used to it.

The counselor sees a discrepancy between what Clarissa is saying and what she is doing. She is actually letting herself slip into a "culture of unemployment."

**Challenge distortions.** Some clients would rather not see the world as it is and therefore distort it in various ways. The distortions are self-serving. For instance:

- Arnie is afraid of his supervisor and therefore sees her as aloof, whereas in reality she is a caring person.
- Edna sees her counselor in some kind of divine role and therefore makes unwarranted demands on him.
- Nancy sees her getting her own way as commitment.

Nancy and Milan come from different cultures. They fought a geat deal in the early years of their marriage, but then things settled down. Now, squabbles have broken out about the best way to bring up their children. Milan is not convinced that counseling is a good idea, so the counselor is talking to Nancy alone. She has forbidden her 12-year-old son to bicycle to school because she doesn't want "his picture to end up on a milk carton." Milan stalked out of the house, yelling back at her, "Why don't you keep him locked in his room!"

NANCY: Milan's just too permissive. Now that they're entering their teenage years, they need more guidance, not less. Let's face it, the world they live in is dangerous.

COUNSELOR: So from your point of view, this is not the time for letting your guard down. . . . I'm making the assumption that Milan is not indifferent to the children's welfare.

NANCY: Of course not! Good grief, he cares as much as I do. We just disagree on how to do it. "Safe, not sorry" is my philosophy.

COUNSELOR: Let's widen the discussion a bit. What other issues do you and Milan disagree on?

NANCY: Well, we used to disagree a lot. But we've put that behind us, it would seem. He leaves a lot of the home decisions to me.

COUNSELOR: I'm not sure whether you've worked things out . . . or if he's given up.

NANCY (curtly): It just happened.

COUNSELOR: You got a bit annoyed when I asked whether Milan was as committed to the kids as you. . . . Since he cares a lot, maybe he's drawing a line in the sand. Maybe he wants to take a stand on this one.

NANCY (heatedly): Well, he's not getting his way on this one! Not on your life! . . . (She catches herself, looks embarrassed, and falls silent.)

COUNSELOR (caringly): Could we try once more to get him to come with you? . . . or maybe have him call me, and I'll try.

The counselor has a hunch that the problem is as much about power and getting one's own way as it is about the kids. Nancy's violent reaction to his suggestion seems to put her in touch with one of her own "little dishonesties."

**Challenge games, tricks, and smoke screens.** If clients are comfortable with their delusions and profit by them, they will obviously try to keep them. If they are rewarded for playing games, inside the counseling sessions or outside, they will continue a game approach to life (see Berne, 1964). Consider some examples:

- Clarence plays the "Yes, but . . ." game: He presents himself as one in need of help and then proceeds to show his helper how ineffective she is.
- Dora makes herself appear helpless and needy when she is with her friends, but when they come to her aid, she is angry with them for treating her like a child.
- Kevin seduces others in one way or another and then becomes indignant when they accept his implied invitations.

The number of games we can play to avoid the work involved in squarely facing the tasks of life is endless. Clients who are fearful of changing will attempt to lay down smoke screens to hide from the helper the ways in which they fail to face up to life. Such clients use communication in order not to communicate (see Beier & Young, 1984). It helps to establish an atmosphere that discourages clients from playing games. An attitude of "Nonsense is challenged here" should pervade the helping sessions.

**Challenge excuses.** Snyder, Higgins, and Stucky (1983; see also Halleck, 1988; Snyder, 1984; Snyder & Higgins, 1988) have examined excuse-making behavior in depth. Excuse making, of course, is universal, part of the fabric of everyday life. Like games and distortions, it has its positive uses in life. Even if it were possible, there is no real reason for setting up a world without myths. On the other hand, excuse making, together with avoidance behavior, probably contributes a great deal to the "psychopathology of the average." Roberto tells the helper that he engaged in benign attempts to sabotage his wife's career because she would get hurt in the Anglo world and because "he was not ready" for the changes in style that his wife's behavior was demanding from him.

We have only skimmed the surface of the games, evasions, tricks, distortions, excuses, rationalizations, and subterfuges resorted to by clients (together with the rest of the population). Skilled helpers are caring and empathic, but they do not let themselves be conned. That helps no one.

## CHALLENGE AND THE SHADOW SIDE OF CLIENTS: SHADOW-SIDE RESPONSES TO CHALLENGE

Even when challenge is a response to a client's plea to be helped to live more effectively, it can precipitate some degree of disorganization in the client. Different writers refer to this experience under different names: "crisis," "disorganization," "a sense of inadequacy," "disequilibrium," and "beneficial uncertainty" (Beier & Young, 1984). As the last of these terms implies, counseling-precipitated crises can be beneficial for the client. Whether they are or not depends, to a great extent, on the skill of the helper.

Even when an invitation to self-challenge or a direct challenge is accurate and delivered caringly, some clients still dodge and weave. Cognitive-dissonance theory (Festinger, 1957) gives us some insight into the dynamics of this. Since dissonance (discomfort, crisis, disequilibrium) is an uncomfortable state, the client will try to get rid of it. According to dissonance theory, there are five typical ways in which people experiencing dissonance attempt to rid themselves of this discomfort. Let's examine them briefly as they apply to being challenged. Since these responses are forms of resistance, review the ways in which resistance can be avoided and managed (Chapter 8) and see how they might apply to the following examples.

**1. Discredit challengers.** The challenger is confronted and discredited. Some attempt is made to point out that he or she is no better than anyone else. In the following example, the client has been discussing her marital problems and has been invited by the helper to take a second look at her behavior.

CLIENT: It's easy for you to sit there and suggest that I be more responsible in my marriage. You've never had to experience the misery in which I live. You've never experienced his brutality. You probably have one of those nice middle-class marriages.

Counterattack is a common strategy for coping with challenge. How do you think this particular instance of counterattack should be handled?

**2. Persuade challengers to change their views.** In this approach challengers are reasoned with. They are urged to see what they have said as misinterpretations and to revise their views. In the following example, the client has been talking about the way she blows up at her husband and has been invited to explore the consequences of this pattern of venting her anger.

CLIENT: I'm not so sure that my anger at home isn't called for. I think that it's a way in which I'm asserting my own identity. If I were to lie down and let others do what they want, I would become a doormat at home. And, as you have helped me see, assertiveness should be part of my style. I think you see me as a fairly reasonable person. I don't get angry here with you because there is no reason to.

Sometimes a client like this will lead an unwary counselor into an argument about the issue in question. How would you respond to this client?

**3. Devalue the issue.** This is a form of rationalization. A client who is being invited to challenge himself about his sarcasm points out that he is rarely sarcastic, that "poking fun at others" is just that, good-natured fun, that everyone does it, and that it is a very minor part of his life not worth spending time on. The client has a right to devalue a topic if it really isn't important. The counselor has to be sensitive enough to discover which issues are important and which are not. How would you handle this client's devaluing the issue?

**4. Seek support elsewhere for the views being challenged.** Some clients leave one counselor and go to another because they feel they aren't being understood. They try to find helpers who will agree with them. This is an extreme way of seeking support of one's own views elsewhere. But a client can remain with the same counselor and still use this strategy by offering evidence that others contest the helper's point of view.

CLIENT: I asked my wife about my sarcasm. She said she doesn't mind it at all. And she thinks that my friends see it as humor and as a part of my style.

This is an indirect way of telling the counselor she is wrong. The counselor might well be wrong, but if the client's sarcasm is really dysfunctional in his interpersonal life, the counselor should find some way of pressing the issue. What would you do in this case?

**5. Cooperate in the helping session, but then do nothing about it outside.** The client can agree with the counselor as a way of dismissing an issue. However, the purpose of challenging is not to get the client's agreement but to develop new perspectives that lead to constructive action. Consider this client.

CLIENT: To tell you the truth, I am pretty lazy and manipulative. And I'm not very clever.

In truth this client was lazy and manipulative and was being both in this instance. Also he was very clever. It was up to the helper to break through this screen of "honesty." What would you do in this case?

## CHALLENGE AND THE SHADOW SIDE
## OF HELPERS

Two shadow-side areas are addressed here—the reluctance of some helpers to challenge clients or invite them to challenge themselves and the blind spots helpers themselves have.

**The "MUM effect."** Initially, some counselor trainees are quite reluctant to help clients challenge themselves. They become victims of what has been called the "MUM effect," the tendency to withhold bad news even when it is in the other's interest to hear it (Rosen & Tesser, 1970, 1971; Tesser & Rosen, 1972; Tesser, Rosen, & Batchelor, 1972; Tesser, Rosen, & Tesser, 1971). In ancient times the person who bore bad news to the king was sometimes killed. That obviously led to a certain reluctance on the part of messengers to bring such news. Bad news—and, by extension, the kind of "bad news" that is involved in any kind of invitation to self-challenge—arouses negative feelings in the challenger, no matter how he or she thinks the receiver will react. If you are comfortable with the supportive dimensions of the helping process but uncomfortable with helping as a social-influence process, you could fall victim to the MUM effect and become less effective than you might be.

**Excuses for not challenging.** Reluctance to challenge is not a bad starting position. In my estimation, it is far better than being too eager to challenge. However, all helping, even the most client-centered, involves social influence. It is important for you to understand your reluctance (or eagerness) to challenge — that is, to challenge yourself on the issue of challenging and on the very notion of helping as a social-influence process. When trainees examine how they feel about challenging others, here are some of the things they discover.

- I am just not used to challenging others. My interpersonal style has had a lot of the live-and-let-live in it. I have misgivings about intruding into other people's lives.
- If I challenge others, then I open myself to being challenged. I may be hurt, or I may find out things about myself that I would rather not know.
- I might find out that I like challenging others and that the floodgates will open and my negative feelings about others will flow out. I have some fears that deep down I am a very angry person.
- I am afraid that I will hurt others, damage them in some way or other. I have been hurt or I have seen others hurt by heavy-handed confrontations.
- I am afraid that I will delve too deeply into others and find that they have problems that I cannot help them handle. The helping process will get out of hand.
- If I challenge others, they will no longer like me. I want my clients to like me.

Finally, being willing to challenge responsibly is one thing; having the skills and wisdom to do so is another. Chapter 10 focuses on the skills and wisdom needed to invite others to take a closer, more constructive look at themselves.

**Helpers' blind spots.** There is an interesting literature on the humanity and flaws of helpers (see Kottler & Blau, 1989; Yalom, 1989) that can be of enormous help to beginners — since prevention is infinitely better than cure — and to old-timers — since you can teach old dogs new tricks. Since helpers are as human as their clients, they too can have blind spots that detract from their ability to help. For instance, in one study (Atkinson and his associates, 1991) counselors were almost unanimous in their preference for a feeling approach to counseling, whereas the majority of male clients preferred either a thinking or an acting orientation. Helpers sometimes prevent clients from moving forward by playing the "insight" game. They help clients develop one insight after another without linking those insights to problem-managing action. Many of the insights end up being oriented to the theories of the helper rather than to the problems of the client. There is an unsurfaced assumption that insight will ultimately cure. On the other hand, some helpers see themselves as challenging when they are, in fact, punishing their clients. Still other helpers don't realize that the client has taken charge of the helping process and is manipulating them.

One of the critical responsibilities of supervisors is to help counselors identify their blind spots and learn from them. Once out of training, skilled helpers use different forums or methodologies to continue this process, especially with difficult cases. They ask themselves, "What am I missing here?" They take counsel with colleagues. Without becoming self-obsessive, they scrutinize and challenge themselves and the role they play in the helping relationship.

## CHAPTER 10

# STEP I-B:
# II. SPECIFIC
# CHALLENGING SKILLS

**ADVANCED EMPATHY: THE MESSAGE BEHIND THE MESSAGE**

Help Clients Make the Implied Explicit

Help Clients Identify Themes in Their Stories

Help Clients Make Connections They May Be Missing

Advanced Empathy as Sharing Hunches with Clients

**INFORMATION SHARING: FROM NEW PERSPECTIVES TO ACTION**

**HELPER SELF-DISCLOSURE**

**IMMEDIACY: DIRECT, MUTUAL TALK**

Types of Immediacy in Helping and Principles for Using Them

Situations Calling for Immediacy

**USING SUGGESTIONS AND RECOMMENDATIONS**

**CONFRONTATION**

**EVALUATION QUESTIONS FOR STEP I-B**

There are any number of ways in which helpers can challenge clients to develop new perspectives, change their internal behavior, and change their external behavior. The following are discussed and illustrated in this chapter: (1) advanced empathy, (2) information sharing, (3) helper self-disclosure, (4) immediacy, (5) suggestions and recommendations, and (6) confrontation. It has already been noted that both probing and summarizing—and even accurate empathy—can challenge clients to rethink their attitudes and behavior. The wisdom just discussed should permeate them all.

## ADVANCED EMPATHY: THE MESSAGE BEHIND THE MESSAGE

Chapter 5 outlined the characteristics of empathy in general and of basic empathy in particular. However, as skilled helpers listen intently to clients, they often see clearly what clients only half see and hint at. In Chapter 4, this was called "listening for the slant." On the listening side, this deeper kind of empathy involves "sensing meanings of which the client is scarcely aware" (Rogers, 1980, p. 142) or, in broader terms, the "story behind the story" (Berger, 1989).

On the responding side, it means sharing your understanding of the message behind the message with the client in the spirit of the wisdom outlined earlier. For instance, Clarence talks about and expresses his anger with his wife, but as he talks, the helper hears not just anger but also *hurt*. It may be that the client can talk with relative ease about his anger but not as easily about his feelings of hurt. In a basic empathic response, clients recognize themselves almost immediately: "Yes, that's what I meant." However, since advanced empathy digs a bit deeper, clients might not immediately recognize themselves in the helper's response. That's what makes advanced empathy, though still empathy, a form of challenge. For instance, the helper said something like this to Clarence: "It's pretty obvious that you really get steamed when she acts like that. But I thought I sensed, mixed in with the anger, a bit of hurt." At that, Clarence's eyes misted up a bit, just a bit, and he looked down, and finally said, "She can still get to me. She certainly can." This appreciably broadened the discussion of the problem situation.

Here are some questions helpers can ask themselves to probe a bit deeper as they listen to clients.

- What is this person only half saying?
- What is this person hinting at?
- What is this person saying in a confused way?
- What covert message is behind the explicit message?

Saying to a client, "You feel angry because your wife has been paying a lot of attention to a younger man—angry and perhaps a bit hurt" can make the client stop short, because the anger is up front, whereas the hurt, though real, is in the background. That's what makes advanced empathy challenging. Note that advanced empathic listening deals with what the client is actually saying or expressing, however confusedly, and not with the helper's interpretations of what the client is saying. Advanced empathy is not an attempt to "psych the client out."

Advanced empathy focuses not just on problems but also on unused or partially used resources. Effective helpers listen for the resources that are buried deeply in clients and often have been forgotten by them. Advanced empathy deals with both the overlooked positive side and the overlooked negative side of the client's experiences, behaviors, and emotions.

In the following example, the client, a soldier who has been thinking seriously about making the army his career, has been talking to a chaplain about his failing to be promoted. He has seen service both in Kuwait and in Bosnia and has performed very well. As he talks, it becomes fairly evident that part of the problem is that he is so quiet and unassuming that it is easy for his superiors to ignore him.

**SOLDIER:** I don't know what's going on. I work hard, but I keep getting passed over when promotion time comes along. I think I work as hard as anyone else, and I work efficiently, but all of my efforts seem to go down the drain. I'm not as flashy as some others, but I'm just as substantial.

**CHAPLAIN A:** You feel it's quite unfair to do the kind of work that merits a promotion and still not get it.

**SOLDIER:** Yeah. . . . I suppose there's nothing I can do but wait it out. (A long silence ensues.)

Chaplain A tries to understand the client from the client's frame of reference. He deals with the client's feelings and the experience underlying those feelings. That is basic empathy. In his response to Chaplain A, the client merely retreats more into himself and chooses a wait-and-see strategy. Let's see a different approach.

**CHAPLAIN B:** It's depressing to put out as much effort as those who get promoted and still get passed by. . . . Tell me more about the not-as-flashy bit. What in your style might make it easy for others not to notice you, even when you're doing a good job?

**SOLDIER:** You mean I'm so quiet I could get lost in the shuffle? Or maybe it's the guys who make more noise, the squeaky wheels my dad called them, who get noticed.

From the context, from the discussion of the problem situation, from the client's manner and tone of voice, Chaplain B picks up a theme that the client states in passing in the phrase "not as flashy." That is, the client is so unassuming that his best efforts go unnoticed. Advanced empathy, then, goes beyond the expressed to the partially expressed and the implied. If helpers are accurate and if their timing is good, they will assist the client to develop a new and useful perspective. In his response to Chaplain B, the client begins to see that his unassuming, nonassertive style may contribute to the problem situation. Once he becomes aware of the self-limiting dimensions of his style, he is in a better position to do something about it. Advanced empathy can take a number of forms. Let's consider some of them.

## Help Clients Make the Implied Explicit

The most basic form of advanced empathy is to help clients give fuller expression to what they are implying. In the following example, the client has been discussing ways of getting back in touch with his wife after a recent divorce, but when he speaks about doing so, he expresses very little enthusiasm.

**CLIENT** (somewhat hesitatingly):  I could wait to hear from her. But I suppose there's nothing wrong with calling her up and asking her how she's getting along.
**COUNSELOR A:**  It seems that it would be all right for you to take the initiative to find out if everything is well with her.
**CLIENT** (somewhat drearily):  Yeah, I suppose I could.

Counselor A's response might have been fine at an earlier stage of the helping process, but it misses the mark here, and the client grinds to a halt.

**COUNSELOR B:**  You've been talking about getting in touch with her, but, unless I'm mistaken, I don't hear a great deal of enthusiasm in your voice.
**CLIENT:**  To be honest, I don't really want to talk to her. But I feel guilty, guilty about the divorce, guilty about seeing her out on her own. I'm taking care of her all over again. And that's one of the reasons we got divorced—I had a need to take care of her and she let me do it. That was the story of our marriage. I don't want to do that anymore.

The goal of Step I-B is to help the client dig deeper in the interest of problem management. Counselor B bases her response not only on the client's immediately preceding remark but also on the entire context of the storytelling process. Her response hits the mark, and the client moves forward. As with basic empathy, there is no such thing as a good advanced empathic response in itself. The question is, Does the response help the client clarify the issue more fully so that he or she might begin to see the need to act differently?

Advanced empathy is part of the social-influence process—it places demands on clients to take a deeper look at themselves. Challenges are based on a basic empathic understanding of the client and are made with genuine care and respect, but they are demands nevertheless.

## Help Clients Identify Themes in Their Stories

When clients tell their stories, certain themes emerge. Thematic material might refer to feelings (such as themes of hurt, of depression, of anxiety), to behavior (such as themes of controlling others, of avoiding intimacy, of blaming others, of overwork), to experiences (such as themes of being a victim, of being seduced, of being punished, of being ignored, of being picked on), or some combination of these. Once you see a self-defeating theme or pattern emerging from your discussions, your task is to communicate your perception to the client in a way that enables the client to check it out.

In the following example, a counseling-psychology trainee is talking with his supervisor. The trainee has four clients. In the past week she has seen each of them for the third time. This dialogue takes place in the middle of a supervisory session.

**SUPERVISOR:** You've had a third session with each of four clients this past week. Even though you're at different stages with each because each started in a different place, you have a feeling, if I understand what you've been saying, that you're "going 'round the mulberry bush" a bit with each.

**TRAINEE:** Yes, I'm grinding my wheels. I don't have a sense of movement.

**SUPERVISOR:** Any thoughts on why not?

**TRAINEE:** Well, they seem willing enough. And I think I've been very good at listening and empathy.

**SUPERVISOR:** Are they enough to get the movement you think is possible? Let's listen to one of the tapes.

They listen to a segment of one of the sessions. The trainee turns off the recorder.

**TRAINEE:** It sounds weird to listen to it. It's all empathy with a few un-huhs. I don't want to be pushy, but this is as far from pushy as you can get.

**SUPERVISOR:** Right. At least in the segment we listened to, no probes and nothing close to summaries or mild challenges.

**TRAINEE:** I've been so preoccupied with not being pushy. Certainly some probes would have given much more focus and direction to the session.

**SUPERVISOR:** Let me play the client as well as I can and see how you might redo the session.

They then spend about fifteen minutes in a role-playing session. The trainee mingles probes with empathy, and the result is quite different. The theme that the supervisor helps the trainee surface is fear of "being pushy," which, as the evidence shows, is an unwarranted fear. Make sure, however, that the themes you discover are based on the client's story and are not just the artifacts of some psychological theory. Advanced empathy works because clients recognize themselves in what you say.

## Help Clients Make Connections They May Be Missing

Clients often reveal experiences, behaviors, and emotions in a hit-or-miss way. The counselor's job, then, is to help them make the kinds of connections that provide action-oriented insights or perspectives.

- Cynaé didn't see that her inability to develop strategies for her chosen goals was related to the fact that she was not really committed to those goals.
- Finnbar didn't relate the trouble he was having with a strong woman supervisor to a bit of sexism in his style.
- Joanna failed to link her lack of influence with fellow team members to her passive-aggressive approach to colleagues.
- Dieter didn't realize that his failed attempts to control his anxiety when working with clients were related to the perfectionistic standards he had set for himself.

The following client, who has a full-time job, is finishing his final two courses for his college degree and is going to get married right after graduation. He talks about being progressively more anxious and tired in recent weeks. Later, he talks about getting ready for his marriage in a few months and about deadlines for turning in papers for current courses. Still later, he talks about his need to succeed, to compete, and to meet the expectations of his parents and grandparents. He has begun to wonder whether all of this is telling him that it is a mistake to get married or that he is an inadequate person. The fact that a recent physical examination showed him to be in good health has actually made things worse.

COUNSELOR: John, there might be a simpler explanation for your growing anxiety and tiredness. One, you are really working very hard at school and at work. Two, competing as hard as you do is bound to take its physical and emotional toll. And three, the emotional drain of getting ready for a marriage is enormous for anyone, including you. I wonder how Superman might manage all this.

JOHN: I thought that the medical exam would have shown some physical basis for being so tired. The idea that I'm trying to juggle too many things never crossed my mind. Ever since I was a kid, I was supposed to handle whatever came along. My grandfather did. My dad did. No complaints.

John had been talking about these four "islands" — a full-time job, finishing school, preparing for marriage, and meeting the expectations of his family — as if they were unrelated to one another. The counselor does two things. First, he helps John make the connections. Second, he provides a bit of information. Things like finishing school and getting ready for marriage score high on the stress scale. High expectations on the part of his family act as a multiplier. They go on to explore ways in which John can manage his stress.

## Advanced Empathy as Sharing Hunches with Clients

From one perspective, advanced empathy means sharing educated hunches or guesses about clients and their overt and covert experiences, behaviors, and feelings that you feel will help them see their problems and concerns more clearly and help them move on to developing new scenarios, setting goals, and acting. Such hunches, of course, must be based on the helper's interactions with the client in which active listening, understanding, empathy, and probing play a large part.

- Hunches can help clients see the bigger picture. In the following example, the counselor is talking to a client who is at odds with his wife's brother: "The problem doesn't seem to be just your attitude toward your brother-in-law anymore; your resentment seems to be spreading to his friends. How do you see it?"

- Hunches can help the clients see what they are expressing indirectly or merely implying. In this example, the counselor is talking to a client who feels that a friend has let her down: "I think I might also be hearing you say that you are more than disappointed—perhaps a bit hurt and angry."

- Hunches can help clients draw logical conclusions from what they are saying. A manager is counseling one of his team members: "From all that you've said about her, it seems that you are also saying that right now you resent having to work with her at all. I know you haven't said that directly. But I'm wondering if you are feeling that way about her."

- Hunches can help clients open up areas they are only hinting at. In this case, a school counselor is talking to a senior in high school: "You've brought up sexual matters a number of times, but you haven't pursued them. My guess is that sex is a pretty important area for you but perhaps pretty touchy, too."

- Hunches can help clients see things they may be overlooking. A counselor is talking to a minister: "I wonder if it's possible that some people take your wit too personally, that they see it, perhaps, as sarcasm rather than humor."

- Hunches can help clients identify themes. In this case, a counselor is talking to a woman who has been abused by her husband: "If I'm not mistaken, you've mentioned in two or three different ways that it is sometimes difficult for you to stick up for your own legitimate rights. For instance, . . ."

- Hunches can help clients take fuller ownership of partially owned experiences, behaviors, and feelings. Example: "You sound as if you have already decided to quit. Or is that overstating the case?"

Hunches should be based on your experience of your clients—their experiences, behaviors, and emotions both within the helping sessions themselves and in their day-to-day lives. Do not base your hunches on "deep" psychological theories. Later on, you will be asked to identify the experiential and behavioral clues on which your hunches are based.

This section ends with a caution or two. First, advanced empathy is not license to draw inferences from clients' history, experiences, or behavior at will (see MacDonald, 1996). Nor is it license to load clients with interpretations that are more deeply rooted in the helper's psychological theories than in the realities of the client's world. Second, advanced empathy is one of the most important of the challenging skills. Of the skills mentioned in this chapter, it is the most frequently used. The following challenging skills, used in moderation, can also be very helpful. But *moderation* is the salient word.

## INFORMATION SHARING: FROM NEW PERSPECTIVES TO ACTION

Sometimes clients are unable to explore their problems fully and proceed to action because they lack information of one kind or another. Information can help clients at any stage or step of the helping process. For instance, in Stage I it helps many clients to know that they are not the first to try to cope with a particular problem. It may also help them to know that moving to Stage II can help them further clarify problems and opportunities. Problem- and opportunity-related information can be of enormous help — for instance, where to look for job opportunities.

The skill or strategy of information sharing is included under challenging skills because it helps clients develop new perspectives on their problems or shows them how to act. It includes both giving information and correcting misinformation. In some cases, the information can prove to be quite confirming and supportive. For instance, a parent who feels responsible following the death of a newborn baby may experience some relief through an understanding of the earmarks of the Sudden Infant Death Syndrome. This information does not "solve" the problem, but the parent's new perspective can help him or her handle self-blame.

The new perspectives clients gain from information sharing can also be both comforting and painful. Consider the following example.

> Adrian was a college student of modest intellectual means. He made it through school because he worked very hard. In his senior year he learned that a number of his friends were going on to graduate school. He, too, applied to a number of graduate programs in psychology. He came to see a counselor in the student services center after being rejected by all the schools to which he had applied. In the interview it soon became clear to the counselor that Adrian thought that many, perhaps even most, college students went on to graduate school. After all, most of his closest friends had been accepted in one graduate school or another. The counselor shared with him the statistics of what could be called the educational pyramid — the decreasing percentage of students attending school at higher levels. Adrian did not realize that just finishing college made him part of an elite group. Nor was he completely aware of the extremely competitive nature of the graduate programs in psychology to which

he had applied. He found much of this relieving but then found himself suddenly faced with what to do now that he was finishing school. Up to this point he had not thought much about it. He felt disconcerted by the sudden need to look at the world of work.

Giving information is especially useful when lack of accurate information either is one of the principal causes of a problem situation or is making an existing problem worse or when information is needed to manage the problem.

In some medical settings, doctors team up with counselors to give clients messages that are hard to hear and to provide them with information needed to make difficult decisions. For instance, Lester, a 54-year-old accountant, has been given a series of diagnostic tests for a heart condition. The doctor and the counselor sit down and talk with him about the findings. Bypass surgery is one option, but it has risks and there is no absolute assurance that the surgery will take care of all his heart problems. Angioplasty is another option, but it, too, has pluses and minuses. The counselor helps Lester cope with the news, process the information, and come to a decision.

There are some cautions helpers should observe in giving information. When information is challenging, or even shocking, be tactful and know how to help the client handle the disequilibrium that comes with the news. Do not overwhelm the client with information. Make sure that the information you provide is clear and relevant to the client's problem situation. Don't let the client go away with a misunderstanding of the information. Be sure not to confuse information giving with advice giving; the latter is seldom useful. Finally, be supportive; help the client process the information.

All these cautions come into play for counselors of those who test positive for the AIDS virus.

> Angie, an unmarried woman who has just given birth to a baby boy, needs to be told that both she and her son are HIV-positive. The implications are enormous. She needs information about the virus, what she needs to do for herself and her son medically, and the implications for her sex life. Obviously the way in which this information is given to her can determine to a great extent how she will handle the immediate crisis and the ensuing lifestyle demands.

It is obvious that Angie's case is not one for an amateur. Experts, particularly those with special training in dealing with AIDS cases, know the range of ways in which clients receive the news that they have tested positive, how to provide support in each case, and how to challenge clients to rally resources and manage the crisis.

At times of major decisions, helpers should not use information as a subtle (or not too subtle) way of pushing their own values. For instance, helpers should not immediately give clients with unwanted pregnancies information about abortion clinics. Conversely, if abortion is contrary to the helper's values, then the helper needs to let the client looking for abortion counseling know that he or she cannot help her in this area.

# HELPER SELF-DISCLOSURE

A third skill of challenging involves the ability of helpers to constructively share some of their own experiences, behaviors, and feelings with clients (Edwards & Murdock, 1994; Hendrick, 1990; Mathews, 1988; Simon, 1988; Stricker & Fisher, 1990; Watkins, 1990; Weiner, 1983). In one sense counselors cannot help but disclose themselves: "The counselor communicates his or her characteristics to the client in every look, movement, emotional response, and sound, as well as with every word" (Strong & Claiborn, 1982, p. 173). This is indirect disclosure. As they attend, listen, and respond, helpers should track the impressions they are making on clients.

Here, however, it is a question of direct self-disclosure. It should be said at the outset that the research in this area does not provide clear-cut guidelines. Many worry about helper self-disclosure because of its possible downside. In some forms of helping, however, direct self-disclosure serves as a form of modeling. Self-help groups such as Alcoholics Anonymous use such modeling extensively. This helps new members get an idea of what to talk about and arouse the courage to do so. It is the group's way of saying, "You can talk here without being judged and getting hurt." Even in one-to-one counseling dealing with alcohol and drug addiction, extensive helper self-disclosure is the norm.

> Beth is a counselor in a drug rehabilitation program. She herself was an addict for a number of years but "kicked the habit" with the help of the agency where she is now a counselor. It is clear to all addicts in the program that the counselors there were once addicts themselves and are not only rehabilitated but also intensely interested in helping others both rid themselves of drugs and develop a kind of lifestyle that helps them stay drug-free. Beth freely shares her experience, both of being a drug user and of her rather agonizing journey to freedom, whenever she thinks that doing so can help a client.

Ex-addicts often make excellent helpers in programs like this. They know from the inside the games addicts play. Sharing their experience is central to their style of counseling and is accepted by their clients. New perspectives are developed, and new possibilities for action are discovered. Such self-disclosure is challenging. It puts pressure on clients to talk about themselves more cogently or in a more focused way.

Helper self-disclosure is challenging for at least two reasons. First, it is a form of intimacy and, for some clients, intimacy is not easy to handle. Helpers need to know precisely why they might be divulging information about themselves. Second, the message to the client is, indirectly, a challenging "You can do it, too," because helper revelations, even when they deal with past failures, usually deal with problem situations that have been overcome.

Research into helper self-disclosure has produced mixed and even contradictory results. Some researchers have discovered that helper self-disclosure can frighten clients or make them see helpers as less well adjusted. Other studies have suggested that helper self-disclosure is appreciated by clients. Some clients see self-disclosing helpers as "down-to-earth" and "honest." Since current research does not tell us a great deal, we need to stick to common sense. In the following example, the helper, Rick, has had a number of sessions with Tim, a client who has had a number of developmental problems and is now taking an overly cautious approach to almost everything in life.

RICK:  In my junior year in high school I was expelled for stealing. I thought that it was the end of the world. My Catholic family took it as the ultimate disgrace. We even moved to a different neighborhood in the city.

TIM:  What did it do to you?

Rick briefly tells his story, a story that includes setbacks — not unlike Tim's — but one that eventually has a successful outcome. Rick does not overdramatize his story. In fact, his story makes it clear that developmental crises are normal. How they are interpreted and managed is the critical issue.

The confused results of research studies tell us that, currently, helper self-disclosure is not a science but an art. Here are some guidelines for using it.

• ***Include helper self-disclosure in the contract.*** In self-help groups and in the counseling of addicts by ex-addicts, helper self-disclosure is an explicit part of the contract. If you don't want your disclosures to surprise your clients, let them know that you may self-disclose. Therefore, a helper might say somewhere toward the beginning of the counseling process something like this: "From time to time I might share with you some of my own life experiences if they can help us move forward."

• ***Make sure that your disclosures are appropriate.*** Sharing yourself is appropriate if it helps clients achieve treatment goals. Don't disclose more than is necessary. Helper self-disclosure that is exhibitionistic or engaged in for effect is obviously inappropriate. Timing is critical. Premature helper self-disclosure can turn clients off. For instance, Goodyear and Shumate (1996) presented evidence indicating that helpers who respond in kind to clients who disclosed feelings of sexual attraction toward them were seen as less expert and less helpful by their clients.

• ***Keep your disclosure selective and focused.*** Don't distract clients with rambling stories about yourself. In the following example, the helper is talking to a first-year grad student in a clinical-psychology program. The client is discouraged and depressed by the amount of work he has to do.

COUNSELOR:  Listening to you brings me right back to my own days in graduate school. I don't think that I was ever busier in my life. I also believe that the most depressing moments of my life took place then. On any number of occasions, I wanted to throw in the towel. I remember once toward the end of my third year when . . .

It may be that selective bits of this counselor's experience in graduate school would be useful in helping the student get a better conceptual and emotional grasp of her problems, but he has wandered off into the kind of reminiscing that meets his needs rather than the client's. Helper self-disclosure is inappropriate if it is too frequent. Some research (Murphy & Strong, 1972) suggested that if helpers disclose themselves too frequently, clients tend to see them as phony and suspect that they have ulterior motives.

- **Do not burden the client.** Do not burden an already overburdened client. One counselor thought that he would help make a client who was sharing some sexual problems more comfortable by sharing some of his own experiences. After all, he saw his sexual development as not too different from the client's. However, the client reacted by saying: "Hey, don't tell me your problems. I'm having a hard enough time dealing with my own." This novice counselor shared too much of himself too soon. He was caught up in his own willingness to disclose rather than its potential usefulness to the client.

- **Remain flexible.** Take each client separately. Adapt your disclosures to differences in clients and situations. In our earlier example, Rick shared his high school experience with Tim relatively early on in their talks as a way of challenging Tim to stop seeing himself as unique. When asked directly, clients say that they want helpers to disclose themselves (see Hendrick, 1988), but this does not mean that every client in every situation wants it or would benefit from it. Self-disclosure on the part of helpers should be a natural part of the helping process, not a gambit.

Once more, moderation in the use of this skill is very important. The full use of it requires a great deal of experience and wisdom.

# IMMEDIACY: DIRECT, MUTUAL TALK

Many, if not most, clients who seek help have trouble with interpersonal relationships. This is either their central concern or part of a wider problem situation. Some of the difficulties clients have in their day-to-day relationships are also reflected in their relationships to helpers. For instance, if they are compliant outside, they are often compliant in the helping process. If they become aggressive and angry with authority figures outside, they often do the same with helpers. Therefore, the client's interpersonal style can be examined, at least in part, through an examination of his or her relationship with the helper. If counseling takes place in a group, then the opportunity is even greater. The package of skills enabling helpers to explore their relationship with their clients or enabling clients to do the same with fellow group members has been called "immediacy" by Robert Carkhuff (see Carkhuff 1969a, 1969b; Carkhuff & Anthony, 1979).

## Types of Immediacy in Helping and Principles for Using Them

Three kinds of immediacy are reviewed here. First, immediacy that focuses on the overall relationship—"How are you and I doing?" Second, immediacy that focuses on some particular event in a session—"What's going on between you and me right now?" Third, self-involving statements. There are three broad principles.

**Review your overall relationship with the client if it adds value.** General-relationship immediacy refers to your ability to discuss with a client where you stand in your overall relationship with him or her. The focus is not on a particular incident but on the way the relationship itself has developed and how it is helping or standing in the way of progress. In the following example, the helper is a 44-year-old woman working as a counselor for a large company. She is talking to a 36-year-old man she has been seeing once every other week for about two months. One of his principal problems is his relationship with his supervisor, who is also a woman.

COUNSELOR: We seem to have developed a good relationship here. I feel we respect each other. I have been able to make demands on you, and you have made demands on me. There has been a great deal of give-and-take in our relationship. You've gotten angry with me, and I've gotten impatient with you at times, but we've worked it out. I'm wondering if you see things the same as I do and, if so, what our relationship has that is missing in your relationship with your supervisor.

CLIENT: Well, for one thing, you listen to me, and I don't think she does. On the other hand, I listen pretty carefully to you, but I don't think I listen to her at all, and she probably knows it. I think she's dumb, and I guess I'm not hiding it from her.

The review of the relationship helps the client focus more specifically on his relationship with his supervisor.

Here is another example. Norman, a 38-year-old trainer in a counselor-training program, is talking to Chen Zeng, 25, one of the trainees.

NORMAN: Chen Zeng, I'm a bit bothered about some of the things that are going on between you and me. When you talk to me, I get the feeling that you are being very careful. You talk slowly — you seem to be choosing your words, sometimes to the point that what you are saying sounds almost prepared. You have never challenged me on anything in the group. When you talk most intimately about yourself, you seem to avoid looking at me. I find myself giving you less feedback than I give others. I've even found myself putting off talking to you about all this. Perhaps some of this is my own imagining, but I want to check it out with you.

CHEN ZENG: I've been afraid to talk about all this, so I keep putting it off, too. I'm glad that you've brought it up. A lot of it has to do with how I relate to people in authority, even though you don't come across as an "authority figure." Still, I hesitate because you don't act the way an authority figure is supposed to act.

In this case, cultural differences are problematic. For Chen Zeng, giving direct feedback to someone in authority is not natural. However, he does go on to talk to Norman about his misgivings. He thinks that Norman's interventions in the training group are too "unorganized" and that he plays favorites. He has not wanted to bring it up because he fears that his position will be jeopardized. But now that Norman has made the overture, he accepts the challenge.

**Address relationship issues as they come up.** Here-and-now immediacy refers to your ability to discuss with clients what is happening between the two of you in the here and now of any given transaction. It is not the entire relationship that is being considered, but only this specific interaction or incident. In the following example, the helper, a 43-year-old woman, is a counselor in a church-related human-services center. Agnes, a 49-year-old woman who was recently widowed, has been talking about her loneliness. Agnes seems to have withdrawn quite a bit, and the interaction has bogged down.

COUNSELOR: I'd like to stop a moment and take a look at what's happening right now between you and me.

AGNES: I'm not sure what you mean.

COUNSELOR: Well, our conversation today started out quite lively, and now it seems rather subdued to me. I've noticed that the muscles in my shoulders have become tense. I sometimes tense up that way when I feel that I might have said something wrong. It could be just me, but I sense that things are a bit strained between us right now.

AGNES (hesitatingly): Well, a little. . . .

Agnes goes on to say how she resented one of the helper's remarks early in the session. She thought that the counselor had intimated that she was lazy. Agnes knows that she isn't lazy. They discuss the incident, clear it up, and move on. The purpose of here-and-now immediacy is to strengthen the working alliance. Research has shown that too much support can actually weaken the working alliance (see Kivlighan, 1990). The relationship needs some fiber. Immediacy is a way of balancing support with challenge (Kivlighan & Schmitz, 1992; Tryon & Kane, 1993).

**Use self-involving statements.** Self-involving statements are present-tense, personal responses to the client (see Robitschek & McCarthy, 1991). They can be positive in tone.

HELPER: I like the way you've begun to show initiative both in discussion and outside. I thought that telling your boss a bit about your past showed guts.

This self-involving remark is also a challenging statement, because the implication is "Keep it up." Clients tend to appreciate positive self-involving statements: "During the initial interview, the support and encouragement offered through the counselor's positive self-involving statements may be especially important because they put clients at ease and allay their anxiety about beginning counseling" (Watkins & Schneider, 1989, p. 345).

Negative self-involving statements are much more directly challenging in tone. Carl Rogers, the dean of client-centered therapy, recounts the following incident.

> I am quite certain even before I stopped carrying individual counseling cases, I was doing more and more of what I would call confrontation. That is, confrontation of the other person with my feelings. . . . For example, I recall a client with whom I began to realize I felt bored every time he came in. I had a hard time staying awake during the hour, and that was not like me at all. Because it was a persisting feeling, I realized I would have to share it with him. I had to confront him with my feeling and that really caused a conflict in his role as a client. . . . So with a good deal of difficulty and some embarrassment, I said to him, "I don't understand it myself, but when you start talking on and on about your problems in what seems to me a flat tone of voice, I find myself getting very bored." This was quite a jolt to him and he looked very unhappy. Then he began to talk about the way he talked and gradually he came to understand one of the reasons for the way he presented himself verbally. He said, "You know, I think the reason I talk in such an uninteresting way is because I don't think I have ever expected anyone to really hear me." . . . We got along much better after that because I could remind him that I heard the same flatness in his voice I used to hear. (See Landreth, 1984, p. 323)

Rogers's self-involving statement, genuine but quite challenging, helped the client move forward. But there is another point of view. Someone once said, "Boredom is a self-indictment." Rogers was bored because he restricted himself to empathy in his interactions with clients. On principle, he did not ordinarily use probing, summaries, and challenging with clients. He was a master at empathy, and some of the results he produced were amazing. In this case, however, he got what he "deserved." This story also points the direction in which Rogers was moving—adding the "spice" of probing and challenging to his interactions with clients—toward the end of his career.

## Situations Calling for Immediacy

Part of skilled helping is knowing when to use any given communication skill. The skill of immediacy can be useful in the following situations:

- **Lack of direction.** When a session is directionless and it seems that no progress is being made: "I feel that we're bogged down right now. Perhaps we could stop a moment and take a look at what we're doing. We might see what we're doing right and what's going wrong."
- **Tension.** When there is tension between helper and client: "We seem to be getting on each other's nerves. It might be helpful to stop a moment and clear the air."
- **Trust.** When trust seems to be an issue: "I see your hesitancy to talk, and I'm not sure whether it's related to me or not. You're talking about pretty sensitive issues. It might still be hard for you to trust me."

- *Diversity.* When diversity, some kind of "social distance," or widely differing interpersonal styles between client and helper seem to be getting in the way: "There are some hints that the fact that I'm black and you're white is making both of us a bit hesitant."
- *Dependency.* When dependency seems to be interfering with the helping process: "You don't seem willing to explore an issue until I give you permission to do so. And I seem to have let myself slip into the role of permission giver."
- *Counterdependency.* When counterdependency seems to be blocking the helping relationship: "It seems that we're letting this session turn into a kind of struggle between you and me. And, if I'm not mistaken, both of us would like to win."
- *Attraction.* When attraction is sidetracking either helper or client: "I think we've liked each other from the start. Now I'm wondering whether that might be getting in the way of the work we're doing here."

Immediacy—both in counseling and in everyday life—is a difficult, demanding skill. It is difficult, first of all, because the helper needs to be aware of what is happening in the relationship without becoming self-preoccupied and without "psyching out" the client. Second, immediacy demands both social intelligence and social competence in all the communication skills discussed to this point. Third, the helper must move beyond the MUM effect and challenge the client even though he or she is reluctant to do so. The helper needs backbone. It is clear that people get into trouble in their day-to-day relationships because they do not have the skills—or perhaps the courage—needed to engage in either general-relationship or here-and-now immediacy. Immediacy within sessions as part of the social-emotional reeducation effort, then, can help clients develop the relationship-enhancing skill of immediacy in real-life situations. The key words in the use of this skill are still *experience, wisdom,* and *moderation.*

## USING SUGGESTIONS AND RECOMMENDATIONS

Don't tell clients what to do. Don't try to take over clients' lives. Let clients make their own decisions. All these imperatives flow from the values of respect and empowerment. Does this mean, however, that suggestions and recommendations are absolutely forbidden? Never say never. It was mentioned earlier that there is a natural tension between helpers' desire to have their clients manage their lives better and respecting their freedom. If helpers build strong, respectful relationships with their clients, then "stronger" interventions sometimes make sense. In this context, suggestions and recommendations can stimulate clients to "get off the dime" and *do* something. Research has shown that clients will generally go along with recommendations from helpers if those recommendations are clearly related to the problem situation, challenge clients' strengths, and are not too difficult (Conoley, Padula, Payton, & Daniels, 1994). Effective helpers can provide suggestions, recommendations, and even directives without robbing clients of their autonomy or their integrity.

Here is a classic example from Cummings's (1979) work with addicts. Substance abusers came to him because they were hurting in many ways. He used every communication skill available to listen to and understand their plight.

> During the first half of the first session the therapist must listen very intently. Then, somewhere in midsession, using all the rigorous training, therapeutic acumen, and the third, fourth, fifth, and sixth ears, the therapist discerns some unresolved wish, some long-gone dream that is still residing deep in that human being, and then the therapist pulls it out and ignites the client with a desire to somehow look at that dream again. This is not easy, because if the right nerve is not touched, the therapist loses the client. (p. 1123)

So Cummings used advanced empathy to understand their plight, their longings, and also their games. The addicts came knowing how to play every game in the book with helpers. But Cummings knew all the games, too. Toward the end of the first session he told them they could have a second session — which they invariably wanted — only when they were "clean." The time of the second session depended on the withdrawal period for the kind of substance they were abusing. They screamed, shouted "foul," tried to play games, but he remained adamant. The directive — "Get clean, then return" — was part of the therapeutic process. And the vast majority did return.

In the hands of the socially intelligent and socially competent helper, the use of suggestions, advice, and directives is an adjunct to the rest of the process. One manager was arguing with a consultant in a coaching and counseling session about changes in his communication style at work. Finally, the consultant said, "Just try it. Then we'll talk about it." The relationship was a good one. Giving the directive was one way of breaking through the manager's dysfunctional communication style. He tried the change, liked it, and worked at incorporating it into his style. Suggestions, advice, and directives need not always be taken literally. They can act as stimuli to get clients to come up with their own package. One client said something like this to her helper: "You told me to let my teenage son have his say instead of constantly interrupting and arguing with him. What I did was make a contract with him. I told him that I would listen carefully to what he said and even summarize it and give it back to him. But he had to do the same for me. Now we have some monologuing going on. And we avoid our usual shouting matches. My hope is to find a way to turn it into dialogue."

In daily human interactions, people feel free to give one another advice. It goes on all the time. But helpers must proceed with caution. Suggestions, advice, and directives are not for novices. It takes a great deal of experience with clients and a great deal of savvy to know when they might work. Experience, wisdom, and moderation.

# CONFRONTATION

What about clients who don't want to develop new perspectives and change both internal and external behavior in the interest of managing problems and developing opportunities? Some clients who don't want to change or don't want to pay the price of changing simply terminate the helping relationship. However, those who stay stretch across a continuum from totally cooperative to extremely reluctant and resistant. Or they may be collaborative on some issues but reluctant when it comes to others. For instance, Hester is quite willing to work on career development but very reluctant to work on improving relationships, even though relationship building is part of the career package. "That's my private world," she says of her relationships.

If inviting clients to challenge themselves is at one end of the continuum, what's at the other? Where does protecting clients' rights to be themselves stop and placing demands on them to live more fully begin? Different helpers would answer those questions differently. Therefore, helpers differ, both theoretically and personally, in their willingness to confront. "Traumatic confrontation" (one wonders about the choice of name) is a cognitive behavior modification technique (Lowenstein, 1993) that involves challenging youths to face up to and change dysfunctional behavior. For example, a 12-year-old boy who had become involved in criminal activity after the disappearance of his father was confronted about his behavior. At first he denied everything, but then decided to face up to the situation.

When all is said and done, there is a place in helping for interventions strong enough to merit the term *confrontation*. Confrontation, as intimated earlier, means challenging clients to develop new perspectives and to change both internal and external behavior even when they show reluctance and resistance to doing so. When helpers confront, they "make the case" for more effective living. Confrontation does not involve "do this or else" ultimatums. More often it is a way of making sure that clients understand what it means *not* to change — that is, making sure they understand the consequences of persisting in dysfunctional patterns of behavior or of refusing to adopt new behaviors.

Both advice giving and confrontation require high levels of social intelligence and social competence on the part of the helper. They are not for everyone and, as suggested earlier, can go wrong in the hands of novices. In most cases, they should be used sparingly. And when helpers do judge that they might well serve the interests of their clients, they should be guided by the values outlined in Chapter 3.

**?**

# Evaluation Questions for Step I-B: The Use of Specific Challenge Skills

How effectively have I developed the communication skills that serve the process of challenging?

- *Advanced empathy.* Sharing hunches with clients about their experiences, behaviors, and feelings to help them move beyond blind spots and develop needed new perspectives
- *Information sharing.* Giving clients needed information or helping them search for it to help them see problem situations in a new light and to provide a basis for action
- *Helper self-disclosure.* Sharing your own experience with clients as a way of modeling nondefensive self-disclosure and helping them move beyond blind spots
- *Immediacy.* Discussing aspects of your relationship with your clients to improve the working alliance
- *Suggestions and recommendations.* Pointing out ways in which clients can more effectively manage problems and develop opportunities, and can engage more productively in the stages and steps of the helping process
- *Confrontation.* Using a solid relationship with the clients to challenge them more forcefully when they show signs of reluctance

## CHAPTER 11

# STEP I-B:
# III. THE WISDOM
# OF CHALLENGING

# GUIDELINES FOR EFFECTIVE CHALLENGING

All challenges should be permeated by the spirit of the client-helper relationship values discussed in Chapter 3; that is, they should be caring (not power games or put-downs), genuine (not tricks or games), and designed to increase rather than decrease the self-responsibility of the client (not expressions of helper control). They should also serve the stages and steps of the helping process, moving it forward for the purpose of constructive change (not an endless search for insight). Clearly, challenging well is not a skill that comes automatically. It needs to be learned and practiced. The following principles constitute some basic guidelines.

## Keep the Goals of Challenging in Mind

Challenge must be integrated into the entire helping process. Keep in mind that the goal is to help clients develop the kinds of alternative perspectives, internal behavior, and external actions needed to get on with the stages and steps of the helping process. To what degree do the new perspectives developed lead to problem-managing and opportunity-developing action? Are the insights and the challenges to action problem- and opportunity-relevant rather than merely dramatic?

## Encourage Self-Challenge

Invite clients to challenge themselves, and give them ample opportunity to do so. You can provide clients with probes and structures that help them engage in self-challenge. In the following excerpt, the counselor is talking to a man who has discussed at length his son's ingratitude. There has been something cathartic about his complaints, but it is time to move on.

**COUNSELOR:** People often have blind spots in their relationships with others, especially in close relationships. Picture your son sitting with some counselor. He is talking about his relationship with you. What's he saying?

**CLIENT:** Well, I don't know. . . . I guess I don't think about that very much. . . . Hmm. . . . He'd probably say . . . well, that he loves me. . . (pauses). And then he might say that since his mother died, I have never really let him be himself. I've done a lot — everything, he would say — to fashion his life. He'd say that he loves me but he always resented that, even though I wouldn't let him say that to me.

**COUNSELOR:** So both love for you and resentment for all that control.

**CLIENT:** And he'd be right. I'm still doing it. Not with him. He won't let me. But with lots of others. Especially in my business.

The counselor provides a structure that enables the client to challenge himself and apply his learnings to other social settings of life. Would that all clients would respond so easily! Alternatively, the counselor might have asked this client to list three things he thinks he does right and three things he thinks he should reconsider in his relationship with his son. The point is to be inventive with the probes and structures you provide to help clients challenge themselves.

## Earn the Right to Challenge

Berenson and Mitchell (1974) maintained that some helpers don't have the right to challenge others, since they do not fulfill certain essential conditions. Here are some of the factors that earn you the right to challenge.

- *Develop a working relationship.* Challenge only if you have spent time and effort building a relationship with your client. If your rapport is poor or you have allowed your relationship with the client to stagnate, then deal with the relationship.

- *Make sure you understand the client.* Effective challenge flows from accurate understanding. Only when you see the world through the client's eyes can you begin to see what he or she is failing to see.

- *Be open to challenge yourself.* Don't challenge unless you are open to being challenged. If you are defensive in the counseling relationship or in your relationship with supervisors or in your everyday life, your clients will follow your lead. Model the kind of nondefensive attitudes and behavior that would benefit your clients.

- *Work on your own life.* How important is constructive change in your own life? Berenson and Mitchell claimed that only people who are striving to live fully according to their value system have the right to challenge others, for only such persons are potential sources of human nourishment for others. In other words, helpers should be striving to develop physically, intellectually, socially, and emotionally.

In summary, ask yourself, "What is there about me that will make clients willing to be challenged by me?"

## Be Tentative but Not Apologetic in the Way You Challenge Clients

Tentative interpretations are generally viewed more positively than absolute interpretations (see Jones & Gelso, 1988). The same challenging message can be delivered in such a way as to invite the cooperation or arouse the resistance of the client. Deliver challenges tentatively, as *hunches* that are open to discussion rather than as accusations. Challenging is not an opportunity to browbeat clients or put them in their place. On the other hand, challenges that are delivered with too many qualifications—either verbally or through the helper's tone of voice—sound apologetic and can be easily dismissed by clients. I was once working in a career-development center. As I listened to one of the clients, it soon became evident that one reason he was getting nowhere was that he engaged in so much self-pity. When I shared this observation with him, I overqualified it. These are not my exact words, but it must have sounded something like this:

**HELPER:** Has it ever, at least in some small way, struck you that one possible reason for not getting ahead, at least as much as you would like, could be that at times you tend to engage in a little bit of self-pity?

I still remember his response. He paused, looked me in the eye, and said, "A little bit of self-pity?" He paused again. I said to myself, "I've been too harsh!" He continued, "I *wallow* in self-pity." We moved on to explore what he might do to decrease his self-pity.

## Challenge Unused Strengths More Than Weaknesses

Berenson and Mitchell (1974) discovered that successful helpers tend to challenge clients' strengths rather than their weaknesses. Individuals who focus on their failures find it difficult to change their behavior. As Bandura (1986) has pointed out, clients who regularly review their shortcomings tend to belittle their achievements, to withhold rewards from themselves when they do achieve, and to live with anxiety. All of this tends to undermine performance (see Bandura, 1986, p. 339). Challenging strengths means pointing out to clients the assets and resources they have but fail to use. In the following example, the helper is talking to a woman in a rape crisis center who is very good at helping others but who is always down on herself.

**COUNSELOR:** Ann, in the group sessions, you provide a great deal of support for the other women. You have an amazing ability to spot a person in trouble and provide an encouraging word. And when one of the women wants to give up, you are the first to challenge her, gently and forcibly at the same time. When you get down on yourself, you don't accept what you provide so freely to others. You are not nearly as kind to yourself as you are to others.

The counselor helps her place a demand on herself to use her rather substantial resources in her own behalf.

Adverse life experiences can be a source of strength. For instance, McMillen, Zuravin, and Rideout (1995) studied adult perceptions of benefit from child sexual abuse. Almost half the adults reported some kind of benefit, including increased knowledge of child sexual abuse, protecting other children from abuse, learning how to protect themselves from others, and developing a strong personality. Counselors, therefore, can help clients "mine" benefits from adverse experiences, putting to practical use the age-old dictum that "good things can come from evil things."

# Build on the Client's Successes

Effective helpers do not urge clients to place too many demands on themselves all at once. Rather, they help clients place reasonable demands on themselves and in the process help them appreciate and celebrate their successes. In the following example, the client is a boy in a detention center who is rather passive and allows the other boys to push him around. Recently, however, he has made a few halfhearted attempts to stick up for his rights in the counseling group. The counselor is talking to him alone after a group meeting.

**COUNSELOR A:** You're still not standing up for your own rights the way you need to. You said a couple of things in there, but you still let them push you around.

This counselor emphasizes the negative and browbeats the client. The following counselor takes a different tack.

**COUNSELOR B:** Here's what I've noticed. In the group meetings you have begun to speak up. You say what you want to say, even though you don't say it really forcefully. And I get the feeling that you feel good about that. You've got power. You need to find ways of using it more effectively.

The second counselor emphasizes the success — its size isn't the issue — and provides some support for further development.

# Be Specific in Your Challenges

Specific challenges hit the mark. Vague challenges get lost. Clients don't know what to do about them. Statements such as "You need to pull yourself together and get on with it" may satisfy some helper need, such as the ventilation of frustration, but they do little for clients. On the other hand, statements such as "How about listing the resources you used when you took charge and completed last month's project; then we can see how they can be applied to the project you're having so much trouble with" are much more specific.

In the following interchange, the helper, Shawon, engages in a bit of immediacy with Adler, the client:

**SHAWON:** Adler, sometimes we seem to hit it off. The sessions go quickly, we get work done, and you seem to be ready almost to race out of here and apply what you learned to your daily life. At other times, like now, we seem to be fencing. I thrust, you sidestep. Something like that. It might help both of us to find out what's going on.

**ADLER** (pauses): You really disarm me when you say things like that.

**SHAWON:** I'm not sure I know what you mean. Like I catch you off guard?

**ADLER:** Yeah. I think when things more or less hum, I'm dealing with you just as you. . . . I start dancing when you're not you. Rather, when I make you something else.

**SHAWON:** So I'm more than just me when things get tricky between us.

**ADLER:** Right. Two things happen. Sometimes, maybe it's me or something you say, but all of a sudden you're every person in authority I've ever known. You're dad, you're teacher, you're minister, you're boss—the whole mess. And then I don't talk to you like you anymore. We might as well stop.

**SHAWON:** And then what we do together is like a charade. . . . Is that how it is now—or has been today?

**ADLER:** No, it's the other you. . . . Every person who has ever been decent to me or taken care of me.

**SHAWON** (pauses): So two very different groups. . . . And two different Shawons.

**ADLER:** To be honest, not two different Shawons, but two different sets of perceptions of you on *my* part. . . . I hate to think that I still go funny around people in authority, but I do. I hate it even more when I think I'm dependent. . . . I can't stand dependent people! Maybe because there's too much of it in me. . . . Hey! We're back on course. You're you again.

**SHAWON** (laughing): And it's a relief to both of us. . . . But there's also a couple of new issues on the table.

**ADLER:** That's all right. Because they *are* on the table. All of a sudden it seems that they don't have to own me.

**SHAWON:** And we can use our relationship, at least in part, to get at them.

**ADLER:** You know, these two things, attitudes or ways I am, whatever, don't rule my life, but I can see that they certainly influence it. They do it here.

Through immediacy, vague terms like fear of authority and dependency needs become tangible.

## Respect Clients' Values

Challenge clients to clarify their values and to make reasonable choices based on them. Be wary of using challenging, even indirectly, to force clients to accept your values. This violates the empowerment value discussed in Chapter 3.

**CLIENT** (a 21-year-old woman who is a junior in college): I have done practically no experimentation with sex. I'm not sure whether it's because I think it's wrong or whether I'm just plain scared.

**COUNSELOR A:** A certain amount of exploration is certainly normal. Perhaps some basic information on contraception would help allay your fears a bit.

This counselor does little to help the client clarify her values. She makes the assumption, by inference, that the client is scared or fearful about pregnancy. She edges toward making some choices for the client.

**COUNSELOR B:** Perhaps you're saying that it's time to find out which it is and what you really want.

This counselor challenges her gently to find out what she really wants. Helpers can assist clients to explore the consequences of the values they hold, but that is not the same as attacking the client's values.

## Deal Honestly, Caringly, and Creatively with Client Defensiveness

Do not be surprised when clients react strongly to being challenged. Follow the principles outlined in Chapter 8 on reluctance and resistance. Help them share and work through their emotions. If they seem not to react externally to what you have said, elicit their reactions. In the following example, the helper has just delivered a brief summary of the problem situation to the client and has gently pointed out the self-destructive nature of some of his behaviors.

**HELPER:** I'm not sure how all this sounds to you.

**CLIENT:** I thought you were on my side. Now you sound like all the others. And I'm paying you to talk like this to me!

Even though the helper was tentative in his challenge, the client still reacts defensively. Here are two different approaches to the client's defensiveness.

**HELPER A:** All I've done is summarize what you have been saying about yourself. Let's look at each point and see if this isn't the case.

This helper takes a defensive, judicial approach. He's about to assemble the evidence. This could well lead to an argument, and an argument is not a dialogue. Helper B backs up a bit.

**HELPER B:** So it sounds harsh and unfair to you. . . . Maybe dumping all this on you is unfair. . . . Let's back up.

This helper backs off without saying that his summary was wrong. He is giving the client some space. It may be that the client needs time to think about what the helper has said. Helper B tries to get into a constructive dialogue with the angered client.

These principles are, of course, guidelines, not absolute prescriptions. In the long run, use your common sense. The more flexible and versatile you are, the more likely you are to be of benefit to your clients.

## LINKING CHALLENGE TO ACTION

More and more theoreticians and practitioners are stressing the need to link insight to problem-managing action (see Westerman, 1989). Wachtel (1989, p. 18) put it succinctly: "There is good reason to think that the really crucial insights are those closely linked to new actions in daily life and, moreover, that insights are as much a product of new experience as their cause." Ishiyama (1990), in a discussion of Morita therapy, talked about challenging as a stimulus for removing attitudinal blocks to constructive action. Certain attitudes stand in the way of problem-managing action—the need to avoid uncomfortable feelings, preoccupation with self, neglect of constructive aspirations, and so forth. These attitudes sap energy that is needed for action. For instance, in Morita therapy clients are encouraged to see anxiety and other unwanted feelings as part of human nature. Helping clients "normalize" such feelings allows them to use their energy for more practical activities. An older woman on a television talk show dealing with the problems of the elderly said, "A good day is one when you wake up and there is no *new* pain. You just gotta get up and do things." "This is the way life is, so move on" is neither a heartless statement nor an acceptance of the status quo.

## Is Challenge Enough to Stimulate Problem-Managing Action?

For some, yes. A few well-placed challenges might be all that some clients need to move to constructive action. It is as if they were looking for someone to challenge them. Once challenged, they do whatever is necessary to manage their problems. Some clients are on the brink of challenging themselves and need only a nudge. Others, once they develop a few new perspectives, are off to the races. Still others have the resources to manage their lives better, but not the will. They know what they need to do but are not doing it. A few nudges in the right direction help them overcome their inertia. I have had many one-session encounters that included a bit of listening, some empathy, and a new perspective that sent the client off on some useful course of action. On the other hand, if helping clients challenge themselves to develop new perspectives leads to one profound insight after another but no action and behavioral change, then once more we are whistling in the wind.

## Self-Efficacy: "I Can, I Will"

Given the many faces of inertia, we cannot assume that clients will take action. In collaboration with them, we need to build action into the helping process right from the start. Understanding the processes of self-efficacy and self-regulation will help us do so. The opposite of passivity is "agency," "assertion or assertiveness" (Galassi & Bruch, 1992), or "self-efficacy" (Bandura, 1977, 1980, 1982, 1986, 1989, 1991; Cervone & Scott, 1995; Lightsey, Jr., 1996; Locke & Latham, 1990; Maddux, 1995). A great deal of research continues to be done on the concept of self-efficacy and its applications to various settings—for instance, education (see Multon, Brown, & Lent, 1991) and health care (O'Leary, 1985).

**The nature of self-efficacy.** Bandura has suggested that people's expectations of themselves have a great deal to do with their willingness to put forth effort to cope with difficulties, the amount of effort they will expend, and their persistence in the face of obstacles. In particular, people tend to take action if two conditions are fulfilled:

1. They see that certain behavior will most likely lead to certain desirable results or accomplishments (outcome expectations).
2. They are reasonably sure that they can successfully engage in such behavior (self-efficacy expectations).

For instance, Yolanda not only believes that participation in a rather painful and demanding physical rehabilitation program following an accident and surgery will literally help her get on her feet again (an outcome expectation), but also believes that she has what it takes to inch her way through the program (a self-efficacy expectation). She therefore enters the program with a very positive attitude. Yves, on the other hand, is not convinced that an uncomfortable chemotherapy program will prevent his cancer from spreading and give him some quality living time (a negative outcome expectation), even though he knows he could endure it, so he says no to the doctors. Xavier is convinced that a series of radiation and chemotherapy treatments would help him (a positive outcome expectation), but he does not feel that he has the courage to go through with them (a negative self-efficacy expectation). He, too, refuses the treatment.

**Helping clients develop self-efficacy.** People's sense of self-efficacy can be strengthened in a variety of ways (see Mager, 1992). Lest self-efficacy be seen as a paradigm that applies only to the weak, let's take the case of a very strong manager, let's call him Nick, who wanted to change his abrasive supervisory style but was doubtful that he could do so. "After all these years, I am what I am," he would say. It would have been silly to merely tell him, "Nick, you can do it; just believe in yourself." It was necessary to help him do a number of things to help strengthen his sense of self-efficacy. That did not make him a wimp by a long shot.

- *Skills.* Make sure that clients have the skills they need to perform desired tasks. Self-efficacy is based on ability. Nick first read about and then attended some skill-building sessions on such "soft" skills as listening, responding with empathy, giving feedback that is softer on the person while being harder on the problem, and constructive challenging. In truth, he had many of these skills, but they lay dormant. These short training experiences put him back in touch with some things he could do but didn't do.
- *Feedback.* Provide feedback that is based on deficiencies in performance, not on the deficiencies of the client. Since I attended many meetings with Nick, I routinely described his performance. This feedback was not about his personality but about his behavior.
- *Success.* Help clients not only act but also see that their behavior produces results. Often success in a small endeavor will give them the courage

to try something more difficult. Some of Nick's peers and team members began to notice changes in his behavior. He was pleased by this, but mentioned it only to me.

• *Modeling.* Help clients see others doing what they are trying to do and then encourage them to try it themselves. I pointed out a few people in the company who were good at these "soft" skills but who were still considered competent, even tough, managers. On occasion he would bring up instances of the soft approach working better than a hard one.

• *Encouragement.* Exhort clients to try, challenge them, and support their efforts. This should never be patronizing. Had I patronized Nick, I would have been dead. My encouragement was, let's say, subtle and indirect.

• *Reducing fear and anxiety.* Help clients overcome their fears. If people are overly fearful that they will fail, they generally do not act. Therefore, procedures that reduce fear and anxiety help heighten the sense of self-efficacy. Deep down, Nick was fearful of two things regarding changing his supervisory style — being less effective in managing the business and making a fool of himself. Business results helped allay the former. He even noticed that two of his team members seemed to become more productive. As to the latter, his behavior outside the office came to the rescue. When we visited plants and field offices, Nick was very upbeat. He was as good at "rallying the troops" as anyone I have ever seen. Discussions about his two different styles helped allay fears that he would make a fool of himself by changing his style.

As a helper, you can do a great deal to help people develop a sense of agency or self-efficacy. First of all, you can help them challenge self-defeating beliefs and attitudes about themselves and substitute realistic beliefs about their ability to act. This includes helping them reduce the kinds of fears and anxieties that keep them from mobilizing their resources. Second, you can help them develop the working knowledge, life skills, and resources they need to succeed. Third, you can help them challenge themselves to take reasonable risks and support them when they do.

Thus, summarizing can lead to new perspectives or alternative frames of reference. In the following example, the client is a 52-year-old man who has revealed and explored a number of problems in living and is concerned about being depressed.

HELPER: Let's take a look at what we've seen so far. You're down — not just a normal slump; this time it's hanging on. You worry about your health, but you check out all right physically, so this seems to be more a symptom than a cause of your depression. There are some unresolved issues in your life. One that you seem to be stressing is the fact that your recent change in jobs has meant that you don't see much of your old friends anymore. Since you're single, you don't find this easy. Another issue — one you find painful and embarrassing — is your struggle to stay young. You don't like facing the fact that you're getting older. A third issue is the way you — to use your own word — "overinvest" yourself in work, so much so that when you finish a long-term project, suddenly your life is empty.

**CLIENT** (pauses): It's painful to hear it all that baldly, but that about sums it up. I've suspected I've got some screwed-up values, but I haven't wanted to stop long enough to take a look at it. Maybe the time has come. I'm hurting enough.

**HELPER:** One way of doing this is by taking a look at what all this would look like if it looked better.

**CLIENT:** That sounds interesting, even hopeful. How would we do that?

The counselor's summary hits home — somewhat painfully — and the client draws his own conclusion. Care should be taken not to overwhelm clients with the contents of the summary. Nor should summaries be used to "build a case" against a client. Helping is not a judicial procedure. Perhaps the foregoing summary would have been more effective if the helper had also summarized some of the client's strengths. That would have provided a more positive context.

# Evaluation Questions for Step I-B:
## The Process and Wisdom of Challenging

How well do I do each of the following as I try to help my clients?

### The Process of Challenging

- Help clients become aware of their blind spots in thinking and acting, and help them develop both new perspectives and more constructive behaviors
- Use challenge seamlessly whenever it is needed in the helping process
- Keep in mind the goals of challenging—that is, helping clients move beyond blinds spots to more effective mind-sets and change both internal and external patterns of behavior that keep them mired in problems and ineffective in developing unused opportunities
- Help clients participate fully in the helping process
- Help clients own their problems
- Help clients state problems as solvable and opportunities as doable
- Help clients correct faulty interpretations of their experiences, actions, and feelings
- Help clients identify and move beyond the predictable dishonesties of life
- Help clients link new insights to problem-managing action
- Help clients develop a sense of self-efficacy
- Help clients generally move beyond discussion and inertia to action

### The Wisdom of Challenging

- Invite clients to challenge themselves
- Earn the right to challenge by
  - developing an effective working alliance with my client
  - working at seeing the client's point of view
  - being open to challenge myself
  - managing problems and developing opportunities in my own life
- Be tactful and tentative in challenging without being insipid or apologetic
- Be specific, developing challenges that hit the mark
- Challenge clients' strengths rather than their weaknesses
- Don't ask clients to do too much too quickly
- Invite clients to clarify and act on their own values, not mine

*(continued)*

*(continued)*

## The Shadow Side of Challenging

- Identify the games my clients attempt to play with me without becoming cynical in the process
- Become comfortable with the social-influence dimension of the helping role
- Incorporate challenge into my counseling style without becoming a confrontation specialist
- Develop the assertiveness needed to overcome the MUM effect
- Challenge the excuses I give myself for failing to challenge clients
- Come to grips with my own imperfections and blind spots both as a helper and as a "private citizen"

## CHAPTER 12

# STEP I-C:
# LEVERAGE: HELPING CLIENTS WORK ON THE RIGHT THINGS

Step I-C involves helping clients choose what issues to work on during the helping process. That involves decision making on the part of the client. However, since decision making permeates the helping process, this chapter starts with an overview of the ups and downs of decision making.

## CLIENTS AS DECISION MAKERS

One of the reasons clients get into trouble in the first place is that they make poor decisions. Clients' stories are riddled with poorly made decisions. We need only to review our own experience to see how often our decisions or our failure to make decisions gets us into trouble. In the helping process itself, there are many decision points. Clients must decide to come to a counseling interview in the first place, to talk about themselves, to return for a second session, to respond to the helper's empathy, probes, and challenges, to choose issues to work on, to determine what they want, to set goals, to develop strategies, to make plans, and to implement those plans. Deciding — or letting the world decide for you — is at the heart of helping as it is at the heart of living.

Decision making in its broadest sense is the same as problem solving. Indeed, this book could be called a decision-making approach to helping. In this chapter, however, the focus is on decision making in a narrower sense — the internal (mental) action of identifying alternatives or options and choosing from among them. It is a commitment to do or to refrain from doing something:

- "I have decided to discuss my career problems but not my sexual concerns."
- "I have decided to ask the courts to remove artificial life support from my comatose wife."
- "I have decided not to undergo chemotherapy."
- "I have decided to move into a retirement home."

The commitment can be to an internal action — "I have decided to get rid of my preoccupation with my ex-wife" — or to an external action — "I have decided to confront my son about his drinking." Decision making in the fullest sense includes the implementation of the decision: "I made a resolution to give up smoking, and I haven't smoked for three years." "I decided that I was being too hard on myself, and I took a week off work and just enjoyed myself."

## Rational Decision Making

Traditionally, decision making has been presented as a rational, linear process involving information gathering, analysis, and choice. Here are the bare essentials of the decision-making process.

**Information gathering.** The first rational task is to gather information related to the particular issue or concern. A patient who must decide whether to have a series of chemotherapy treatments needs some essential information. What are the treatments like? What will they accomplish? What are the side effects? What are the consequences of not having them? What would another doctor say? And so forth. And there is a whole range of ways in which she might gather this information — reading, talking to doctors, talking to patients.

**Analysis.** The next rational step is processing the information. This includes analyzing, thinking about, working with, discussing, meditating on, and immersing oneself in the information. Just as there are many ways of gathering information, so there are many ways of processing it. Effective information processing leads to a clarification and an understanding of the range of possible choices. "Now, let's see, what are the advantages and disadvantages of each of these choices?" is a way of analyzing information. Effective analysis assumes that decision makers have criteria, whether objective or subjective, for comparing alternatives. For instance, a patient wants to determine whether the weeks or months of life she will gain through a series of chemotherapy treatments will be relatively comfortable or miserable.

**Making a choice.** Finally, decision makers need to make a choice — that is, commit themselves to some internal or external action that is based on the analysis: "After thinking about it, I have decided to sue for custody of the children." As indicated earlier, the fullness of the choice includes an action: "I had my lawyer file the custody papers this morning." There are also rational "rules" that can be used to make a decision. For instance, one rule, stated as a question, deals with the consequences of the decision: "Will it get me everything I want or just part?"

Counselors help clients engage in rational decision making; that is, they help clients gather information, analyze what they find, and then base action decisions on the analysis. Although that indeed does happen, it is not the full story.

# The Shadow Side of Decision Making: Choices in Everyday Life

In actuality, decision making in everyday life, and in counseling, is not the straightforward rational process just outlined (see Cosier & Schwenk, 1990; Etzioni, 1989; Heppner, 1989; Kaye, 1992; March, 1982; Russo & Schoemaker, 1992; Schoemaker & Russo, 1990; Stroh & Miller, 1993; Whyte, 1991). Rather, it is an ambiguous, highly complicated process. That is, it has a "shadow side." For instance, Gati, Krausz, and Osipow (1996) discuss the messiness associated with making career decisions and list ten ways in which such decisions can be flawed.

Headlee and Kalogjera (1988) saw some of the roots of this shadow side in childhood. Some children are allowed too much choice, and others are given too few choices. Moreover, in the early years distortions of choice evolve because of racial, ethnic, sexual, religious, and other prejudices. By the time the child becomes an adult, these distortions are ingrained in the decision-making process and are not reflected on. In everyday life the decision making is often confused, covert, difficult to describe, unsystematic, and, at times, quite irrational. A shadow-side analysis of decision making as it is actually practiced reveals a less-than-rational application of its three parts.

**Information gathering.** Information gathering should lead to a clear definition of the matter to be decided. A client trying to decide whether to pursue a divorce needs information about that entire process. However, information gathering is practically never straightforward. Decision makers, for whatever reason, are often complacent and engage in an inadequate search; they get too much, too little, inaccurate, or misleading information; the search for information and the information itself are clouded with emotion. The client trying to decide whether to proceed with therapy may have already made up his or her mind and therefore may not be open to confirming or disconfirming information. Since full, unambiguous information is never available, all decisions are at risk. In fact, there is no such thing as completely objective information. All information, especially in decision making, is received by the decision maker and takes on a subjective cast. In view of all this, Ackoff (1974) called human problem solving "mess management" (p. 21).

> Eloise wanted to make a decision about whether to marry her companion or not. One of the obstacles was conflicting careers. She didn't know whether she'd be in the same career five years from now; neither did he. She knew little about his past and thought that it didn't matter. She liked him now. He knew that she was a nonpracticing Catholic but knew little about how her Catholicism affected her or how it would affect them in the future, especially if they had children. Since religion was not currently an issue, he did not explore it.

There were many other things they did not know about themselves and each other. They eventually did marry, but the marriage lasted less than a year.

Granted, clients' stories are never complete, and information will always be partial and open to distortion. Yet, though counselors cannot help clients make information gathering perfect, they can help them make it adequate for problem management and opportunity development.

**Processing the information.** Since it is impossible to separate the decision from the decision maker, the processing of information is as complex as the person making the decision. Factors affecting the analyzing of information include clients' feelings and emotions, their values-in-use, which often differ from their espoused values, their assumptions about "the way things work," and their level of motivation. There is no such thing as full, objective processing of gathered information. Poorly gathered information is often subjected to further mistreatment. Clients, because of their biases, focus on bits and pieces of the information they have gathered rather than seeing the full picture. Furthermore, few clients have the time or the patience to spell out all possible choices related to the issue at hand, together with the pros and cons of each. Therefore, some say, most decisions are based not on evidence but on taste: "I like it. It sounds good."

> Jamie was in a high-risk category for AIDS because of occasional drug use and sexual promiscuity. When he was busted once for drug use, he had to attend a couple of sessions on AIDS awareness. He listened to all the information, but he processed it poorly. These were problems for "other people." He engaged in risky behavior "only occasionally." He was sure that his sexual partners were "clean." One or two "mistakes" were not going to do him in. He knew others that engaged in much riskier behavior than he and "nothing had happened to them." He'd be "more careful," though it was not clear what that meant as far as his behavior was concerned. He was in good health and "healthy people can take a lot."

Jamie distorted information and rationalized away most of the risk of his current lifestyle.

Up to a point, counselors can help clients overcome inertia and biases and tackle the work of analysis. If a client says his values have changed but he still automatically makes decisions based on his former values, then he can be challenged to get his new values into his decision making. One client, trying to make a decision about a career change, kept moving toward options in the helping professions even though he had become quite interested in business. There was something in him that kept saying, "You have to choose a helping profession. Otherwise you will be a traitor." The counselor helped him see his bias. He first became a consultant, then a manager, then a senior manager. But in the end he had to salve his conscience by noting both to himself and to others that "running a successful business is an important contribution to society."

**Choice and execution.** A host of things can happen at the point of decision — the point of commitment — and at the implementation stage to make decision making an unpredictable process. Decision makers sometimes

- skip the analysis stage and move quickly to choice.
- ignore the analysis and base the decision on something else entirely; the analysis was nothing but a sham, because the decision criteria, however covert, were already in place.
- engage in what Janis and Mann (1977) called "defensive avoidance." That is, they procrastinate, attempt to shift responsibility, or rationalize delaying a choice.

- confuse confidence in decision making with competence.
- panic and seize upon a hastily contrived solution that gives promise of immediate relief. The choice may work in the short term but have negative long-term consequences.
- are swayed by a course of action that is most salient at the time or by one that comes highly recommended, even though it is not right for them.
- let enthusiasm and other emotions govern their choices.
- announce a choice, to themselves or to others, but then do nothing about it.
- translate the decision into action only halfheartedly.
- decide one thing but do another.

The fact that choices do not necessarily make life easier, as Goslin (1985) pointed out so well, but more difficult for self and others explains a great deal of the shadow side of decision making. It is clear that counselors cannot help clients avoid all the pitfalls involved in making decisions, but they can help clients minimize them.

In summary, then, pure-form rational, linear decision making has probably never been the norm in human affairs. Decision making goes on at more than one level. There is, as it were, the rational decision-making process in the foreground and an emotional decision-making process in the background. Gelatt (1989) proposed an approach to decision making that factors in these realities: "What is appropriate now is a decision and counseling framework that helps clients deal with change and ambiguity, accept uncertainty and inconsistency, and utilize the nonrational and intuitive side of thinking and choosing" (p. 252). Positive uncertainty means, paradoxically, being positive (comfortable and confident) in the face of uncertainty (ambiguity and doubt)—feeling both uncertain about the future and positive about the uncertainty.

Stages II and III provide methodologies clients can use to make decisions, explore their consequences, and act on them.

# THE GOALS OF STEP I-C

Since Step I-C is both a stage-specific task and a process that permeates all the stages of the helping process, its goals reflect both. Furthermore, even though the following goals can be distinguished conceptually, in practice they intermingle.

• *Determine whether the client is ready to invest in constructive change.* Change requires work on the part of clients. If they do not have the incentives to do the work, they will begin and then drop out. This is an expensive proposition. In this sense, then, all counseling requires helping clients identify and evaluate incentives. For instance, Beth is an intelligent empty-nester whose husband travels a great deal. She has plenty of incentives to deal with her malaise and to identify and develop opportunities. On the other hand, Helmut and Gretchen are mildly dissatisfied with their marriage and seem to be looking for a "psychological pill" that will magically make things better. There seem to be few incentives for the work required to reinvent the marriage.

• *Help clients become more effective decision makers.* This is obviously a process goal; that is, it relates to the entire helping process. Indeed, since self-responsibility is a key helping value, helping clients not only make good decisions but also become better decision makers is not an amenity but a necessity.

• *Help clients screen problems and opportunities.* Clients need to be helped to judge whether helping or therapy is what they need at this point. Not everyone needs therapy, and not every problem merits the kind of attention that counseling and therapy provide. The first stage-specific goal, then, is to decide whether or not to continue. If the story being told has little substance, then a decision may be made to terminate the relationship or probe for more substantive issues.

• *Help clients work on the right things.* Since all the concerns in a complex problem-opportunity situation cannot be dealt with at once, another goal is to help clients establish some priorities. Even problems that are told as single-issue stories — "I'd like to get rid of these headaches" — often become, when examined, multiple-issue problem situations. For instance, the headaches are a symptom of overwork, poor interpersonal relationships, self-esteem problems, financial concerns, and an uncontrolled temper. Effective counselors help clients work on problems and opportunities that will make a difference. This is the search for leverage. The helper asks, "What can I do to help this client get the most out of his or her investment?" Once a client chooses a problem for attention, then, of course, the helper uses all the communication skills outlined earlier to help that client explore the problem and develop new perspectives on it.

• *Help clients stay focused throughout the helping process.* A broader goal is to help clients focus on the right things — not just the right issues but also the right kind of relationship with the helper, the right goals, the right strategies for achieving their goals, and the right actions both within the helping sessions and in day-to-day life. Step I-C in this sense is not a step but a process that deals with the "economics" of helping. Helpers need to ask themselves, "Am I adding value through each of my interactions with this client?" Clients need to be helped to ask themselves, "Am I spending my time well? Do the decisions I am making have the potential of adding value to my life?" We now take a closer look at two of these goals — screening and working on the right things.

# Screening — The Initial Search for Leverage

Helping is expensive both financially and psychologically. It should not be undertaken lightly. Relatively little is said in the literature about screening — that is, about deciding whether any given problem situation or opportunity deserves attention. The reasons are obvious. Helpers-to-be are rightly urged to take their clients and their clients' concerns seriously. They are also urged to adopt an optimistic attitude, an attitude of hope, about their clients. Finally, they are schooled to take their profession seriously and are convinced that their services can make a difference in the lives of clients. For those and other reasons, the first impulse of the average counselor is to try to help clients no matter what the problem situation might be.

There is something very laudable about that. It is rewarding to see helpers prize people and express interest in their concerns. It is rewarding to see helpers put aside the almost instinctive tendency to evaluate and judge others and to offer their services to clients just because they are human beings. However, like other professions, helping can suffer from the "law of the instrument." A child, given a hammer, soon discovers that almost everything needs hammering. Helpers, once equipped with the models, methods, and skills of the helping process, can see all human problems as needing their attention. In fact, in many cases counseling may be a useful intervention and yet be a luxury whose expense cannot be justified. The problem-severity formula discussed earlier is a useful tool for screening.

Under the term *differential therapeutics*, Frances, Clarkin, and Perry (1984) discussed ways of fitting different kinds of treatment to different kinds of clients. They also discussed the conditions under which "no treatment" is the best option. In the no-treatment category they included clients who have a history of treatment failure or who seem to get worse from treatment, such as

- criminals trying to avoid or diminish punishment by claiming to be suffering from psychiatric conditions — "We may do a disservice to society, the legal system, the offenders, and ourselves if we are too willing to treat problems for which no effective treatment is available" (p. 227);
- patients with malingering or fictitious illness;
- chronic nonresponders to treatment;
- clients likely to improve on their own;
- healthy clients with minor chronic problems;
- reluctant and resistant clients who refuse treatment.

Although a decision needs to be taken in each case, and although some might dispute some of the categories proposed, the possibility of no treatment deserves serious attention.

The no-treatment option can do a number of useful things: interrupt helping sessions that are going nowhere or that are actually destructive; keep both client and helper from wasting time, effort, and money; delay help until the client is ready to do the work required for constructive change; consolidate gains from previous treatments; provide clients with an opportunity to discover that they can do without treatment; keep helpers and clients from playing games with themselves and one another; and provide motivation for the client to find help in his or her own daily life. However, a no-treatment decision on the part of helping professionals is countercultural and therefore difficult to make.

It goes without saying that screening clients' stories, or helping clients screen their own stories, can be done in a heavy-handed way. Statements such as the following are not useful:

- "Your concerns are actually not that serious."
- "You should be able to work that through without help."
- "I don't have time for problems as simple as that."

Whether such sentiments are expressed or implied, they obviously indicate a lack of respect and constitute a caricature of the screening process. Helpers are not alone in grappling with this problem. Doctors face clients day in and day out with problems that run from the life-threatening to the inconsequential. Statistics suggest that more than half of the people who come to doctors have nothing physically wrong with them. Doctors consequently have to find ways to screen patients' complaints. I am sure that the best find ways to do so that preserve the dignity of their patients.

Experienced helpers, because they are empathic, stay close to the experience of clients and develop hypotheses about both the substance of a client's problems and the client's commitment and then test those hypotheses in a humane way. If clients' problems seem inconsequential, they probe for more substantive issues. If clients seem reluctant, resistant, and unwilling to work, they challenge clients' attitudes and help them work through their resistance. In both cases, they realize that there may come a time, and it may come fairly quickly, to judge that further effort is uncalled for because of lack of results. It is better, however, to help clients make such a decision themselves or challenge them to do so. In the end, the helper might have to call a halt, but his or her way of doing so should reflect basic counseling values.

## LEVERAGE: WORKING ON ISSUES THAT MAKE A DIFFERENCE

Clients often need help to get a handle on complex problem situations. A 41-year-old depressed man with a failing marriage, a boring, run-of-the-mill job, deteriorating interpersonal relationships, health concerns, and a drinking problem cannot work on everything at once. Priorities need to be set. Put bluntly, the question is, Where is the biggest payoff? Where should the limited resources of both client and helper be invested? If the problem situation has many dimensions, the helper and the client are faced with the problem of where to start.

> Andrea, a woman in her mid-30s, is referred to a neighborhood mental-health clinic by a social worker. During her first visit, she pours out a story of woe both historical and current—brutal parents, teenage drug abuse, a poor marriage, unemployment, poverty, and the like. Andrea is so taken up with getting it all out that the helper can do little more than sit and listen.

Where is Andrea to start? How can the time, effort, and money invested in helping provide a reasonable return to her? In the broadest sense, what are the economics of helping?

## Some Principles of Leverage

The following principles of leverage—a reasonable return on the investment of the client's, the helper's, and third-party resources—serve as guidelines for choosing issues to be worked on. These principles overlap; more than one may apply at the same time.

- If there is a crisis, first help the client manage the crisis.
- Begin with the problem that seems to be causing pain for the client.
- Begin with issues the client sees as important.
- Begin with some manageable subproblem of a larger problem situation.
- Begin with a problem that, if handled, will lead to some kind of general improvement in the client's condition.
- Focus on a problem for which the benefits will outweigh the costs.

Underlying all these principles is an attempt to make clients' initial experiencing of the helping process rewarding so that they will have the incentives they need to continue to work. Examples of the use and abuse of these principles follow.

**If there is a crisis, first help the client manage the crisis.**  Although crisis intervention is sometimes seen as a special form of counseling (Baldwin, 1980; Janosik, 1984), it can also be seen as a rapid application of the three stages of the helping process to the most distressing aspects of a crisis situation.

> *Principle violated:* Zachary, a student near the end of the second year of a four-year doctoral program in counseling, gets drunk one night and is accused of sexual harassment by a student whom he met at a party. Knowing that he has never

sexually harassed anyone, he seeks the counsel of a faculty member whom he trusts. The faculty member asks him many questions about his past, his relationship with women, how he feels about the program, and so on. Zachary becomes more and more agitated and then explodes: "Why are you asking me all these silly questions?" He stalks out and goes to a fellow student's house.

*Principle used*: Seeing his agitation, his friend says, "Good grief, Zach, you look terrible! Come in. What's going on?" He listens to Zachary's account of what has happened, interrupting very little, merely putting in a word here and there to let his friend know he is with him. He sits with Zachary when he falls silent or cries a bit, and then slowly and reassuringly "talks Zach down," engaging in an easy dialogue that gives his friend an opportunity gradually to find his composure again. Zach's student friend has a friend in the university's student services department. They call him up, go over, and have a counseling and strategy session on the next steps in dealing with the harassment charges.

The friend's instincts are much better than those of the faculty member. He does what he can to defuse the immediate crisis and helps Zach take the next crisis-management step.

**Begin with the problem that seems to be causing pain.** Clients often come for help because they are hurting even though they are not in crisis. Their hurt, then, becomes a point of leverage. Their pain also makes them vulnerable. This means they are open to influence from helpers. If it is evident that they are open to influence because of their pain, seize the opportunity, but move cautiously. Their pain may also make them demanding. They can't understand why you cannot help them get rid of it immediately. This kind of impatience may put you off, but it, too, needs to be understood. Such clients are like patients in the emergency room, each one seeing himself or herself as needing immediate attention. Their demands for immediate relief may well signal a self-centeredness that is part of their character and therefore part of the broader problem situation. It may be that their pain is, in your eyes, self-inflicted, and ultimately you may have to challenge them on it. But pain, whether self-inflicted or not, is still pain. Part of your respect for clients is your respect for them as vulnerable.

*Principle violated*: Rob, a man in his mid-20s, comes to a counselor in great distress because his wife has just left him. The counselor sees quickly that Rob is an impulsive, self-centered person with whom it would be difficult to live. The counselor immediately challenges him to take a look at his interpersonal style and the ways in which he alienates others. Rob seems to listen, but he does not return.

*Principle used*: Rob goes to a second counselor, who also sees a number of clues indicating self-centeredness and a lack of maturity and discipline. However, she listens carefully to his story, even though it is one-sided. Instead of adding to his pain by making him come to grips with his selfishness, she focuses on the future. Of course, Rob thinks that his wife's return is the most important part of a better future. Since that is the case, the counselor also helps him describe in some detail what life would be like once they got back together. His hurting provides the incentive for working with the counselor on how he needs to change to get what he wants.

The helper's focusing on a better future responds to Rob's need. This future, then, becomes the stepping-stone to challenge. What is Rob willing to do to create the future he says he wants?

**Begin with issues the client sees as important and is willing to work on.** The frame of reference of the client is a point of leverage. Given the client's story, you may think that he or she has not chosen the most important issues for initial consideration. However, helping clients work on issues that are important to them sends an important message: "Your interests are important to me."

> *Principle violated*: A woman comes to a counselor complaining about her relationship with her boss. She believes that he is sexist. After listening to her story, the counselor believes that she probably has some leftover developmental issues with her father that affect her attitude toward older men. He pursues this line of thinking with her. The client is confused and feels put down. When she does not return for a second interview, the counselor says to himself that she is not motivated to do something about her problem.

> *Principle used*: The woman seeks out a lawyer who deals with equal-opportunity cases. The lawyer listens carefully to her story and probes for missing detail. Then he gives her a snapshot of what such cases involve if they go to litigation. Against that background he helps her determine what she really wants. Is it more respect? more pay? a promotion? a different kind of boss? a better use of her talents? Once she names her preferences, he discusses with her options for getting what she wants. Litigation is not one of them.

It may well be that a client's immediate frame of reference needs broadening. This does not mean, however, that the helper should substitute his or her agenda for the client's concerns.

**Begin with some manageable subproblem of a larger problem situation.** Large, complicated problem situations often remain vague and unmanageable. Dividing a problem into manageable bits can provide leverage. Most larger problems can be broken down into smaller, more manageable subproblems.

> *Principle violated*: Aaron and Ruth, in their mid-50s, have a 25-year-old son living at home who has been diagnosed as schizophrenic. Aaron is a manager in a manufacturing concern that is downsizing. His wife has a history of panic attacks and chronic anxiety. All these problems have placed a great deal of strain on the marriage. They feel guilty about their son's illness. The son has become quite abusive at home and has been stigmatized by people in the neighborhood for his "odd" behavior. Aaron and Ruth have also been stigmatized for "bringing him up wrong." They have been seeing a counselor who specializes in a

"systems" approach to such problems. They are confused by his "everything is related to everything else" approach. They are looking for relief but are exposed to more and more complexity. They finally come to the conclusion that they do not have the internal resources to deal with the enormity of the problem situation and drop out.

*Principle used:* A couple of weeks pass before they screw up their courage to see a psychiatrist. Although she understands the systemic complexity of the problem situation, she also understands their need for some respite. She first sees the son and prescribes some antipsychotic medication. His odd and abusive behavior is greatly reduced. Even though Aaron and his wife are not actively religious, she also arranges a meeting with a rabbi from the community known for his ecumenical activism. The rabbi puts them in touch with an ecumenical group of Jews and Christians who are committed to developing a "neighborhood that cares." Involvement with the group helps diminish their sense of stigma. They still have many concerns, but they now have some relief and better access to both internal and community resources to help them with those concerns.

The psychiatrist targets two manageable subproblems—the son's behavior and the couple's sense of isolation from the community.

**Move as quickly as possible to a problem that, if handled, will lead to some kind of general improvement.** Some problems, when addressed, yield results beyond what might be expected. This is the spread effect.

*Principle violated:* Jeff, a single carpenter in his late 20s, comes to a community mental-health center with a variety of complaints, including insomnia, light drug use, feelings of alienation from his family, a variety of psychosomatic complaints, and temptations toward exhibitionism. He also has an intense fear of dogs, something that occasionally affects his work. The counselor sees this last problem as one that can be managed through the application of behavior modification methodologies. He and Jeff spend a fair amount of time in the desensitization of the phobia. Jeff's fear of dogs abates quite a bit and many of his symptoms diminish. But gradually the old symptoms reemerge. His phobia, though significant, is not related closely enough to his primary concerns to involve any kind of major spread effect.

*Principle used:* Jeff's major problems reemerge. One day he becomes disoriented and bangs a number of cars near his job site with his hammer. He is overheard saying, "I'll get even with you." He is admitted briefly to a general hospital with a psychiatric ward. The immediate crisis is managed quickly and effectively. During his brief stay, he talks with a psychiatric social worker. He feels good about the interaction, and they agree to have a few sessions after he is discharged. In their talks, the focus turns to his isolation. This has a great deal to do with his lack of self-esteem—"Who would want to be my friend?" He has managed this problem by staying away from close relationships with both men and women. They discuss ways in which he can begin to socialize. As he begins to get involved with others, his symptoms begin to disappear.

Guilt about intimacy and the lack of intimacy are high-leverage issues in Jeff's life. Dealing with intimacy and the associated guilt takes care of many of his other complaints.

**Focus on a problem for which the benefits will outweigh the costs.** This is not an excuse for not tackling difficult problems. If you demand a great deal of work from both yourself and the client, then basic laws of behavior suggest that there will be some kind of reasonable payoff for both of you.

> *Principle violated*: Margaret discovers to her horror that Hector, her husband, has been found to be HIV-positive. Tests reveal that she and her recently born son have not contracted the disease. This helps cushion the shock. But she has difficulty with her relationship with Hector. He claims that he picked up the virus from a "dirty needle." But she didn't even know that he had ever used drugs. The helper's approach focused on "reconstruction of the marital relationship" together with elements of "personality reconstruction" for both of them. He told her that some of this would be painful because it meant looking at areas of their lives that they had never reviewed or discussed. She wanted some practical help in reorienting herself to her husband and to family life. After three sessions she did not return.

> *Principle used*: Margaret still searches for help. The doctor who is treating Hector suggests a self-help group for spouses, children, and companions of HIV-positive patients. In the sessions Margaret learns a great deal about how to relate to someone who is HIV-positive. The meetings are very practical. The fact that Hector is not in the group helps. She begins to understand herself and her needs better. In the security of the group, she explores mistakes she had made in relating to Hector. Although she does not "reconstruct" either her relationship with Hector or her own personality, she does learn how to live more creatively with both herself and him. She realizes that there will probably be further anguish, but she also sees that she is better prepared to face that future.

Reconstructing both relationships and personality, even if possible, is a very costly — and chancy — proposition. It is not at all clear that Margaret needs such reconstruction. The cost-benefit ratio is out of balance. Once more it is a question of a helper more committed to his theories than to the needs of his clients. Margaret gets the help she needs in the group and in sessions she and Hector have with a caring doctor.

## A No-Formula Approach

These approaches to leverage — and others you may think of – are principles, not formulas. Keeping the outcome in mind is important — helping the client get focused and the effective and efficient use of time and other resources for the purpose of problem management and opportunity development. The principles of leverage, used judiciously and in combination, can help counselors shape the helping process to suit the needs and resources of individual clients. However, helpers must avoid using these very same principles to water down or retard the helping process. Ineffective helpers

- begin with the framework of the client, but never get beyond it;
- fail to challenge clients to consider significant issues they are avoiding;
- allow the client's pain and discomfort to mask the roots of his or her problems;
- fail to help clients build on their successes;
- continue to deal with small, manageable problems and fail to help clients face more demanding problems in living;
- fail to help clients generalize learning in one area of life (for instance, self-discipline in a fitness program) to other, more difficult areas of living (for instance, self-control in closer interpersonal relationships).

In short, ineffective helpers are afraid of making reasonable demands of clients or of influencing clients to make reasonable demands of themselves.

## FOCUS AND LEVERAGE: THE LAZARUS TECHNIQUE

Arnold Lazarus (Rogers, Shostrom, & Lazarus, 1977), in a film on his multi-modal approach to therapy (Lazarus, 1976, 1981), uses a focusing technique I find useful. He asks the client to use just one word to describe her problem. She searches around a bit and then offers a word. For instance, a client might say, "Cloudy." Then he asks the client to put the word in a simple sentence that would describe her problem. The client might say, "My mind is cloudy and I can't think straight." The helper then moves from a simple sentence to a more extended description of the issue. This client says something like this:

> When I say that my mind is cloudy and I can't think straight I mean that this is the reaction to what my boss told me last week. I have been working very hard. And learning a lot. Over the past two years I've received two awards for special projects. And all along I've gotten good ratings. But then the recession hit and they were getting rid of all sorts of people. I was let go because the unit I was in was eliminated. But if I was supposedly so good, why didn't they find a way of keeping me? . . . I was shell-shocked. . . . I've been very confused. . . . A fog . . . I was loyal . . . It's hit me very hard.

I'm not suggesting that you begin all sessions or any session that way, but it is a simple way to bring a session into focus. This methodology can be used at any stage or step of the helping process. For instance, clients can be asked to use one word to describe what they want.

BOX 12-1

## Leverage Questions for Step I-C

- What problem or opportunity should I really be working on?
- Which issue, if faced, would make a substantial difference in my life?
- Which problem or opportunity has the greatest payoff value?
- Which issue do I have both the will and the courage to work on?
- Which problem, if managed, will take care of other problems?
- Which opportunity, if developed, will help me deal with critical problems?
- What is the best place for me to start?
- If I need to start slowly, where should I start?
- If I need a boost or a quick win, which problem or opportunity should I work on?

**HELPER:** Now that you have explored some of your concerns and against this background, tell me in just one word or short phrase what you want.

**CLIENT:** Let's see . . . My family.

**HELPER:** All right, your family. Now put that in a sentence. Describe what you want in a short sentence.

**CLIENT:** I want my family and my family life back.

This client had achieved financial success in a business venture rather quickly. Success went to his head. He left his wife and three children and began living with a "flashier" woman. Now, some six years later, the shine was off the apple. He was successful but unhappy. He wanted his family back. He wanted family life as he once knew it again. That for him was a leverage point. How realistic it was and how it could be achieved was another issue. But the Lazarus technique provided focus. Box 12-1 points out leverage questions that you can help clients ask themselves.

## STEP I-C AND ACTION

Like every other step of the helping model, Step I-C can act as a stimulus to client action. Helping clients find leverage means, concomitantly, helping them discover incentives for acting inside the helping sessions and outside. Helping clients identify and deal with high-leverage issues helps them move toward the little actions that precede the formal plan for constructive change. One client, a man in his early 30s, discussed all sorts of problems and unused opportunities with a counselor. He skipped from one topic to another, usually settling on the issue that had caused him the most trouble the previous week. The counselor always listened attentively. At the beginning of about the third session she said, "I have a hypothesis I'd like to explore with you." She went on to name what she saw as a thread that wove its way through most of the issues he discussed—a reluctance on his part to commit himself. He was thunderstruck, because he instantly recognized the truth. In the next few weeks he began to explore his problems and concerns from this perspective. Time after time he saw himself withdrawing when he should have been committing himself, whether to a project or to a person. Gradually he began to commit himself in little ways. For instance, when an intimate woman friend of his began talking about where their relationship was going, he said, "There is something in me that wants to change the subject or run away. But no. Let's begin talking about the future and what future we might have together." At least he committed himself to opening a discussion about the relationship if not to the relationship itself. Discussing issues that make a difference can galvanize some clients into using resources that have lain dormant for years.

# Evaluation Questions for Step I-C

How well am I doing the following?

- Observing clients' decision-making styles and helping them not only to make good decisions but also to become better decision makers
- Helping clients focus on issues that have payoff potential for them
- Maintaining a sense of movement and direction in the helping process
- Avoiding unnecessarily extending the problem identification and exploration stage
- Moving to other stages of the helping process as clients' needs dictate
- Encouraging clients to act on what they are learning

# STAGE II:
# HELPING CLIENTS
# DETERMINE WHAT THEY
# NEED AND WANT

In many ways Stages II and III together with the
Action Arrow are the most important parts of the
helping model. It is here that counselors help clients
develop and implement programs for constructive
change. The payoff for identifying and clarifying both
problem situations and unused opportunities lies in
doing something about them. The skills helpers need
to help clients to do precisely that—engage in
constructive change—are reviewed and illustrated in
Stages II and III. In Stages II and III, counselors help
clients ask and answer the following two
commonsense but critical questions.

*What do you want?*

and

*What do you have to do to get what you want?*

Part Four deals with helping clients discover and commit themselves to what they both need and want. Part Five deals with developing and implementing strategies for a better future.

Stage II focuses on this better future, the client's preferred scenario. Problems can make clients feel hemmed in and closed off. To a greater or lesser extent they have no future, or the future they have looks troubled. But, as Gelatt (1989) noted, "The future does not exist and cannot be predicted. It must be imagined and invented" (p. 255). The steps of Stage II outline three ways in which helpers can be with their clients with a view to exploring and developing this better future.

**Step II-A:** Helping clients identify *possibilities* for a better future. What do you want? What do you need? What are some of the possibilities?

**Step II-B:** Helping clients craft a *change agenda*. Given the possibilities, what do you really want? What are your choices?

**Step II-C:** Helping clients discover incentives for *commitment* to their change agenda. What are you willing to pay for what you want?

Figure II highlights these three steps of the helping process. Without minimizing in any way what helpers can help their clients accomplish through Stage I interventions—that is, problem and opportunity clarification; the development of new, more constructive perspectives of self, others, and the world; the choice of high-leverage issues to work on—the real power of helping lies in helping clients *do something* about the problem situations and unused opportunities discussed in Stage I.

O'Hanlon and Weiner-Davis (1989, p. 6) claimed that a trend "away from explanations, problems, and pathology, and toward solutions, competence, and capabilities" was emerging in the helping professions. Although I hope that this is, as they suggested, a "megatrend" in helping, I am not convinced. An earlier study showed that clients are interested in solutions to their problems and feeling better, whereas many helpers are concerned about the origin of problems and transforming them through insight (Llewelyn, 1988). Too many approaches to helping still focus on Stage I activities. Too many helper training programs still emphasize, or overemphasize, the skills of Stage I. Intensive discussion of problem situations is often based on a "working through" mentality, whereas action approaches are based on the assumption that many problems need to be "transcended" rather than worked through. At any rate, the goal of helping, as stated in Chapter 1, is "problems managed," not "problems understood."

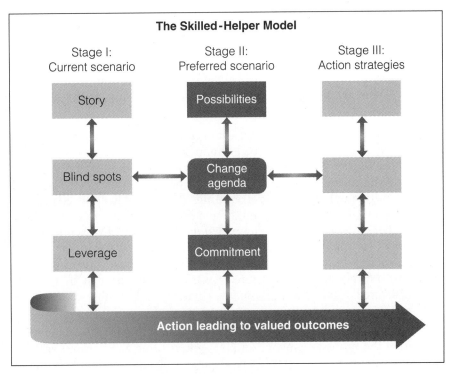

**FIGURE II**
The Helping Model — Stage II

# THE POWER OF GOAL SETTING

Goal setting, whether it is called that or not, is part of everyday life. We all do it all the time.

> Why do we formulate goals? Well, if we didn't have goals, we wouldn't do anything. No one cooks a meal, reads a book, or writes a letter without having a reason, or several reasons, for doing so. We want to accomplish some end with our activity or we want to prevent or avoid some end. We want to make something the way it "should" be or we want to prevent something that is already as it should be from changing. *These desires are beacons for our actions; they tell us which way to go.* When formalized into goals, they play an important role in problem solving. (Dorner, 1996, p. 49) [italics mine]

Even not setting overt goals is a form of goal setting. It means that we choose to continue to pursue the covert default set of goals already in place, however enhancing or limiting they may be. We don't like the sagging muscles and flab we see in the mirror. But not deciding to get into better shape is a decision to continue to drift. Since goals are all-pervasive, it makes sense to make them work for us rather than against us.

# Goal Setting as a Tool of Empowerment

Goals at their best mobilize clients' resources for the purpose of action; that is, they get people moving. They also provide channels for wise action; that is, they get clients headed in the right direction. According to Locke and Latham (1984), helping clients set goals empowers them in four ways.

- *Goals focus clients' attention and action.* A counselor at a refugee center in London described Simon, a victim of torture in a Middle Eastern country, to her supervisor as aimless and minimally cooperative in exploring the meaning of his brutal experience. Her supervisor suggested that she help Simon explore a better future. The counselor started one session by asking, "Simon, if you could have one thing you don't have, what would it be?" Simon came back immediately, "A friend." During the rest of the session, he was totally focused. What was uppermost in his mind was not the torture but the fact that he was so lonely in a foreign country. When he did talk about the torture, it was to express his fear that torture had "disfigured" him, if not physically, then psychologically, thus making him unattractive to others.

- *Goals mobilize clients' energy and effort.* Clients who seem lethargic during the problem-exploration phase often come to life when asked to discuss possibilities for a better future. A patient in a long-term rehabilitation program who had been listless and uncooperative said to her counselor after a visit from her minister, "I've decided that God and God's creation and not pain will be the center of my life. This is what I want." That was the beginning of a new commitment to the arduous program. She collaborated more fully in exercises that helped her manage her pain. Clients with goals are less likely to engage in aimless behavior. Goal setting is not just a "head" exercise. Many clients begin engaging in constructive change after setting even broad or rudimentary goals.

- *Goals motivate clients to search for strategies to accomplish them.* Setting goals, a Stage II task, leads naturally into a search for means to accomplish them, a Stage III task. Lonnie, a woman in her 70s who had been described by her friends as "going downhill fast," decided, after a heart-problem scare that proved to be a false alarm, that she wanted to live as fully as possible until she died. She searched out ingenious ways of redeveloping a social life, including remodeling her house and taking in two young women from a local college as boarders.

- *Goals stated in specific terms increase persistence.* Not only are clients with goals energized to do something, but they also tend to work harder and longer. An AIDS patient who said that he wanted to be reintegrated into his extended family managed, against all odds, to recover from five hospitalizations to achieve what he wanted. He did everything he could to buy the time he needed. Clients with clear and realistic goals don't give up as easily as clients with vague goals or with no goals at all.

One study (Payne, Robbins, & Dougherty, 1991) showed that high-goal-directed retirees were more outgoing, involved, resourceful, and persistent in their social settings than low-goal-directed retirees, who were more self-critical, dissatisfied, sulky, and self-centered. People with a sense of direction don't waste time in wishful thinking. Rather, they translate wishes into specific outcomes toward which they can work. Picture a continuum. At one end is the aimless person; at the other, a person with a sense of direction. Your clients may come from any point on the continuum. They may be at different points with respect to different issues — mature in seizing opportunities for education, for instance, but aimless in developing sexual maturity. All of us are "marginal" at one time or another. Helping clients establish and commit themselves to problem-managing goals is the principal task of Stage II.

Setting goals, however informally, provides clients with a sense of direction. People with a sense of direction have a sense of purpose; live lives that are going somewhere; have self-enhancing patterns of behavior in place; focus on results, outcomes, and accomplishments; don't mistake aimless action for accomplishments; have a defined rather than an aimless lifestyle. Locke and Latham (1990) pulled together years of research on the motivational value of setting goals. Although the motivational value of goal setting is incontrovertible, the challenge for counselors is to help clients do it well.

## CHAPTER 13

# STEP II-A: WHAT DO YOU NEED AND WANT? POSSIBILITIES FOR A BETTER FUTURE

POSSIBILITIES FOR A BETTER FUTURE

SKILLS FOR IDENTIFYING POSSIBILITIES FOR A BETTER FUTURE
Creativity and Helping
Divergent Thinking
Brainstorming: A Tool for Divergent Thinking
Future-Oriented Probes
Models as a Source of Possibilities

POSSIBILITIES FOR A BETTER DEATH: A CASE

A FAMILY CASE

EVALUATION QUESTIONS FOR STEP II-A

## POSSIBILITIES FOR A BETTER FUTURE

The goal of Step II-A is to help clients develop a sense of direction by exploring possibilities for a better future. I once was sitting at the counter of a late-night diner when a young man sat down next to me. The conversation drifted to the problems he was having with a friend of his. I listened for a while and then asked, "Well, if your relationship was just what you wanted it to be, what would it look like?" It took him a bit to get started, but eventually he drew a picture of the kind of relationship he could live with. Then he stopped, looked at me, and said, "You must be a professional." I believe he thought that because this was the first time in his life that anyone had ever asked him to describe the possibilities of a better future.

**The power of possibilities.** Too often the exploration and clarification of problem situations are followed, almost immediately, by the search for "solutions." *Solutions* is an ambiguous word. It can refer to the series of actions that will lead to the resolution of the problem situation or the development of an opportunity. But it can also refer to what will be in *place* once those actions are completed. There is great power in visualizing outcomes, just as there is a danger in formulating action strategies before getting a clear idea of desired outcomes. Stage II is about results, outcomes, accomplishments. Stage III is about strategies and plans for delivering those outcomes.

**Possible selves.** Clients come to helpers because they are stuck. Counseling is a process of helping clients get "unstuck" and develop a sense of direction. Consider the case of Ernesto. He was very young but very stuck for a variety of sociocultural and emotional reasons.

> A counselor first met Ernesto in the emergency room of a large urban hospital. He was throwing up blood into a pan. He was a member of a street gang, and this was the third time he had been beaten up in the last year. He had been so severely beaten this time that it was likely that he would suffer permanent physical damage. Ernesto's style of life was doing him in, but it was the only one he knew. He was in need of a new way of living, a new scenario, a new way of participating in city life. This time he was hurting enough to consider the possibility of some kind of change.

Markus and Nurius (1986) used the term *possible selves* to represent "individuals' ideas of what they might become, what they would like to become, and what they are afraid of becoming" (p. 954). The counselor worked with Ernesto, not by helping him explore the complex sociocultural and emotional reasons he was in this fix, but principally by helping him explore his "possible selves" to discover a different purpose in life, a different direction, a different lifestyle. Step II-A is about "possible selves."

# SKILLS FOR IDENTIFYING POSSIBILITIES FOR A BETTER FUTURE

At its best, counseling helps clients move from problem-centered mode to "discovery" mode. Discovery mode involves creativity and divergent thinking.

## Creativity and Helping

One of the myths of creativity is that some people are creative and others are not. Clients, like the rest of us, can be more creative than they are. It is a question of finding ways to help them be so. Stages II and III help clients tap into their dormant creativity. A review of the requirements for creativity (see Cole & Sarnoff, 1980; Robertshaw, Mecca, & Rerick, 1978, pp. 118–120) shows, by implication, that people in trouble often fail to use whatever creative resources they might have. The creative person is characterized by

- *optimism and confidence*, whereas clients are often depressed and feel powerless.
- *acceptance of ambiguity and uncertainty*, whereas clients may feel tortured by ambiguity and uncertainty and want to escape from them as quickly as possible.
- *a wide range of interests*, whereas clients may be people with a narrow range of interests or whose normal interests have been severely narrowed by anxiety and pain.
- *flexibility*, whereas clients may have become rigid in their approach to themselves, others, and the social settings of life.
- *tolerance of complexity*, whereas clients are often confused and looking for simplicity and simple solutions.
- *verbal fluency*, whereas clients are often unable to articulate their problems, much less their goals and ways of accomplishing them.
- *curiosity*, whereas clients may not have developed a searching approach to life or may have been hurt by being too venturesome.
- *drive and persistence*, whereas clients may be all too ready to give up.
- *independence*, whereas clients may be quite dependent or counterdependent.
- *nonconformity or reasonable risk taking*, whereas clients may have a history of being very conservative and conformist or may get into trouble with others and with society precisely because of their particular brand of nonconformity.

A review of some of the principal obstacles to creativity brings further problems to the surface. Innovation is hindered by

- *fear*, and clients are often quite fearful and anxious.
- *fixed habits*, and clients may have self-defeating habits or patterns of behavior that may be deeply ingrained.
- *dependence on authority*, and clients may come to helpers looking for the "right answers" or be quite counterdependent (the other side of the dependence coin) and fight efforts to be helped with "Yes, but" and other games.
- *perfectionism*, and clients may come to helpers precisely because they are hounded by this problem and can accept only ideal or perfect solutions.

It is easy to say that imagination and creativity are most useful in Stages II and III, but it is another thing to help clients stimulate their own, perhaps dormant creative potential.

## Divergent Thinking

Many people habitually take a convergent-thinking approach to problem solving; that is, they look for the "one right answer." Such thinking has its uses, of course. However, many of the problem situations of life are too complex to be handled by convergent thinking. Such thinking limits the ways in which people use their own and environmental resources.

Divergent thinking, on the other hand — thinking "outside the boxes" — assumes that there is always more than one answer. De Bono (1992) calls it "lateral thinking." In helping, that means more than one way to manage a problem or develop an opportunity. Unfortunately, divergent thinking, as helpful as it can be, is not always rewarded in our culture and sometimes is even punished. For instance, students who think divergently can be thorns in the sides of teachers. Some teachers feel comfortable only when they ask questions in such a way as to elicit the "one right answer." When students who think divergently give answers that are different from the ones expected, even though their responses might be quite useful (perhaps more useful than the expected responses), they may be ignored, corrected, or punished. Students learn that divergent thinking is not rewarded, at least not in school, and they may generalize their experience and end up thinking that it is simply not a useful form of behavior. Consider the following case.

> Quentin wanted to be a doctor, so he enrolled in the pre-med program at school. He did well but not well enough to get into medical school. When he received the last notice of refusal, he said to himself, "Well, that's it for me and the world of medicine. Now what will I do?" When he graduated, he took a job in his brother-in-law's business. He became a manager and did fairly well financially, but he never experienced much career satisfaction. He was glad that his marriage was good and his home life rewarding, because he derived little satisfaction from his work.

Not much divergent thinking went into handling this problem situation. No one asked Quentin what he really wanted. For Quentin, becoming a doctor was the "one right career." He didn't give serious thought to any other career related to the field of medicine, even though there are dozens of interesting and challenging jobs in the field of health care.

The case of Caroline, who also wanted to become a doctor but failed to get into medical school, is quite different from that of Quentin.

> Caroline thought to herself, "Medicine still interests me; I'd like to do something in the health field." With the help of a medical career counselor, she reviewed the possibilities. Even though she was in pre-med, she had never realized that there were so many careers in the field of medicine. She decided to take whatever courses and practicum experiences she needed to become a nurse. Then, while working in a clinic in the hills of Appalachia, where she found the experience invaluable, she managed to get an M.A. in family-practice nursing by attending a nearby state university part-time. She chose this specialty because she thought that it would enable her to be closely associated with delivery of a broad range of services to patients and would also enable her to have more responsibility for the delivery of those services.
>
> When Caroline graduated, she entered private practice with a doctor as a nurse practitioner in a small midwestern town. Since the doctor divided his time among three small clinics, Caroline had a great deal of responsibility in the clinic where she practiced. She also taught a course in family-practice nursing at a nearby state school and conducted workshops in holistic approaches to preventive medical self-care. Still not satisfied, she began and finished a doctoral program in practical nursing. She taught at a state university and continued her practice. Needless to say, her persistence paid off with an extremely high degree of career satisfaction.

A successful professional career in health care remained Caroline's aim throughout. A great deal of divergent thinking and creativity went into the elaboration of that aim into specific goals and into coming up with the courses of action to accomplish them. But for every success story, there are many more failures. Quentin's case is the norm, not Caroline's. For many, divergent thinking is uncomfortable.

## Brainstorming: A Tool for Divergent Thinking

One excellent way of helping clients think divergently and more creatively is brainstorming. Brainstorming is a simple idea-stimulation technique for exploring the elements of complex situations. Brainstorming in Stages II and III is a tool for helping clients develop possibilities for a better future and ways of accomplishing goals.

There are certain rules that help make this technique work: suspend judgment, produce as many ideas as possible, use one idea as a takeoff point for others, get rid of normal constraints to thinking, and produce even more ideas by clarifying items on the list. Here, then, are the rules.

- *Suspend your own judgment, and help clients suspend theirs.* In the brainstorming phase, do not let clients criticize the ideas they are generating and do not criticize these possibilities yourself. There is some evidence that this rule is especially effective when the problem situation has been clarified and defined and goals have not yet been set. In the following example, a woman whose children are grown and married is looking for ways of putting meaning into her life.

CLIENT: One possibility is that I could become a volunteer, but the very word makes me sound a bit pathetic.

HELPER: Add it to the list. Remember, we'll discuss and critique them later.

Having clients suspend judgment is one way of handling the tendency on the part of some to play a "Yes, but" game with themselves. That is, they come up with a good idea and then immediately show why it isn't really a good idea, as in the preceding example. By the same token, don't let yourself say such things as "Explain what you mean," "I like that idea," "This one is useful," "I'm not sure about that idea," or "How would that work?" Premature approval and criticism cut down on creativity. A marriage counselor was helping a couple brainstorm possibilities for a better future. When Nina said, "We will stop bringing up past hurts," Tip replied, "That's your major weapon when we fight. You'll never be able to give that up." The helper said, "Add it to the list. We'll look at the realism of these possibilities later on."

- *Encourage clients to come up with as many possibilities as possible.* The principle is that quantity ultimately breeds quality. Some of the best ideas come along later in the brainstorming process. Cutting the process short can be self-defeating. In the following example, a man in a sex-addiction program has been brainstorming activities that might replace his preoccupation with sex.

CLIENT: Maybe that's enough. We can start putting it all together.

HELPER: It didn't sound like you were running out of ideas.

CLIENT: I'm not. It's actually fun. It's almost liberating.

HELPER: Well, let's keep on having fun for a while.

CLIENT (pausing): Ha! I could become a monk.

Later on, the counselor, focusing on this "possibility," asked, "What would a modern-day monk who's not even a Catholic look like?" This helped the client explore the concept of sexual responsibility from a completely different perspective and to rethink the place of religion in his life. And so, within reason, the more ideas the better. Helping clients identify many possibilities for a better future increases the quality of the possibilities that are eventually chosen and turned into goals.

- *Help clients use one idea as a takeoff point for another.* This is called piggybacking. Without criticizing the client's productivity, encourage him or her both to develop strategies already generated and to combine different ideas to form new possibilities. In the following example, a client suffering from chronic pain is trying to come up with possibilities for a better future.

CLIENT: Well, if there is no way to get rid of all the pain, then I picture myself living a full life without pain at its center.

HELPER: Expand that a bit for me.

CLIENT: The papers are filled with stories of people who have been living with pain for years. When they're interviewed, they always look miserable. They're like me. But every once in a while there is a story about someone who has learned how to live creatively with pain. Very often they are involved in some sort of cause which takes up their energies. They don't have time to be preoccupied with pain.

When one client with multiple sclerosis said, "I'll have a friend or two with whom I can share my frustrations as they build up," the helper asked, "What would that look like?" The client replied, "It would not be just a poor-me thing. It would be a give-and-take relationship. We'd be sharing both joys and pains like other people do."

•  *Help clients let themselves go and develop some "wild" possibilities.* When clients seem to be "drying up" or when the possibilities being generated are quite pedestrian, I often say, "Okay, now draw a line under the items on your list and write the word *wild* under the line. Now let's see if we can develop some really wild ways of getting your goals accomplished." Later it is easier to cut suggested strategies down to size than to expand them. The wildest possibilities often have within them at least a kernel of an idea that will work. In the following example, an older single man who is lonely is exploring possibilities for a better future.

CLIENT: I can't think of anything else. And what I've come up with isn't very exciting.

HELPER: How about getting a bit wild? You know, some crazy possibilities.

CLIENT: Well, let me think. . . . I'd start a commune and would be living in it. . . . And . . .

Clients often need permission to let themselves go even in harmless ways. They repress good ideas because they might sound foolish. Helpers need to create an atmosphere where such apparently foolish ideas will be not only accepted but also encouraged. Help clients come up with conservative possibilities, liberal possibilities, radical possibilities, and even outrageous possibilities.

# Future-Oriented Probes

One way of helping clients invent the future is to ask them, or get them to ask themselves, future-oriented questions related to their current unmanaged problems or undeveloped opportunities. The following questions are different ways of helping clients answer the questions "What do you want? What do you need?" They all deal with needs and wants in terms of what will be *in place* after the clients act.

- *What would this problem situation look like if you were managing it better?* Ken, a college student who has been a "loner," has been talking about his general dissatisfaction with his life. A couple of the things he said: "I'd be having fewer anxiety attacks. And I'd be spending more time with people rather than by myself."

- *What changes in your present lifestyle would make sense?* Cindy, who described herself as a "bored homemaker," replied, "I would not be drinking as much. I'd be getting more exercise. I would not sit around and watch the soaps all day. I'd have something meaningful to do."

- *What would you be doing differently with the people in your life?* Lon, a graduate student at a university near his parents' home, realized that he had not yet developed the kind of autonomy suited to his age. He mentioned these possibilities: "I would not be letting my mother make my decisions for me. I'd be sharing an apartment with one or two friends."

- *What patterns of behavior would be in place that are not currently in place?* Bridget, a depressed resident in a nursing home, had this suggestion: "I'd be engaging in more of the activities offered here in the nursing home."

- *What current patterns of behavior would be eliminated?* Bridget added, "I would not be putting myself down for incontinence I cannot control. I would not be complaining all the time. It gets me and everyone else down!"

- *What would you have that you don't have now?* Sissy, a single woman who has lived in a housing project for 11 years, said, "I'd have a place to live that's not rat-infested. I'd have some friends. I wouldn't be so miserable all the time."

- *What accomplishments would be in place that are not in place now?* Ryan, a divorced man in his mid-30s, said, "I'd have my degree in practical nursing. I'd be doing some part-time teaching. I'd have someone to marry."

- *What would this opportunity look like if you developed it?* Enid, a woman with a great deal of talent who has been given one modest promotion in her company but who feels like a second-class citizen, had this to say: "In two years I'll be an officer of this company or have a very good job in another firm."

It is a mistake to suppose that clients will automatically gush with answers. Ask the kinds of questions just listed, or encourage them to ask themselves the questions, but then help them answer them. Many clients don't know how to use their innate creativity. Thinking divergently is not part of their mental lifestyle. You have to work with clients to help them produce some creative output. Some clients are reluctant to name possibilities for a better future because they sense that this will bring more responsibility. They will have to move into action mode.

# Models as a Source of Possibilities

Some clients can see future possibilities better when they see them embodied in others. You can help clients brainstorm possibilities for a better future by helping them to identify models. By models I don't mean superstars or people who do things perfectly. That would be self-defeating. In the next example, a marriage counselor is talking with a middle-aged, childless couple. They are bored with their marriage. When he asked them, "What would your marriage look like if it looked a little better?" he could see that they were stuck.

COUNSELOR: Maybe the question would be easier to answer if you reviewed some of your married relatives, friends, or acquaintances.

WIFE: None of them have super marriages. (Husband nods in agreement.)

COUNSELOR: No, I don't mean super marriages. I'm looking for things you could put in your marriage that would make it a little better.

WIFE: Well, Fred and Lisa are not like us. They don't always have to be doing everything together.

HUSBAND: Who says we have to be doing everything together? I thought that was your idea.

WIFE: Well, we always *are* together. If we weren't always together, we wouldn't be in each other's hair all the time.

Even though it was a somewhat torturous process, these two people were able to come up with a range of possibilities for a better marriage. The counselor had them write them down so they wouldn't lose them. At this point the purpose was not to get the clients to commit themselves to these possibilities but to review them.

In the following case, the client finds herself making discoveries by observing people she had not identified as models at all.

Fran, a somewhat withdrawn college junior, realizes that when it comes to interpersonal competence, she is not ready for the business world she intends to enter when she graduates. She wants to do something about her interpersonal style and a few nagging personal problems. She sees a counselor in the Office of Student Services. After a couple of discussions with him, she joins a "lifestyle" group on campus that includes some training in interpersonal skills. Even though she expands her horizons a bit from what the members of the group say about their experiences, behaviors, and feelings, she tells her counselor that she learns even more by watching her fellow group members in action. She sees behaviors that she would like to incorporate in her own style. A number of times she says to herself in the group, "Ah, there's something I never thought of." Without becoming a slavish imitator, she begins to incorporate some of the patterns she sees in others into her own style.

Models or exemplars can help clients name what they want more specifically. Models can be found anywhere: among the client's relatives, friends, and associates, in books, on television, in history, in movies. Counselors can help clients identify models, choose those dimensions of others that are relevant, and translate what they see into realistic possibilities for themselves.

## POSSIBILITIES FOR A BETTER DEATH: A CASE

Tom was HIV-positive and it was clear that he was getting sicker. But he wanted to "get some things done" before he died. Tom's action orientation helped a great deal. Over the course of a few months, a counselor helped him to name some of the things he wanted before he died or on his journey toward death. Tom came up with the following possibilities:

- He wanted someone with a religious orientation, such as a minister, with whom he could occasionally talk about the "bigger" issues of life and death.
- He wanted to die with a sense of meaning.
- He wanted to be a member of some kind of community, maybe a self-help group of fellow AIDS victims, people who did not fear him.
- He wanted fewer financial worries.
- He wanted one or two intimates with whom he could share the ups and downs of daily life.
- He wanted to be engaged in some kind of productive work, whether paid or not, as long as he could.
- He wanted a decent place to live, maybe with others.
- He wanted access to decent medical attention from a medical staff who would not treat him like a new-age leper.
- He wanted to manage bouts of anxiety and depression better.
- He wanted to be reintegrated with his family.
- He wanted especially some kind of reconciliation with his father.
- He wanted to make peace with one or two of his closest friends who abandoned him when they learned he had AIDS.
- He wanted to die in his home town.

Tom didn't name all these possibilities at once, nor did he necessarily use this language. The counselor helped him stitch together a set of goals from these possibilities (Stage II) and ways of accomplishing them (Stage III). This help contributed substantially to Tom's quality of life, even under very difficult circumstances. Box 13-1 outlines the kinds of questions you can help clients ask themselves to discover possibilities for a better future.

---

BOX 13-1
## Questions for Exploring Possibilities

- What are my most critical needs and wants?
- What are some possibilities for a better future?
- What outcomes or accomplishments would take care of my most pressing problems?
- What would my life look like if I were to develop a couple of key opportunities?
- What should my life look like a year from now?
- What should I put in place that is currently not in place?
- What are some wild possibilities for making my life better?

---

# A FAMILY CASE

This case is more complex because it involves a family. Not only does the family as a unit have its wants and needs, but also each of the individual members has his or her own. Therefore, it is even more imperative to review possibilities for a better future so that competing needs can be reconciled.

Lane, the 15-year-old son of Troy and Rhonda Washington, was hospitalized with what was diagnosed as an "acute schizophrenic attack." He had two older brothers, both teenagers, and two younger sisters, one 10 and one 12, all living at home. The Washingtons lived in a large city. Although both parents worked, their combined income still left them pinching pennies. They also ran into a host of problems associated with their son's hospitalization—the need to arrange ongoing help and care for Lane, financial burdens, behavioral problems among the other siblings, marital conflict, and stigma in the community ("They're a funny family with a crazy son"; "What kind of parents are they?"). To make things worse, they did not think the psychiatrist and the psychologist they met at the hospital took the time to understand their concerns. They felt that the helpers were trying to push Lane back out into the community; in their eyes, the hospital was "trying to get rid of him." "They give him some pills and then give him back to you" was their complaint. No one explained to them that short-term hospitalization was meant to guard the civil rights of patients and avoid the negative effects of longer-term institutionalization.

When Lane was discharged, his parents were told that he might have a relapse, but they were not told what to do about it. They faced the prospect of caring for Lane in a climate of stigma without adequate information, services, or relief. Feeling abandoned, they were very angry with the mental-health establishment. They had no idea what they should do to respond to Lane's illness or to the range of family problems that had been precipitated by the episode. By chance the Washingtons met someone who had worked for the National Alliance for the Mentally Ill (NAMI), an advocacy and education organization. This person referred them to an agency that provided support and help.

What does the future hold for such a family? With help, what kind of future can be fashioned? Social workers at the agency helped the Washingtons identify both needs and wants in seven areas (see Bernheim, 1989).

- *The home environment.* The Washingtons needed an environment in which the needs of all the family members are balanced. They didn't want their home be an extension of the hospital. They wanted Lane taken care of, but they wanted to attend to the needs of the other children and to their own needs as well.

- *Care outside the home.* They wanted a comprehensive therapeutic program for Lane. They needed to review possible services, identify relevant services, and arrange access to those services. They needed to find a way of paying for all this.

- *Care inside the home.* They wanted all family members to know how to cope with Lane's residual symptoms. He might be withdrawn or aggressive, but they needed to know how to relate to him and help him handle behavioral problems.

- *Prevention.* Family members needed to be able to spot early warning symptoms of impending relapse. They also needed to know what to do when they saw those signs, including such things as contacting the clinic or, in the case of more severe problems, arranging for an ambulance or getting help from the police.

- *Family stress.* They needed to know how to cope with the increased stress all of this would entail. They needed forums for working out their problems. They wanted to avoid family blowups, and when blowups occurred, they wanted to manage them without damage to the fabric of the family.

- *Stigma.* They wanted to understand and be able to cope with whatever stigma might be attached to Lane's illness. For instance, when taunted for having a "crazy brother," the children needed to know what to do and what not to do. Family members needed to know whom to tell, what to say, how to respond to inquiries, and how to deal with blame and insults.

- *Limitation of grief.* They needed to know how to manage the normal guilt, anger, frustration, fear, and grief that go with problem situations like this.

Bernheim's schema constituted a useful checklist for stimulating thinking about possibilities for a better future. The Washingtons first needed to be helped to develop these possibilities. The next step would be to help them set priorities and establish goals to be accomplished. That is the work of Step II-B.

## Evaluation Questions for Step II-A

- How at home am I in working with my own imagination?
- In what ways can I apply the concept of "possible selves" to myself?
- What problems do I experience as I try to help clients use their imaginations?
- Against the background of problem situations and unused opportunities, how well do I help clients focus on what they want?
- To what degree do I prize divergent thinking and creativity in myself and others?
- How effectively do I use empathy, a variety of probes, and challenge to help clients brainstorm what they want?
- Besides direct questions and other probes, what kinds of strategies do I use to help clients brainstorm what they want?
- How effectively do I help clients identify models and exemplars that can help them clarify what they want?
- How easily do I move back and forth in the helping model, especially in establishing a "dialogue" between Stage I and Stage II?
- How well do I help clients act on what they are learning?

## CHAPTER 14

# STEP II-B:
# WHAT DO YOU REALLY WANT? MOVING FROM POSSIBILITIES TO CHOICES

# FROM POSSIBILITIES TO CHOICES

Once possibilities for a better future have been developed, clients need to make some choices; that is, they need to choose one or more of those possibilities and turn them into a *program for constructive change*. Step II-A is, in many ways, about *creativity*, getting rid of boundaries, thinking beyond one's limited horizon, moving outside the box. Step II-B is about *innovation* — that is, turning possibilities into a practical program for change. If implemented, a goal constitutes the "solution" for the client's problem. Consider the following case.

> Bea, an African American woman, had been arrested when she went on a rampage in a bank and broke several windows. She had exploded with anger because she felt that she had been denied a loan mainly because she was black and a single mother. In discussing the incident with her minister, she came to see that she had a great deal of pent-up anger. She also realized that venting her anger as she had done in the bank led to a range of negative consequences. But she still harbored a great deal of anger about social injustice. To complicate the picture, she tended to take it out on those around her, including her friends and her two children. The minister helped her look at four possibilities — venting her anger, repressing her anger, channeling her anger, and simply giving up. Giving up was not in her makeup. Merely venting her anger seemed to do little but make her more angry. Repressing her anger, she reasoned, was just another way of giving up, and that was demeaning. The "channeling" option needed to be explored. In the end, Bea joined a political activist group involved in community organizing. She learned that she could channel her anger without giving up her values or her intensity. The impact on her life was quite positive. She also discovered that she was good at influencing others and getting things done. She felt very good about herself.

Since goals can be highly motivational, helping clients set realistic goals is one of the most important steps in the helping process.

# HELPING CLIENTS SHAPE THEIR GOALS

Practical goals do not usually leap out fully formed. They need to be shaped. Effective counselors add value by using the communication skills outlined earlier to help clients choose, craft, shape, and develop their goals. Goals are specific statements about what clients want and need. The goals that either emerge through client-helper dialogue or that are explicitly set by clients in fashioning problem-managing and opportunity-developing programs for constructive change are more likely to be *workable* if they have certain characteristics. They need to be

- stated as *outcomes* rather than activities,
- *specific* enough to be verifiable and to drive action,

- *challenging* and substantive,
- both venturesome and *prudent,*
- *realistic* in regard to resources and control,
- *sustainable* over a reasonable time period,
- *flexible* without being wishy-washy,
- congruent with the *client's values,* and
- set in a reasonable *time frame.*

In keeping with our no-formula approach, just how this package of goal characteristics will look in practice will differ from client to client. From a practical point of view, these characteristics can be seen as "tools" that counselors can use to help clients shape their goals. They can also be used to reshape and evaluate goals that have been set. Ineffective helpers will get lost in the details of these characteristics. Effective helpers will keep them in the back of their mind and, in a second-nature manner, turn them into helpful "sculpting" probes at the right time. These characteristics, then, take on life through the following flexible principles.

## Help Clients State What They Need and Want as Outcomes or Accomplishments

The goal of counseling, as emphasized again and again, is neither discussing nor planning nor engaging in activities but problem-managing *results.* "I want to start doing some exercise" is an activity rather than an outcome. "Within six months I will be running three miles in less than thirty minutes at least four times a week" is an outcome, a pattern of behavior in place. If a client says, "My goal is to get some training in interpersonal communication skills," then she is stating her goal as a set of activities rather than as an accomplishment. If she says that she wants to become a better listener as a wife and mother, then she is stating her goal as an accomplishment, even though "better listener" needs further clarification. Goals stated as outcomes provide *direction* for clients. They are targets.

You can help clients develop this "past-participle" approach — drinking *stopped,* number of marital fights *decreased,* anger habitually *controlled* — to describing what they need and want. Stating goals as outcomes or accomplishments is not just a question of language. Helping clients state goals as accomplishments rather than activities helps them avoid directionless and imprudent action. If a woman with breast cancer says that she thinks she should join a self-help group, she should be helped to see what she wants to get out of such a group. Joining a group and participating in it are activities. Goals, at their best, are expressions of what clients need and want. Clients who know what they want are more likely than those who don't to work not just harder but also smarter.

Consider the case of Dillard, a former Marine suffering from posttraumatic shock disorder.

Dillard was involved in Desert Storm. Three of his buddies were killed by "friendly fire." Afterward he began acting in strange ways, wandering around at times in a daze. He was sent home and given a medical discharge. Although he seemed to recover, he lived an aimless life. He went to college but dropped out during the first semester. He became rather reclusive but, as a friend who referred him to the helper said, "Dillard never really showed odd behavior." He moved in and out of a number of low-paying jobs. He also became less careful about his person. "You know, I used to be very careful about the way I dressed. Kind of proud of myself in the marine tradition. Don't get me wrong; I'm not a bum and don't smell or anything, but I'm not myself." The whole direction of Dillard's life was wrong; he was headed for serious trouble. He was bothered by thoughts about the war and had taken to sleeping whenever he felt like it, day or night, "just to make it all stop."

Ed, Dillard's counselor, had a good relationship with Dillard. He helped Dillard tell his story and challenged some of his self-defeating thinking. He went on to help Dillard focus on what he wanted from life. They moved back and forth between Stage I and Stage II, between problems and possibilities for a better future. Eventually, Dillard began talking about his real needs and wants — that is, what he needed to accomplish to "get back to his old self."

**DILLARD:** I've got to stop hiding in my hole. I'm going to get out and see people more. I'm going to stop feeling so damn sorry for myself. Who wants to be with a nothing!

**COUNSELOR:** What will Dillard's life look like a year or two from now?

**DILLARD:** One thing for sure. He will be seeing women again. He might not be married, yet, but he will probably have a special girlfriend. And she will see him as an ordinary guy.

Here Dillard talks about changes as patterns of behavior that will be in place. He is painting a picture of what he wants to be. The helper's probe reinforces this outcome approach.

## Help Clients Move from Broad Aims to Clear and Specific Goals

Specific rather than general goals tend to drive behavior. Specific goals will, of course, be different for each client. Therefore, broad goals need to be translated into more specific goals and tailored to the needs and abilities of each client. Skilled helpers use probes and challenges to help clients say what they really want to accomplish.

Dillard said that he wanted to become "more disciplined." His counselor helped him make that more specific.

**COUNSELOR:** What areas do you want to focus on?

**DILLARD:** Well, if I'm going to put more order in my life, I need to look at the times I sleep. I've been going to bed whenever I feel like it and getting up whenever I feel like it. It was the only way I could get rid of those thoughts and the anxiety. But I'm not nearly as anxious as I used to be. Things are calming down.

COUNSELOR: So more disciplined means a more regular sleep schedule because there's no particular reason now for not having one.

DILLARD: Yeah, now sleeping whenever I want is just a bad habit. And I can't get things done if I'm asleep.

Dillard goes on to translate "more disciplined" into other problem-managing needs and wants related to school, work, and care of his person. Greater discipline will have a decidedly positive impact on his life. It will make him a doer rather than an onlooker.

Counselors often add value by helping clients move from vague desires to quite specific goals. The movement might be from good intentions through broad aims to quite specific goals.

- *Good intentions.* "I need to do something about this" is a statement of intent. However, even though good intentions are a good start, they need to be translated into aims and goals. In the following example, the client, Jon, has been discussing his relationship with his wife and children. The counselor has been helping him see that his "commitment to work" is perceived negatively by his family. Jon is open to challenge and is a fast learner.

JON: Boy, this session has been an eye-opener for me. I've really been blind. My wife and kids don't see my investment — rather, my overinvestment — in work as something I'm doing for them. I've been fooling myself, telling myself that I'm working hard to get them the good things in life. In fact, I'm spending most of my time at work because I like it. My work is mainly for me. It's time for me to realign some of my priorities.

The last statement is a good intention, an indication on Jon's part that he wants to do something about a problem now that he sees it more clearly. It may be that Jon will now go out and put a different pattern of behavior in place without further help from the counselor. Or he may need some help in realigning his priorities.

- *Broad aims.* A broad aim is more than a good intention. It has content; that is, it identifies the area in which the client wants to work and makes some general statement about that area. Let's return to the example of Jon and his overinvestment in work.

JON: I don't think I'm spending so much time at work in order to run away from family life. But family life is deteriorating because I'm just not around enough. I must spend more time with my wife and kids. Actually, it's not just a case of must. I want to.

Jon moves from a declaration of intent to an aim or a broad goal, spending more time at home. But he still has not created a picture of what that will look like.

- *Specific goals.* To help Jon move toward greater specificity, the counselor uses such probes as "Tell me what 'spending more time at home' will look like."

**JON:** I'm going to consistently spend three out of four weekends a month at home. During the week I'll work no more than two evenings.

**COUNSELOR:** So you'll be at home a lot more. Tell me what you'll be doing with all this time.

Notice how much more specific Jon's statement is than "I'm going to spend more time with my family." He sets a goal as a specific pattern of behavior he wants to put in place. But his goal as stated deals with quantity, not quality. The counselor's probe is really a challenge. It's not just the *amount* of time Jon is going to spend with his family but also the kinds of things he will be doing. Quality time, some call it. This warrants discussion because maybe the family wants a relaxed rather than an intense Jon at home.

Helping clients move from good intentions to more and more specific goals is a shaping process. Consider the example of a couple whose marriage has degenerated into constant bickering, especially about finances.

- *Good intention.* "We want to straighten out our marriage."
- *Broad aim.* "We want to handle our decisions about finances in a much more constructive way."
- *Specific goal.* "We try to solve our problems about family finances by fighting and arguing. We'd like to reduce the number of fights we have and begin making mutual decisions about money. We yell instead of talking things out. We need to set up a month-by-month budget. Otherwise we'll be arguing about money we don't even have. We'll have a trial budget ready the next time we meet with you."

Declarations of intent, broad goals, and specific goals can all drive constructive behavior, but specific goals have the best chance. Is it possible to get clients to be too specific about their goals? Yes, if they get lost in the planning details and lose their spontaneity.

If the goal is clear enough, the client will be able to determine progress toward the goal. For many clients, being able to measure progress is an important incentive. If goals are stated too broadly, it is difficult to determine both progress and accomplishment. "I want to have a better relationship with my wife" is a very broad goal, difficult to verify. "I want to spend more time with her" comes closer, but it is still not clear what "more time" means.

It is not always necessary to count things to determine whether a goal has been reached, though sometimes counting is helpful. Helping is about living more fully, not about accounting activities. At a minimum, however, desired outcomes need to be capable of being verified in some way. For instance, a couple might say something like "Our relationship is better, not because we've found ways of spending more time together, but because the quality of our time together has improved. We listen more carefully, we talk about more personal concerns, we are more relaxed, and we make more mutual decisions about issues that affect us both, such as finances." The couple's accomplishment is one they can verify because they have spelled out what they mean by quality.

# Help Clients Establish Goals That Make a Difference

Outcomes and accomplishments are meaningless if they do not have the required *impact* on the client's life. The goals clients choose should have substance to them — that is, some decided impact on their lives. First of all, to have substance, a goal must be substantially related to managing the original problem situation or developing some opportunity.

> Vitorio ran the family business. His son, Anthony, worked for his father, mainly in sales. After spending a few years learning the business and getting an MBA part-time at a local university, Anthony wanted more responsibility and authority. His father never thought that he was "ready." They began arguing quite a bit, and their relationship suffered from it. Finally, a friend of the family persuaded them to spend time with a consultant-counselor who worked with small family businesses. He spent relatively little time listening to their problems. After all, he had seen this same problem over and over again: the reluctance and conservatism of the father, the pushiness of the son.
>
> Vitorio wanted the business to stay on the rails. Anthony wanted to be the company's marketer, to move it into new territory. After a number of discussions with the consultant-counselor, they settled on this scenario: A "marketing department" headed by Anthony would be created. He could divide his time between sales and marketing as he saw fit, provided that he maintained the current level of sales. Vitorio agreed not to interfere. They would meet once a month with the consultant-counselor to discuss problems and progress. Vitorio insisted that the consultant's fee come from increased sales. After some initial turmoil, the bickering decreased dramatically. Anthony easily found new customers, although they demanded modifications in the product line, which Vitorio reluctantly approved. Both sales and margins increased to the point that another person was needed in sales.

Not all issues in family businesses are handled as easily. In fact, a few years later Anthony left the business and founded his own. But the goal package they worked out made quite a difference both in the father-son relationship and in the business.

Second, goals have substance to the degree that they help clients "stretch" themselves. There is a good deal of research (Locke & Latham, 1984, 1990) suggesting that, other things being equal, harder goals can be more motivating than easier goals.

> Extensive research . . . has established that, within reasonable limits, the . . . more challenging the goal, the better the resulting performance. . . . People try harder to attain the hard goal. They exert more effort. . . . In short, people become motivated in proportion to the level of challenge with which they are faced. . . . Even goals that cannot be fully reached will lead to high effort levels, provided that partial success can be achieved and is rewarded. (Locke & Latham, 1984, pp. 21, 26)

I met an AIDS patient who was, in the beginning, full of self-loathing and despair. Eventually, however, over time he painted a new scenario in which he saw himself not as a victim of his own lifestyle but as a helper to other AIDS patients. Until close to the time of his death, he worked hard, within the limits of his physical disabilities, seeking out other AIDS sufferers, getting them to join self-help groups, and generally helping them to manage an impossible situation in a more humane way. When he was near death, he said that the last two years, though at times they were very bitter, were among the best years of his life. He had set his goals high, but they proved to be quite realistic. On the other hand, "difficult" should not mean "impossible."

## Help Clients Set Goals That Are Prudent

Although the helping model described in this book encourages a bias toward action on the part of clients, action needs to be both directional and wise. Discussing and setting goals should contribute to both direction and wisdom. The following case begins poorly but ends well.

> Harry was a sophomore in college who was admitted to a state mental hospital because of some bizarre behavior at the university. He was one of the disc jockeys for the university radio station. He came to the notice of college officials one day when he put on an attention-getting performance that included rather lengthy dramatizations of grandiose religious themes. In the hospital it was soon discovered that this quite pleasant, likable young man was actually a loner. Everyone who knew him at the university thought that he had many friends, but in fact he did not. The campus was large, and his lack of friends went unnoticed.
>
> Harry was soon released from the hospital but returned weekly for therapy. At one point he talked about his relationships with women. Once it became clear to him that his meetings with women were perfunctory and almost always took place in groups—he had thought he had a rather full social life with women—Harry launched a full program of getting involved with the opposite sex. His efforts ended in disaster, however, because Harry had some basic sexual and communication problems. He also had serious doubts about his own worth and therefore found it difficult to make a gift of himself to others. He ended up in the hospital again.
>
> The counselor helped Harry get over his sense of failure by emphasizing what Harry could learn from the "disaster." With the therapist's help, Harry returned to the problem-clarification and new-perspectives part of the helping process and then established more realistic short-term goals regarding getting back "into community." The direction was the same—establishing a realistic social life—but the goals were now more prudent because they were "bite-size." Harry attended socials at a local church where a church volunteer provided support and guidance.

Harry's leaping from problem clarification to action without taking time to discuss possibilities and set a direction toward reasonable goals was part of the problem rather than part of the solution. His lack of success in establishing solid relationships with women actually helped him see his problem with women more clearly. There are two kinds of prudence—playing it safe is one, doing the wise thing is the other. Helping is venturesome. It is about wise choices rather than playing it safe.

## Help Clients Formulate Realistic Goals

Setting stretch goals can help clients energize themselves. They rise to the challenge. On the other hand, goals set too high can do more harm than good. Locke and Latham (1984, p. 39) put it succinctly: "Nothing breeds success like success. Conversely, nothing causes feelings of despair like perpetual failure. A primary purpose of goal setting is to increase the motivation level of the individual. But goal setting can have precisely the opposite effect if it produces a yardstick that constantly makes the individual feel inadequate." A goal is realistic if the client has access to the *resources* needed to accomplish it, the goal is under the client's control, and external circumstances do not prevent its accomplishment.

**Resources: Help clients choose goals for which the resources are available.** It does little good to help clients develop specific, substantive, and verifiable goals if the resources needed for their accomplishment are not available. Consider the case of Rory, who, because of downsizing in his company, has had to take a demotion. He now wants to leave the company and become a consultant.

> *Insufficient resources:* Rory does not have the assertiveness, marketing savvy, industry expertise, or interpersonal style needed to become an effective consultant. Even if he did, he does not have the financial resources needed to tide him over while he developed a business.

> *Sufficient resources:* Challenged by the outplacement counselor, Rory changes his focus. Graphic design is an avocation of his. He is not good enough to take a technical position in the company's design department, but he does apply for a supervisory role in that department. He is good with people, very good at scheduling and planning, and knows enough about graphic design to discuss issues meaningfully with the members of the department.

Rory combines his managerial skills with his interest in graphic design to move in a quite different direction. The move is challenging, but it can have a substantial impact on his work life. For instance, the opportunity to hone his design skills will open up further career possibilities.

**Control: Help clients choose goals that are under their control.** Sometimes clients defeat their own purposes by setting goals that are not under their control. For instance, it is common for people to believe that their problems would be solved if only other people would not act the way they do. In most cases, however, we do not have any direct control over the ways others act. Consider the following example.

Tony, a 16-year-old boy, felt that he was the victim of his parents' inability to relate to each other. Each tried to use him in the struggle, and at times he felt like a Ping-Pong ball. A counselor helped him see that he could probably do little to control his parents' behavior but that he might be able to do quite a bit to control his reactions to his parents' attempts to use him. For instance, when his parents started to fight, he could simply leave instead of trying to "help." If either tried to enlist him as an ally, he could say that he had no way of knowing who was right. Tony also worked at creating a good social life outside the home. That helped him weather the tensions he experienced there.

Tony needed a new way of managing his interactions with his parents to minimize their attempts to use him as a pawn in their own interpersonal game. Goals are not under clients' control if they are blocked by external forces that they cannot influence. "To live in a free country" may be an unrealistic goal for a person living in a totalitarian state because he cannot change internal politics, nor can he change emigration laws in his own country or immigration laws in other countries. "To live as freely as possible in a totalitarian state" might well be an aim that could be translated into realistic goals.

## Help Clients Set Goals That Can Be Sustained

Clients need to commit themselves to goals that have staying power. One separated couple said that they wanted to get back together again. They did so only to separate and get divorced within six months. Their goal of getting back together again was achievable but not sustainable. Perhaps they should have asked themselves, "What do we need to do not only to get back together but also to stay together? What would our marriage have to look like to become and remain workable?" In discretionary-change situations, the issue of sustainability needs to be visited early on.

Many Alcoholics Anonymous–like programs work because of their one-day-at-a-time approach. The goal of being drug-free has to be sustained only over a single day. The next day is a new era. In a previous example, Vitorio and Anthony's arrangement had enough staying power to produce good results in the short term. It also allowed them to reset their relationship and to improve the business. The goal was not designed to produce to-the-end-of-our-lives results. In the end, Anthony's aspirations were bigger than the family business.

## Help Clients Choose Goals That Have Some Flexibility

In many cases, goals have to be adapted to changing realities. Therefore, there might be some trade-offs between goal specificity and goal flexibility in uncertain situations. Napoleon noted this when he said, "He will not go far who knows from the first where he is going." Sometimes making goals too specific or too rigid does not allow clients to take advantage of emerging opportunities.

Even though he liked the work and even the company, Jessie felt like a second-class citizen. He thought that his supervisor gave him most of the dirty work and that there was an undercurrent of prejudice against Hispanics in his department. Friends of his working in other departments had no trouble with prejudice. He wanted to quit and get another job, one that would pay the same relatively good wages he is now earning. A counselor challenged Jessie's choice, especially in view of the fact that that particular region of the country was experiencing an economic downturn and there were few jobs available. He helped Jessie choose an interim goal that was more flexible and more directly related to coping with his present situation. The interim goal was to use his time preparing himself for a better job, whether inside the company or outside. In six months to a year he would be better prepared for any career eventuality. Jessie began volunteering for special assignments and worked extra hours for no pay to learn new skills. He felt good about what he was learning and more easily ignored the prejudice.

Counseling is a living, organic process. Just as organisms adapt to their changing environments, the choices clients make need to be adapted to their changing circumstances.

## Help Clients Choose Goals Consistent with Their Values

Although helping is a process of social influence, it remains ethical only if it respects, within reason, the values of the client. Values are criteria we use to make decisions. Helpers may challenge clients to reexamine their values but they should not encourage clients to perform actions that are not in keeping with their values. For instance, the son of Antonio and Consuela Garza is in a coma in the hospital after an automobile accident. He needs a life-support system to remain alive. His parents are experiencing a great deal of uncertainty, pain, and anxiety. They have been told that there is practically no chance that their son will ever come out of the coma. The counselor should not urge them to terminate the life-support system if that action is counter to their values. However, the counselor can help them explore and clarify the values involved. In this case, the counselor suggests that they discuss their decision with their clergyman. In doing so, they find out that the termination of the life-support system would not be against the tenets of their religion. Now they are free to explore other values that relate to their decision.

Some problems involve a client's trying to pursue contradictory goals or values. Dillard, the ex-Marine, wanted to get an education, but he also wanted to make a decent living as soon as possible. The former goal would put him in debt, but failing to get a college education would lessen his chances of securing the kind of job he wanted. The counselor helped him identify and use his values to consider some trade-offs. Dillard chose to work part-time and go to school part-time. He chose a job in an office instead of a job in construction. Even though the latter paid better, it would be much more exhausting and would leave him with little energy for school.

# Help Clients Establish Realistic Time Frames for the Accomplishment of Goals

Goals that are to be accomplished "sometime" probably won't be accomplished at all. Therefore, helping clients put some time frames in their goals can add value. Greenberg (1986) talked about immediate, intermediate, and final outcomes.

- *Immediate* outcomes are changes in attitudes and behaviors evident in the helping sessions themselves. For Janette, the helping sessions constitute a safe forum for her to become more assertive.
- *Intermediate* outcomes are changes in attitudes and behaviors that lead to further change. It takes Janette a while to transfer her assertiveness skills both to the workplace and to her social life.
- *Final* outcomes refer to the completion of the overall program for constructive change through which problems are managed and opportunities developed. It takes more than two years for Janette to become assertive in a consistent day-to-day way.

Jensen, a 22-year-old on probation for shoplifting, was seeing a counselor as part of a court-mandated program. An immediate need in his case was overcoming his resistance to his court-appointed counselor and developing a working alliance. Because of the counselor's skill and her unapologetic caring attitude that had some toughness in it, he quickly came to see her as "on his side." An intermediate outcome was attitudinal. Brainwashed by what he saw on television, he thought that America owed him some of its affluence and that personal effort had little to do with it. The counselor helped him see that his entitlement attitude was unrealistic and that hard work played a key role in most payoffs. There were two significant final outcomes in Jensen's case. First, he made it through the probation period free of any further shoplifting attempts. Second, he acquired and kept a job that helped him pay his debt to the retailer.

Taussig (1987) talked about the usefulness of setting and executing minigoals early in the helping process. Consider the case of Gaston.

> Gaston, a 16-year-old school dropout and loner, was arrested for arson. Though he lived in the inner city and came from a single-parent household, it was difficult to discover just why he had turned to arson. He had torched a few structures that seemed relatively safe to burn. No one was injured. Was his behavior a cry for help? Social rage expressed in vandalism? Just a way of getting some kicks? The social worker assigned to the case found these questions too speculative to be of much help. Instead of looking for the root causes of Gaston's malaise, she tried to help him set some simple goals that appealed to him and that could be accomplished relatively quickly. One goal was social support. The counselor helped Gaston join a social club at a local church. A second goal was having a role model. Gaston struck up a friendship with one of the

BOX 14-1

# Questions for Shaping Goals

- Is the goal stated in outcome or results language?
- Is the goal specific enough to drive behavior? How will I know when I have accomplished it?
- If I accomplish this goal, will it make a difference? Will it really help manage the problems and opportunities I have identified?
- Does this goal have "bite" while remaining prudent?
- Is it doable?
- Can I sustain this goal over the long haul?
- Does this goal have some flexibility?
- Is this goal in keeping with my values?
- Have I set a realistic time frame for the accomplishment of the goal?

more active members of the club, a dropout who had gotten a high school equivalency degree. He also received some special attention from one of the adult monitors of the club. This was the first time he had experienced the presence of a strong adult male in his life. A third goal was broadening his view of the world. A group of seminarians who worked in both the black and the white communities invited Gaston and a couple of the other boys to help them in a white parish. This was the first time he had been out of the ghetto, and the experience helped him push back the walls a bit. The accomplishment of these minigoals fulfilled some of Gaston's needs and helped him experience himself, others, and the world about him in much more constructive ways.

It is not suggested here that goal setting is a facile answer to intractable social problems. But the achievement of sequenced minigoals can go a long way toward making a dent in intractable problems.

There is no such thing as a set time frame for every client. Some goals need to be accomplished now, some soon; others are short-term goals; still others are long-term. Consider the case of a priest who had been unjustly accused of child molestation.

- **A "now" goal.** Some immediate relief from debilitating anxiety attacks and keeping his equilibrium during the investigation and court procedures
- **A "soon" goal.** Obtaining the right kind of legal aid
- **A short-term goal.** Winning the court case
- **A long-term goal.** Reestablishing his credibility in the community and learning how to live with those who continue to suspect him

There is no particular formula for helping all clients choose the right mix of goals at the right time and in the right sequence. Although helping is based on problem-management principles, it remains an art.

It is not always necessary, then, to make sure that each goal in a client's program for constructive change has all the characteristics outlined in this chapter. For some clients, identifying broad goals is enough. For others, some help in formulating more specific goals is called for. The principle is clear: Help clients develop goals that have some probability of success. In one case, this may mean helping a client deal with clarity; in another, with substance; in still another, with realism, values, or time frame. Box 14-1 outlines some questions that you can help clients ask themselves to choose goals from among possibilities.

## NEEDS VERSUS WANTS

In some cases, what clients want and what they need coincide. The lonely person wants a better social life and needs some kind of community to live a full human life. In other cases, clients might not want what they need. The alcoholic may need a life of total abstention but wants to drink moderately. Brainstorming possibilities for a better future should focus on the package of needs and wants that makes sense for this particular client. Consider the case of Irv.

> Irv, a 41-year-old entrepreneur, collapsed one day at work. He had not had a physical in years. He was shocked to learn that he had both a mild heart condition and multiple sclerosis. His future was uncertain. The father of one of his wife's friends had had multiple sclerosis but had lived and worked well into his 70s. But no one knew what the course of the disease would be. Since he had made his living by developing and then selling small businesses, he wanted to continue to do this, but it was too physically demanding. What he needed was a less physically demanding work schedule. Working 60–70 hours per week, even though he loved it, was no longer in the cards. Furthermore, he had always plowed the money he received from selling one business into starting up another. But now he needed to think of the future financial well-being of his wife and three children. Up to this point, his philosophy had been that the future would take care of itself. It was very wrenching for him to move from a lifestyle he wanted to one he needed.

Involuntary clients often need to be challenged to look beyond their wants to their needs. One woman who voluntarily led a homeless life was attacked and severely beaten on the street. She still wanted the freedom that came with her lifestyle. A counselor suggested that there were ways of keeping her freedom even if she were to adopt a different lifestyle. When challenged to consider the kinds of freedom she wanted, she admitted that freedom from responsibility was at the core. It was her choice to live the way she wanted, but the counselor helped her explore the consequences of her choices.

In the following case, the client, dogged by depression, was ultimately able to integrate what he wanted with what he needed.

> Milos had come to the United States as a political refugee. The last few months in his native land had been terrifying. He had been jailed and beaten and got out just before another crackdown. Once the initial euphoria of having escaped had subsided, he spent months feeling confused and disorganized. He tried to live as he had in his own country, but the culture here was too invasive. He thought he should feel grateful, and yet he felt hostile. After two years of misery, he began seeing a counselor. He had resisted getting help, since "back home" he had been "his own man."
>
> In discussing these issues with a counselor, it gradually dawned on him that he *wanted* to reestablish links with his native land but that he *needed* to integrate himself into the life of his host country. He saw that the accomplishment of this twofold goal would be very freeing. He began finding out how other immigrants who had been here longer than he had accomplished this goal. He spent time in the immigrant community, which differed from the refugee community. In the immigrant community, there was a long history of keeping links to the homeland culture alive. But the immigrants had also adapted to the best in their adopted country. The friends he made became role models for him. The more active he became there, the more his depression lifted.

Had Milos focused only on what he wanted or what he needed, he would have remained unhappy.

## EMERGING GOALS

It is not always a question of setting goals. Rather, goals can naturally *emerge* through the client-helper dialogue. The skills and methods of Stage I, used effectively, do not lead exclusively to the clarification of problems and opportunities. Often, possible goals and action strategies come tumbling out, too. Clients, once they are helped to clarify a problem situation through a combination of probing, empathy, and challenge, begin to see more clearly what they want and what they have to do to manage the problem. Some clients must first act in some way before they find out just what they want to do. Once goals begin to emerge, counselors can help clients clarify them and find ways of implementing them. On the other hand, if "emerge" means that clients wait around until "something comes up" or if it means that clients try many different solutions in the hope that one of them will work, then emergence can be a self-defeating process.

Explicit goal setting, however, is not to be underrated. Some clients are looking for this kind of direction right from the start. If that is the case, then helping clients set goals, even quite specific goals, is a way of meeting their needs. Taussig (1987) showed that clients respond positively to goal setting even when problems are discussed and goals set very early in the counseling process. The best approach seems to be a client-centered, "no one right formula" approach. Although all clients need focus and direction in managing problems and developing opportunities, what focus and direction will look like will differ from client to client.

## ADAPTIVE VERSUS STRETCH GOALS

Difficult or "stretch" goals are often the most motivational but not in every case. Some clients choose to make significant changes in their lives. Others take a more modest approach. Wheeler and Janis (1980, p. 98) cautioned against the search for the "absolute best" goal all the time. It is not realistic: "Sometimes it is more reasonable to choose a satisfactory alternative than to continue searching for the absolute best. The time, energy, and expense of finding the best possible choice may outweigh the improvement in the choice." Consider the following case.

> Joyce, a flight attendant nearing middle age, centered most of her nonflying life on her aging mother. Her mother had been pampered by her now-deceased husband and her three children and allowed to have her way all her life. She now played the role of the tyrannical old woman who constantly feels neglected and who can never be satisfied. Though Joyce knew that she could live much more independently without abandoning her mother, she found it very difficult to make choices for herself. Guilt stood in the way of any change in her relationship with her mother. She even said that being a virtual slave to her mother's whims was not as bad as the guilt she experienced when she stood up to her mother or "neglected" her.
>
> The counselor helped Joyce experiment with a few new ways of dealing with her mother. For instance, Joyce went on a two-week trip with friends even though her mother objected, saying that it was ill-timed. Although the experiments were successful in that no harm was done to Joyce's mother and Joyce did not experience excessive guilt, counseling did not help her restructure her relationship with her mother in any substantial way. The experiments did give her a sense of greater freedom. For instance, she felt freer to say no to this or that demand of her mother. This provided enough slack, it seems, to make Joyce's life more livable.

In this case, counseling helped the client fashion a life that was "a little bit better," though not as good as the counselor thought it could be. When asked, "What do you want?" Joyce had in effect replied, "I want a bit more slack and freedom, but I do not want to abandon my mother." Joyce's "new" scenario did not differ dramatically from the old. But perhaps it was enough for her. It was a case of choosing a satisfactory alternative rather than the best.

Helping is the art of the possible. Although most clients can do more to control their lives than they actually do, that does not mean that clients must substitute taking charge of everything for passivity and dependency. Take the case of Debbie. She learned, to her chagrin, that she and her husband had more areas of incompatibility than she had imagined. Though she knew that he liked his work, she had no idea that it was his passion. Work came before almost everything else. However, she did not think that she could make him change. A counselor helped her find ways of coping. Instead of feeling sorry for herself, she developed friends and interests outside the home, activities in which he did not participate. In her mind, this was not the perfect solution. But it certainly made life fuller and more livable.

Leahey and Wallace (1988, p. 216) offered the following example of a client in adaptive mode.

> "For the last five years, I've thought of myself as a person with low self-esteem and have read self-help books, gone to therapists, and put things off until I felt I had good self-esteem. I just need to get on with my life, and I can do that with excellent self-esteem or poor self-esteem. Treatment isn't really necessary. Being a person with enough self-esteem to handle situations is good enough for me."

The following client also redefines the problem situation, seeing it now from a different angle.

> "I would say that I am completely cured. . . . I can still pinpoint these conditions which I had thought to be symptoms. . . . These worries and anxieties make me prepare thoroughly for the daily work I have to do. They prevent me from being careless. They are expressions of the desire to grow and to develop." (See Weisz, Rothbaum, & Blackburn, 1984, p. 964.)

Some helpers, reviewing these last two examples, would be disappointed. Others would see them as legitimate examples of adapting to rather than changing reality. In both cases, however, the clients have acted—that is, done something about the way they think about themselves and their problems. Clients, like the rest of us, have limitations. Like us, they are in many ways not captains of their own fates. Some would say that what Debbie did amounted to giving up. Others would see it as a case of choosing adaptive goals.

## ACTION BIAS AS A METAGOAL

Although clients set goals that are directly related to their problem situations, there are also metagoals, superordinate goals that would make them more effective in pursuing the goals they set and in leading fuller lives. The metagoal of helping clients become more effective in problem management and opportunity development was mentioned in Chapter 1. Another metagoal is to help clients become more effective "agents" in life—doers rather than mere reactors, preventers rather than fixers, initiators rather than followers. This is especially the case if the client has become mired in a problem situation because of a bias against rather than for action.

Frank was liked by his superiors for two reasons. First, he was competent—he got things done. Second, he did whatever they wanted him to do. They moved him from job to job when it suited them. He never complained. However, as he matured and began to think more of his future, he realized that there was a great deal of truth in the adage "If you're not in charge of your own career, no one is." After a session with a career counselor, he outlined the kind of career he wanted and presented it to his superiors. He pointed out to them how this would serve both the company's interests and his own. At first they were nonplussed, but then they agreed. Later, when they wanted to break the implicit contract developed in this interchange, he stood his ground.

The doer is more likely to pursue stretch rather than adaptive goals in managing problems. The doer is also more likely to move beyond problem management to opportunity development.

# Evaluation Questions for Step II-B

- To what degree am I helping clients choose specific goals from among a number of preferred-scenario possibilities?
- How well do I challenge clients to translate good intentions into broad goals and broad goals into specific, actionable goals?
- To what extent do the goals set by each client have the characteristics outlined in Box 14-1?
- How effectively do I help clients establish goals that take into consideration both needs and wants?
- To what degree do I help clients become aware of goals that are naturally emerging from the helping process?
- How effectively do I help clients choose the right mix of adaptive and stretch goals?
- How well do I help clients explore the consequences of the goals they are setting?
- How do I help clients make a bias toward action one of their goals?

## CHAPTER 15

# STEP II-C: COMMITMENT—WHAT ARE YOU WILLING TO PAY FOR WHAT YOU WANT?

As mentioned earlier, Step II-C is not really a step in the true sense of the term but a dimension of the goal-setting process. Clients can formulate goals without being willing to pay for them. Once clients state what they want and set goals, the battle is joined, as it were. It is as if the client's "old self" or old lifestyle begins vying for resources with the client's potential "new self" or new lifestyle. On a more positive note, history is full of examples of people whose strength of will to accomplish some goal has enabled them to do seemingly impossible things.

> A woman with two sons in their 20s was dying of cancer. The doctors thought she could go at any time. However, one day she told the doctor that she wanted to live to see her older son get married in six months' time. The doctor talked vaguely about "trusting in God" and "playing the cards she had been dealt." Against all odds, the woman lived to see her son get married. Her doctor was at the wedding. During the reception, he went up to her and said, "Well, you got what you wanted. Despite the way things are going, you must be deeply satisfied." She looked at him wryly and said, "But, Doctor, my second son will get married someday."

Although the job of counselors is not to encourage clients to heroic efforts, counselors should not undersell clients, either. Clients are capable of turning possibilities into agendas and agendas into causes to which they commit themselves.

In this step, which is usually intermingled with the other two steps of Stage II, counselors help their clients pose and answer such questions as

- Why should I pursue this goal?
- Is it worth it?
- Is this where I want to invest my limited resources of time, money, and energy?
- What competes for my attention?
- What are the incentives for pursuing this agenda?
- How strong are competing agendas?

Again, there is no formula. Some clients, once they establish goals, race to accomplish them. At the other end of the spectrum are clients who, once they decide on goals, stop dead in the water. Furthermore, the same client might speed toward the accomplishment of one goal and drag her feet on another. The job of the counselor is to help clients face up to their commitments.

## HELPING CLIENTS COMMIT THEMSELVES

There is a difference between initial commitment to a goal and an ongoing commitment to a strategy or plan to accomplish the goal. The proof of initial commitment is some movement to action. For instance, one client who chose as a goal a less abrasive interpersonal style began to engage in an "examination of conscience" each evening to review what his interactions with people had been like that day. In doing so, he discovered, somewhat painfully, that in some of his interactions he moved beyond abrasiveness to contempt. That forced him back to a deeper analysis of the problem situation and the blind spots associated with it. Being dismissive of people he did not like who were "not important" had become ingrained in his interpersonal lifestyle.

There is a range of things you can do to help clients in their initial commitment to goals and the kind of action that is a sign of that commitment. Counselors can help clients by helping them make goals appealing, by helping them enhance their sense of ownership, and by helping them deal with competing agendas.

# HELP CLIENTS SET GOALS THAT DON'T COST MORE THAN THEY ARE WORTH

Here we revisit the "economics" of helping. Cost-effectiveness could have been included in the characteristics of workable goals outlined in the previous chapter, but it is considered here instead because of its close relationship to commitment. Some goals that can be accomplished carry too high a cost in relation to the payoff. It may sound overly technical to ask whether any given goal is "cost-effective," but the principle remains important. Skilled counselors help clients budget rather than squander their resources.

> Eunice discovered that she had a terminal illness. In talking with several doctors, she found out that she would be able to prolong her life a bit through a combination of surgery, radiation treatment, and chemotherapy. However, no one suggested that these would lead to a cure. She also found out what each form of treatment and each combination would cost, not so much in monetary terms, but in added anxiety and pain. Ultimately she decided against all three, since no combination of them promised much for the quality of the life that was being prolonged. Instead, with the help of a doctor who was an expert in hospice care, she developed a scenario that would ease both her anxiety and her physical pain as much as possible.

It goes without saying that another patient might have made a different decision. Costs and payoffs are relative. Some clients might value an extra month of life no matter what the cost.

Since it is often impossible to determine the cost-benefit ratio of any particular goal, counselors can add value by helping clients understand the *consequences* of choosing a particular goal. For instance, if a client sets her sights on a routine job with minimally adequate pay, this outcome might well take care of some of her immediate needs but prove to be a poor choice in the long run. Helping clients foresee the consequences of their choices may not be easy. Another woman with cancer felt she was no longer able to cope with the sickness and depression that came with her chemotherapy treatments. She decided abruptly one day to end the treatment, saying that she didn't care what happened. No one helped her explore the consequences of her decision. Eventually, when her health deteriorated, she had second thoughts about the treatments, saying, "There are still a number of things I must do before I die." But it was too late. Some challenge on the part of a helper might have helped her make a better decision.

The balance-sheet methodology outlined in the Appendix is a tool you can use selectively to help clients weigh costs against benefits both in choosing goals and in choosing programs to implement goals. The balance sheet is also used in Chapter 17 to help clients choose best-fit strategies for accomplishing their goals.

## Help Clients Set Appealing Goals

Just because goals are cost-effective does not mean that they are appealing to the client. Setting appealing goals is common sense, but it is not always easy to do. For instance, for many if not most addicts, a drug-free life is not immediately appealing, to say the least.

> Ed tries to help Dillard work through his resistance to giving up prescription drugs. He listens and is empathic; he also challenges the way Dillard has come to think about drugs and his dependency on them. One day Ed says something about "giving up the crutch and walking straight." In a flash Dillard sees himself not as addicted to drugs but as a person with a crippling disability. Some of the soldiers of Desert Storm had lost a leg. He knew that they longed for the days when they could be fitted with a prosthesis and throw their crutches away. The image of "throwing away the crutch" and "walking straight" proved to be very appealing to Dillard.

A goal is appealing if there are incentives for pursuing it. Counselors need to help clients in their search for incentives throughout the helping process. Ordinarily, negative goals — giving something up — need to be translated into positive goals — getting something. It was much easier for Dillard to commit himself to returning to school than to giving up prescription drugs, because school represented something he was getting. Images of himself with a degree and of holding some kind of professional job were solid incentives. The picture of him throwing away the crutch proved to be an important incentive in cutting down on drug use.

## Help Clients Own the Goals They Set

It is essential that the goals chosen be the client's rather than the helper's goals or someone else's. Various kinds of probes can be used to help clients discover what they want to do to manage some dimension of a problem situation more effectively. For instance, Carl Rogers, in a film of a counseling session (Rogers, Perls, & Ellis, 1965), is asked by a woman what she should do about her relationship with her daughter. He says to her, "I think you've been telling me all along what you want to do." She knew what she wanted the relationship to look like, but she was asking for his approval. If he had given it, the goal would, to some degree, have become his goal instead of hers. At another time he asks, "What is it that you want me to tell you to do?" This question puts the responsibility for goal setting where it belongs—on the shoulders of the client.

> Cynthia was dealing with a lawyer because of an impending divorce. They had talked about what was to be done with the children, but no decision had been reached. One day she came in and said that she had decided on mutual custody. She wanted to work out such details as which residence, hers or her husband's, would be the children's principal one and so forth. The lawyer asked her how she had reached a decision. She said that she had been talking to her husband's parents—she was still on good terms with them—and that they had suggested this arrangement. The lawyer challenged Cynthia to take a closer look at her decision. "Let's start from zero," he said, "and you tell me what arrangements you want." He did not want to help her carry out a decision that was not her own.

Choosing goals suggested by others enables clients to blame others if they fail to reach the goals.

Commitment to goals can take different forms—compliance, buy-in, and ownership. The least useful is *mere compliance*. "Well, I guess I'll have to change some of my habits if I want to keep my marriage afloat" does not augur well for sustaining changes in behavior. But it may be better than nothing. *Buy-in* is a level up from compliance. "Yes, these changes are essential if we are to have a marriage that makes sense for both of us. We say we want to preserve our marriage, but now we have to prove it to ourselves." This client has moved beyond mere compliance. But sometimes buy-in alone does not provide enough staying power because it depends too much on reason. "This is logical" is far different from "This is what I really want!" *Ownership* is a higher form of commitment. It means that the client can say, "This goal is not someone else's, it's not just a good idea; it is mine, it is what I want to do." Consider the following case.

> A counselor worked with a manager whose superiors had intimated that he would not be moving much further in his career unless he changed his style in dealing with the members of his team and other key people with whom he worked within the organization. At first the manager resisted setting any goals. "What they want is a lot of hogwash. It won't do anything to make the business better," was his initial response. One day, when asked whether accomplishing what "they" wanted would cost him too much, he pondered a few moments and then said, "No, not really." That got him started.

With a bit of help from the counselor, he identified a few areas of his managerial style that could well be "polished up." He had moved beyond resistance to the compliance level. Within a few months he got much more into the swing of things. Given the favorable response to his changed behavior he had gotten from the people who reported to him, he was able to say, "Well, I now see that this makes sense. But I'm doing it because it has a positive effect on the people in the department. It's the right thing to do." Buy-in had arrived. A year later he moved up another notch. He became much more proactive in finding ways to improve his style. He delegated more, gave people feedback, asked for feedback, held a couple of managerial retreats, joined a human-resource task force, and routinely rewarded his direct reports for their successes. Now he began to say such things as "This is actually fun." Ownership had arrived. The people in his department began to see him as one of the best executives in the company. This process took over two years.

The manager did not have a personality transformation. He did not change his opinion of some of his superiors and was right in pointing out that they didn't observe their own rules. But he did change his behavior because he gradually discovered meaningful incentives to do so.

The use of contracts to structure the helping process itself was discussed in Chapter 3. Self-contracts—that is, contracts that clients make with themselves—can also help clients commit themselves to new courses of action. Although contracts are promises clients make to themselves to behave in certain ways and to attain certain goals, they are also ways of making goals more focused. It is not only the expressed or implied promise that helps but also the explicitness of the commitment. Consider the following example in which one of Dora's sons disappears without a trace.

About a month after one of Dora's two young sons disappeared, she began to grow listless and depressed. She was separated from her husband at the time the boy disappeared. By the time she saw a counselor a few months later, a pattern of depressed behavior was quite pronounced. Although her conversations with the counselor helped ease her feelings of guilt—she stopped engaging in self-blaming rituals—she remained listless. She shunned relatives and friends, kept to herself at work, and even distanced herself emotionally from her other son. She resisted developing images of a better future, because the only better future she would allow herself to imagine was one in which her son had returned.

Some strong challenging from Dora's sister-in-law, who visited her from time to time, helped jar her loose from her preoccupation with her own misery. "You're trying to solve one hurt, the loss of Bobby, by hurting Timmy and hurting yourself. I can't imagine in a thousand years that that's what Bobby would want!" her sister-in-law screamed at her one night. Afterward Dora and the counselor discussed a "recommitment" to Timmy, to herself, and to their home. Through a series of contracts, she began to reintroduce patterns of behavior that had been characteristic of her before the tragedy. For instance, she contracted to opening her life up to relatives and friends once more, to creating a much more positive atmosphere at home, to encouraging Timmy to have his friends over, and so forth. Contracts worked for Dora because, as she said to the counselor, "I'm a person of my word."

When Dora first began implementing these goals, she felt she was just going through the motions. However, what she was really doing was acting herself into a new mode of thinking. Contracts helped Dora in both her initial commitment to a goal and her movement to action. In counseling, contracts are not legal documents but human instruments to be used if they are helpful. They often provide both the structure and the incentives some clients need.

## Help Clients Deal with Competing Agendas

Clients often set goals and formulate programs for constructive change without taking into account competing agendas—other things in their lives that soak up time and energy, such as job, family, and leisure pursuits. The world is filled with distractions. For instance, one manager wanted to change his overly critical interpersonal style, but crises in the economy and a divorce set up competing agendas and sapped his resources. None of the goals of his interpersonal-style agenda was accomplished. Programs for constructive change often involve a rearrangement of priorities. If a client is to be a full partner in the reinvention of his marriage, then he cannot spend as much time with the boys. Or the unemployed white-collar worker might have to put aside some parts of her social life as she works part-time to support herself and gives herself to a search for full-time work. One executive, out of work for over nine months, was pulled up short in a self-help group when one of his peers said, "You're still living your old life and just playing at getting a job."

This is not to say that competing agendas are frivolous. The woman who wants to expand her horizons by getting involved in social settings outside the home has to figure out how to handle the tasks at home. The single parent who wants a promotion at work needs to balance her new responsibilities with involvement with her children. A counselor who had worked with a two-career couple as they made a decision to have a child helped them think of competing agendas once the pregnancy started. A year after the baby was born, they saw the counselor again for a couple of sessions to work on some issues that had come up. However, they started the session by saying, "Are we glad that you talked about competing agendas when we were struggling with the decision to become parents! In our discussions with one another, we have gone back time and time again and reviewed what we said and what we decided. It helped stabilize us for the last two years."

Even self-contracts have a shadow side. There is no such thing as a perfect contract. Most people don't think through the consequences of all the provisions of a contract, whether it be marriage, employment, or self-contracts designed to enhance a client's commitment to goals. And even people of goodwill unknowingly add covert codicils to contracts they make with themselves and others—"I'll pursue this goal until it begins to cost me money" or "I won't be abusive unless she pushes me to the wall." The codicils are buried deep in the decision-making process and only gradually make their way to the surface. Box 15-1 indicates the kinds of questions you can help clients ask themselves about their commitment to their change agendas.

BOX 15-1

# Questions on Client Commitment

You can help clients ask themselves these kinds of questions as they struggle with committing themselves to a program of constructive change:

- What is my state of readiness for change in this area at this time?
- How badly do I want what I say I want?
- How hard am I willing to work?
- To what degree am I choosing this goal freely?
- How highly do I rate the personal appeal of this goal?
- How do I know I have the courage to work on this?
- What's pushing me to choose this goal?
- What incentives do I have for pursuing this change agenda?
- What rewards can I expect if I work on this agenda?
- If this goal is in any way being imposed by others, what am I doing to make it my own?
- What difficulties am I experiencing in committing myself to this goal?
- In what way is it possible that my commitment is not a true commitment?
- What can I do to get rid of the disincentives and overcome the obstacles?
- What can I do to increase my commitment?
- In what ways can the goal be reformulated to make it more appealing?
- To what degree is the timing for pursuing this goal poor?
- What do I have to do to stay committed?
- What resources can help me?

# STAGE II AND ACTION

The work of Step II-A — developing possibilities for a better future — is just what some clients need. It frees them from thinking solely about problem situations and unused resources and enables them to begin fashioning a better future. Once they identify some of their wants and needs, they move into action. Other clients move into action too quickly. The focus on the future liberates them from the past, and the first few possibilities are very attractive.

They need the kind of focus and direction provided by Step II-B. Weighing alternatives and shaping goals help them move to action that is both directed and prudent. For still other clients, the search for incentives for commitment is the trigger for action. Once they see what's in it "for me" — a kind of upbeat and productive selfishness, if you will — they are ready to work.

## THE SHADOW SIDE OF GOAL SETTING

Despite the advantages of goal setting outlined in the last three chapters, some helpers and clients seem to conspire to avoid goal setting as an explicit process. It is puzzling to see counselors helping clients explore problem situations and unused opportunities and then stopping short of asking them what they want to replace what they have. As Bandura (1990, p. xii) put it, "Despite this unprecedented level of empirical support [for the advantages of goal setting], goal theory has not been accorded the prominence it deserves in mainstream psychology." Years ago, the same concern was expressed differently. A U.S. developmental psychologist was talking to a Russian developmental psychologist. The Russian said, "It seems to me that American researchers are constantly seeking to explain how a child came to be what he is. We in the USSR are striving to discover how he can become what he not yet is" (see Bronfenbrenner, 1977, p. 258). One of the main reasons that counselors do not help clients develop realistic goals is that we do not train them to do so.

There are other reasons. First of all, to some clients goal setting sounds very rational, perhaps too rational. Because the real-life process of problem management and opportunity development is messier, both helpers and clients object to this overly rational approach. There is a dilemma. On the one hand, many clients need or would benefit by a rigorous application of the problem-management process. On the other, they resist its rationality and discipline. They find it alien. Second, goal setting means that clients have to move out of the relatively safe harbor of discussing problem situations and of exploring the possible roots of those problems in the past and move into the uncharted waters of the future. This may be uncomfortable for client and helper alike. Third, clients who set goals and commit themselves to them move beyond the victim-of-my-problems game. Victimhood and self-responsibility make poor bedfellows.

Fourth, goal setting involves clients' placing demands on themselves, making decisions, committing themselves, and moving to action. If I say, "This is what I want," then, at least logically, I must also say, "Here is what I am going to do to get it. I know the price and I am willing to pay it." Since this demands work and pain, clients will not always be grateful for this kind of "help." Fifth, goals, though liberating in many respects, also hem clients in. If a woman chooses a career, then it might not be possible for her to have the kind of marriage she would like. If a man commits himself to one woman, then he can no longer play the field. There is some truth in the ironic statement

"There is only one thing worse than not getting what you want, and that's getting what you want." The responsibilities accompanying getting what you want—a drug-free life, a renewed marriage, custody of the children, a promotion, the peace and quiet of retirement, freedom from an abusing husband—often cause a new set of problems. It is one thing for parents to decide to give their children more freedom; it is another thing for them to watch them use that freedom. Finally, there is a phenomenon called post-decisional depression. Once choices are made, clients begin to have second thoughts that often keep them from acting on their decisions.

Effective helpers know what lurks in the shadows of goal setting both for themselves and for their clients and are prepared to manage their own part of it and help clients manage theirs. The answer to all of this lies in helpers' being trained in the entire problem-management process and in their sharing a picture of the entire helping process with the client. Then goal setting, described in the client's language, will be a natural part of the process. Artful helpers weave goal setting, under whatever name, into the flow of helping. They do so by moving easily back and forth among the stages and steps of the helping process.

# Evaluation Questions for Step II-C

- What do I need to do to help clients commit themselves to a better future?
- What do I do to make sure that the goals the client is setting are really his or her goals and not mine or those of a third party?
- How effectively do I help clients examine the benefits of goals they are choosing as measured against the costs?
- In what ways do I help clients focus on the appealing dimensions of the goals being set?
- How effectively do I perceive and deal with the misgivings clients have about the goals they are formulating?
- To what degree do I help clients enter into self-contracts with respect to the accomplishment of goals?
- What am I doing to help clients identify, explore, and manage competing agendas?
- What do I do to help clients move to initial goal-accomplishing action?

## PART FIVE

# STAGE III:
# HELPING CLIENTS WORK
# FOR WHAT THEY NEED
# AND WANT

In a book called *True Success*, Tom Morris (1994) lays down the conditions for achieving success. They include

- determining what you want—that is, a goal or a set of goals "powerfully imagined,"
- focus and concentration in preparation and planning,
- the confidence or belief in oneself to see the goal through,
- a commitment of emotional energy,
- being consistent, stubborn, and persistent in the pursuit of the goal,
- the kind of integrity that inspires trust and gets people pulling for you,
- a capacity to enjoy the process of getting there.

The role of the counselor is to help clients engage in all these internal and external behaviors in the interest of goal accomplishment.

## ACTIVITIES VERSUS OUTCOMES AND IMPACT

Whereas Stage II is about outcomes—goals or accomplishments "powerfully imagined"—Stage III is about the activities—the work—needed to produce those outcomes. Clients, when helped to explore what is going wrong in their lives, often ask, "Well, what should I do about it?" That is, they focus on solutions as *activities*. But, as we shall see, activities, though essential, are valuable only to the degree that they produce *outcomes*. An outcome is a goal, something a client wants or needs. Outcomes, also essential, are valuable only to the degree that they have a constructive *impact* on the life of the client. The distinction between activities, outcomes, and impact is seen in the following example.

> Enid, a 40-year-old single woman, is making a great deal of progress in controlling her drinking through her involvement with an AA program. She engages in certain *activities*; for instance, she attends AA meetings, follows the twelve steps, stays away from situations that would tempt her to drink, and calls fellow AA members when she feels depressed or when the temptation to drink is pushing her hard. The *outcome* is that she has been sober for over seven months. She feels that this is quite an accomplishment. The *impact* of all this is very rewarding. She feels better about herself, and she has had both the energy and the enthusiasm to do things that she has not done in years—developing a circle of friends, getting interested in church activities, and doing a bit of travel.
>
> But Enid is also struggling with a troubled relationship with a man. In fact, her drinking was, in part, an ineffective way of avoiding the problems in the relationship. She knows that she no longer wants to tolerate the psychological abuse she has been getting from her male friend, but she fears the vacuum she

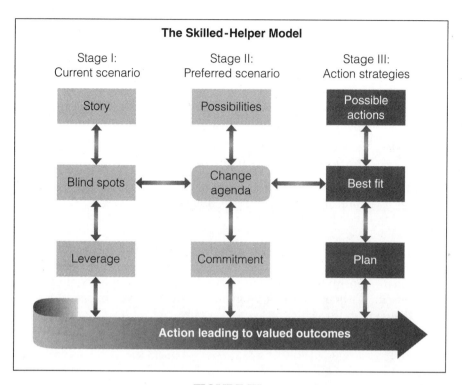

**FIGURE III**
**The Helping Model — Stage III**

will create by cutting off the relationship. She is, therefore, trying to determine what she wants, realizing that ending the relationship might turn out to be the best option.

She has engaged in a number of *activities*. For instance, she has become much more assertive in the relationship. She now cuts off contact whenever her companion becomes abusive. She does not let him make all the decisions in the relationship. But the relationship remains troubled. Even though she is doing many things, there is no satisfactory *outcome*. She has not yet determined what the outcome should be; that is, she has not determined what kind of relationship she would like and if it is possible to have such a relationship with this man. Nor has she determined to end the relationship.

Finally, after one seriously abusive episode, she tells him that she is ending the relationship. She does what she has to do to sever all ties with him (*activities*), and the *outcome* is that the relationship ends and stays ended. The *impact* is that she feels liberated but lonely. The helping process needs to be recycled to help her with this new problem.

In Stage II, clients are helped to determine what they need and want — *outcomes*. In Stage III, they are helped to figure out what they need to do to get what they want — *activities*. However, skilled counselors help clients monitor the *impact* that all of this is to have on their lives. In a fundamental sense, helping is all about outcomes and their impact.

Stage III has three steps, in our usual definition of *step*. They are all aimed at action on the part of the client.

**Step III-A:**  Help clients develop possible strategies for accomplishing their goals. What do I need to do to get what I want?

**Step III-B:**  Help clients choose strategies tailored to their preferences and resources. What actions are best for me?

**Step III-C:**  Help clients formulate actionable plans. What should my campaign for constructive change look like? What do I need to do first? Second?

Stage III, highlighted in Figure III, adds the final pieces to a client's program for constructive change. Stage III deals with the "game plan." However, these three "steps" constitute planning for action and should not be confused with action itself. Without action, a program for constructive change is nothing more than a wish list.

## CHAPTER 16

# STEP III-A: STRATEGIES FOR ACTION—WHAT DO I NEED TO DO TO GET WHAT I NEED AND WANT?

Strategies are actions that help clients accomplish their goals. Step III-A, developing a range of possible strategies to accomplish goals, is a powerful exercise. Clients who feel hemmed in by their problems and unsure of the viability of their goals are liberated through this process. Another way clients can ask themselves the III-A question is, "How do I bridge the gap between what I have and what I need and want?"

Strategy is the art of identifying and choosing realistic courses of action for achieving goals and doing so under adverse conditions, such as war. The problem situations in which clients are immersed constitute adverse conditions; clients often are at war with themselves and the world around them. Helping clients develop strategies to achieve goals can be a most thoughtful, humane, and fruitful way of being with them. This step in the counseling process is another that helpers sometimes avoid because it is too "technological." They do their clients a disservice. Clients with goals but no clear idea of how to accomplish them still need help.

## How Many Ways Are There to Get What I Need and Want?

Once more it is a question of helping clients stimulate their imaginations and engage in divergent thinking. Most clients do not instinctively seek different routes to goals and then choose from among them.

### Help Clients Brainstorm Strategies for Accomplishing Goals

Brainstorming, discussed in Chapter 13, plays an important part in strategy development. The quality and the efficacy of an action plan tend to be better if the program is chosen from among a number of possibilities. Consider the case of Karen, who has come to realize that heavy drinking is ruining her life. Her goal was to stop drinking. She felt that it simply would not be enough to cut down; she had to stop. To her the way forward seemed simple enough: Whereas before she drank, now she wouldn't. Because of the novelty of not drinking, she was successful for a few days; then she fell off the wagon. This happened a number of times until she finally realized that she could use some help. Stopping drinking, at least for her, was not as simple as it seemed.

A counselor at a city alcohol and drug treatment center helps her explore a number of techniques that could be used in an alcohol-management program. Together they come up with the following possibilities:

- Just stop cold turkey and get on with life.
- Join Alcoholics Anonymous.
- Move someplace declared "dry" by local government.

- Take a drug that causes nausea if followed by alcohol.
- Replace drinking with other rewarding behaviors.
- Join some self-help group other than Alcoholics Anonymous.
- Get rid of all liquor in the house.
- Take the "pledge" not to drink; to make it more binding, take it in front of a minister.
- Join a residential hospital detoxification program.
- Avoid friends who drink heavily.
- Change other social patterns; for instance, find places other than bars and cocktail lounges to socialize.
- Try hypnosis to reduce the drive to drink.
- Use behavior modification techniques to develop an aversion for alcohol; for instance, pair painful but safe electric shocks with drinking or even thoughts about drinking.
- Change self-defeating patterns of self-talk, such as "I have to have a drink" or "One drink won't hurt me."
- Become a volunteer to help others stop drinking.
- Read books and view films on the dangers of alcohol.
- Stay in counseling as a way of getting support and challenge for stopping.
- Share intentions to stop drinking with family and close friends.
- Spend a week with an acquaintance who does a great deal of work in the city with alcoholics, and go on his rounds with him.
- Walk around skid row meditatively.
- Have a discussion with members of the family about the impact drinking has on them.
- Discover things to eat that might help reduce the craving for alcohol.
- Get a hobby or an avocation that captures the imagination.
- Substitute a range of self-enhancing activities for drinking.

This list contained many more items than Karen would have thought of had she not been stimulated by the counselor to take a census of possible strategies. One of the reasons that clients are clients is that they are not very creative in looking for ways of getting what they want. Once goals are established, getting them accomplished is not just a matter of hard work. It is also a matter of imagination.

If a client is having a difficult time coming up with strategies, the helper can "prime the pump" by offering a few suggestions. Driscoll (1984) put it well.

> Alternatives are best sought cooperatively, by inviting our clients to puzzle through with us what is or is not a more practical way to do things. But we

must be willing to introduce the more practical alternatives ourselves, for clients are often unable to do so on their own. Clients who could see for themselves the more effective alternatives would be well on their way to using them. That clients do not act more expediently already is in itself a good indication that they do not know how to do so. (p. 167)

Although the helper may need to suggest alternatives, he or she can do so in such a way that the principal responsibility for evaluating and choosing possible strategies stays with the client. For instance, there is the "prompt and fade" technique. The counselor can say, "Here are some possibilities. . . . Let's review them and see whether any of them make sense to you or suggest further possibilities." Or, "Here are some of the things that people with this kind of problem situation have tried. . . . How do they sound to you?" The "fade" part of this technique keeps it from being advice giving; prompts are stated in such a way that the client has to work with the suggestion.

Elton, a graduate student in counseling psychology, is plagued with perfectionism. Although he is an excellent student, he worries about getting things right. After he writes a paper or practices counseling, he agonizes over what he could have done better. The kind of behavior puts him on edge when he practices counseling with his fellow trainees. They tell him that his "edge" makes them uncomfortable and interferes with the flow of the helping process. One student told him, "You make me feel as if I'm not doing the right things as a client." He realizes that "less is more"; that is, becoming less preoccupied with the details of helping will make him a more effective helper. His goal is to become more relaxed in the helping sessions, free his mind of the "imperatives" to be perfect, and learn from mistakes rather than expending an excessive amount of effort trying to avoid them. He and his supervisor are talking about ways he can free himself of these inhibiting imperatives.

SUPERVISOR: What kinds of things can you do to become more relaxed?

ELTON: I need to focus my attention on the client and the client's goals instead of being preoccupied with myself.

SUPERVISOR: So a basic shift in your orientation right from the beginning will help.

ELTON: Right. . . . And this means getting rid of a few inhibiting beliefs.

SUPERVISOR: Such as . . .

ELTON: That technical perfection doing the helping model is more important than the relationship with the client. I've gotten lost in the details of the model and have forgotten that I'm a human being with another human being.

SUPERVISOR: So rehumanizing the helping process in your own mind will help? . . . Any other internal behaviors need changing?

ELTON: Another belief is that I have to be the best in the class. That's my history, at least in academic subjects. Being as effective as I can be in helping a client has nothing to do with competing with my fellow students. Competing is a distraction. I know it's in my bones, but now it's rather high-schoolish.

**SUPERVISOR:** Okay, so the academic-game mentality doesn't work here . . .

**ELTON** (interrupting): That's precisely it. Even the practicing we do with one another is real life, not a game. You know that a lot of us talk about real issues when we practice.

**SUPERVISOR:** You've been talking about getting your attitudes right and the impact that can have on helping sessions. Are there any external behaviors that might also help?

**ELTON** (pauses): I'm hesitating because it strikes me how I'm in my head too much, always figuring *me* out. . . . On a much more practical basis, I like what Jerry and Philomena do. Before each session with their "clients," they spend five or ten minutes reviewing just where the client is in the overall helping process and determining what they might do in the next session to add value and move things forward. That puts the focus where it belongs, on the client.

**SUPERVISOR:** So a mini-prep for each session can help you get out of your world and into the client's.

**ELTON:** Also in debriefing the training videos we make each week. . . . I now see that I always start by looking at my behavior instead of what's happening with the client. . . . And I need to share what we've been talking about with my training partner.

**SUPERVISOR:** You haven't brought up the perfectionism issue either in the practice sessions or in our lifestyle group meetings.

**ELTON** (hesitating): Well, it's pretty pervasive in my life. . . . I guess I haven't brought it up because I'd rather be seen as competent, not perfectionistic. . . . Well, the cat is out of the bag with you, so I guess it makes sense to put it on my lifestyle group agenda.

This dialogue, which includes empathy, probes, and challenge on the part of the supervisor, produces a number of strategies that Elton can use to develop a more client-focused mentality. He ends by saying that all these can be reinforced through his interactions with his training partner.

## Develop Frameworks for Stimulating Clients' Thinking about Strategies

How can helpers find the right probes to help clients develop a range of strategies? Simple frameworks can help. Consider the following case.

> Jackson has terminal cancer. He has been in and out of the hospital several times over the past few months, and he knows that he probably will not live more than a year. He would like the year to be as full as possible, and yet he wants to be realistic. He hates being in the hospital, especially a large hospital, where it is so easy to be anonymous. One of his goals is to die outside the hospital. He would like to die as benignly as possible and retain possession of his faculties as long as possible. How is he to achieve these goals?

Probes and prompts can be used to discover possible strategies by investigating possible resources in the client's life, including people, models, communities, places, things, organizations, programs, and personal resources.

- *Individuals.* What individuals might help clients achieve their goals? Jackson gets the name of a local doctor who specializes in the treatment of chronic cancer-related pain. The doctor teaches people how to use a variety of techniques to manage pain. Jackson says that perhaps his wife and daughter can learn how to give simple injections to help him control the pain. Also, he thinks that talking every once in a while with a friend whose wife died of cancer, a man he respects and trusts, will help him find the courage he needs.

- *Models and exemplars.* Does the client know people who are presently doing what he or she wants to do? One of Jackson's fellow workers died of cancer at home. Jackson visited him there a couple of times. That's what gave him the idea of dying at home, or at least outside the hospital. He noticed that his friend never allowed himself poor-me talk. He refused to see dying as anything but part of living. This touched Jackson deeply at the time, and now reflecting on that experience may help him develop realistic attitudes, too.

- *Communities.* What communities of people are there through which clients might identify strategies for implementing their goals? Even though Jackson has not been a regular churchgoer, he does know that the parish within which he resides has some resources for helping those in need. A brief investigation reveals that the parish has developed a relatively sophisticated approach to visiting the sick. Parishioners are carefully selected and then trained for this program. They visit sick people in the hospital, in hospices, and at home and provide a number of services.

- *Places.* Are there particular places that might help? Jackson immediately thinks of Lourdes, the shrine to which Catholic believers flock with all sorts of human problems. He doesn't expect miracles, but he feels that he might experience life more deeply there. It's a bit wild, but why not a pilgrimage? He still has the time and also enough money to do it.

- *Things.* What things exist that can help clients achieve their goals? Jackson has read about the use of combinations of drugs to help stave off pain and the side effects of chemotherapy. He has heard that certain kinds of electric stimulation can ward off chronic pain. He explores all these possibilities with his doctor and even arranges for second and third opinions.

- *Organizations.* Are there any organizations that help people with this kind of problem? Jackson knows that there are mutual-help groups composed of cancer patients. He has heard of one at the hospital and believes that there are others in the community. He learns that there are such things as hospices for those terminally ill with cancer.

BOX 16-1
# Questions for Developing Strategies

- Now that I know what I want, what do I need to do?
- Now that I know my destination, what are the different routes for getting there?
- What actions will get me to where I want to go?
- Now that I know the gaps between what I have and what I want and need, what do I need to do to bridge those gaps?
- How many ways are there to accomplish my goals?
- How do I get started?
- What can I do right away?
- What do I need to do later?

- *Programs.* Are there any ready-made programs for people in the client's position? A hospice for the terminally ill has just been established in the city. They have three programs. One helps people who are terminally ill stay in the community as long as they can. A second makes provision for part-time residents. The third provides a residential program for those who can spend little or no time in the community. The goals of these programs are practically the same as Jackson's. Box 16-1 outlines some questions that you can help clients ask themselves to develop strategies for accomplishing goals.

## WHAT RESOURCES DO I NEED TO GET WHAT I WANT? THE FOCUS ON SOCIAL SUPPORT

Step III-A can also be seen as helping clients get the resources, both internal and environmental, they need to pursue goals. Many clients do not know how to mobilize needed resources. One of the most important resources is social support (see Basic Behavioral Science Task Force of the National Advisory Mental Health Council, 1996, p. 628).

> Social support has . . . been examined as a predictor of the course of mental illness. In about 75% of studies with clinically depressed patients, social-support factors increased the initial success of treatment and helped patients maintain their treatment gains. Similarly, studies of people with schizophrenia or alcoholism revealed that higher levels of social support are correlated with fewer relapses, less frequent hospitalizations, and success and maintenance of treatment gains.

Of course, studies like this reinforce what, it is to be hoped, we already know through common sense. Which of us has not been helped through difficult times by family and friends?

When it comes to social support, there are two categories of clients. First, there are those who lead an impoverished social life. The objective with this group is to help them find social resources, to get back into community in some productive way. However, a challenging study by Bankoff (1994) provided evidence that most clients are not socially isolated. In this study most clients had close friends from whom they received nurturing support and some kind of ongoing romantic relationship. The objective with this group is to help them tap into those resources in a way that helps them manage problem situations more effectively. Indeed, the National Advisory Mental Health Council study just mentioned showed that people who are highly distressed and therefore most in need of social support may be the least likely to receive it because their expressions of distress drive away potential supporters. Which of us has not avoided at one time or another a distressed friend or colleague? Therefore, distressed clients can be helped to learn how to modulate their expressions of distress. Who wants to help the whiner? On the other hand, potential supporters can learn how to deal with distressed friends and colleagues, even whiners.

The Task Force study suggested two general strategies for fostering social support—helping clients mobilize or increase support from existing social networks and "grafting" new ties onto impoverished social networks. Both of these come into play in the following case.

> Casey, a bachelor whose job involved frequent travel literally around the world, fell ill. He had many friends, but they were spread around the world. Because he was neither married nor in a marriagelike relationship, he had no primary care-giver in his life. He received excellent medical care, but his psyche fared poorly.
>
> Once out of the hospital, he recuperated slowly, mainly because he was not getting the social support he needed. In desperation he had a few sessions with a counselor, sessions that proved to be quite helpful. The counselor challenged him to "ask for help" from his local friends. He had underplayed his illness with them because he didn't want to be a "burden." He discovered that his friends were more than ready to help. But since their time was limited, he "grafted" onto his rather sparse hometown social network some very caring people from the local church. He was fearful that he would be deluged with piety, but instead he found people like himself, and they knew how much or how little care to give. Finally, he hired a couple of students from a local university to do word processing and run errands for him from time to time. Intrigued by his world travels, they also provided some social support.

As the Task Force authors note, it's important not only that people be available to provide support but also that those needing support perceive that it is available. This may mean, as in Casey's case, working with the client's attitudes and openness to receive support.

Eventually, all clients have to make it without the help of a counselor. Therefore, effective helpers right from the beginning try to help them explore the social-support dimensions of problem situations. At the Action Arrow stage, questions like the following are appropriate: "Who might help you do this? Who's going to challenge you when you want to give up? With whom can you share these kinds of concerns? Who's going to give you a pat on the back when you accomplish your goal?"

Although social support is often key, it is not the only resource clients need to pursue their goals. Effective helpers build some kind of resource census into the helping process.

## WHAT WORKING KNOWLEDGE AND SKILLS WILL HELP ME GET WHAT I NEED AND WANT?

It often happens that people get into trouble or fail to get out of it because they lack the needed life skills or coping skills to cope with problem situations. These are the resources they lack. If this is the case, then helping clients find ways of learning the life skills they need to cope more effectively is an important broad strategy. Indeed, the use of skills training as part of therapy — what Carkhuff years ago (1971) called "training as treatment" — might be an essential for some clients. Helping clients choose strategies that they don't have the skills to implement is compounding rather than solving the problem. Helping sessions cannot make up for all deficits in life skills, but they can target the specific skills needed by clients to accomplish their goals. A constant question throughout the helping process should be, What kinds of skills does this client need to get where he or she wants to go? Consider the following case.

> Jerzy and Zelda fell in love. They married and enjoyed a relatively trouble-free honeymoon period of about two years. Eventually, however, the problems that inevitably arise from living together in such intimacy asserted themselves. They found, for instance, that they counted too heavily on positive feelings for each other and now, in their absence, could not "communicate" about finances, sex, and values. They lacked certain critical interpersonal communication skills. Furthermore, they lacked understanding of each other's developmental needs. Jerzy had little working knowledge of the developmental demands of a 20-year-old woman; Zelda had little working knowledge of the kinds of cultural blueprints that were operative in the lifestyle of her 26-year-old husband. The relationship began to deteriorate. Since they had few problem-solving skills, they didn't know how to handle their situation.

In the case of this young couple, it seems reasonable to assume that helping will necessarily include both education and training as essential elements of an action plan. Lack of requisite interpersonal communication and other life skills is often at the heart of marriage breakdowns. One marriage counselor I know does marriage counseling in groups, usually of four couples. Training in communication skills is part of the process. He separates men from women and trains them in attending, listening, and empathy. For practice he begins by pairing a woman with a woman and a man with a man. Next he has a man and a woman, but not spouses, practice the skills together. Finally, spouses are paired, taught a simple version of the problem-management process outlined in this book, and then helped to use the skills they have learned to engage in problem solving with each other. In sum, he equips them in two sets of life skills — interpersonal communication and problem solving — that they have not picked up along the way. Strangely, although the society in which we live values these skill sets intellectually, it does little to make sure that its members acquire them. The acquisition of certain key life skills is left to chance.

The literature is filled with programs designed to equip clients with the working knowledge and skills they need to manage problems and lead a fuller life. Some of them focus on specific problems. For instance, Deffenbacher and his associates (Deffenbacher, Thwaites, Wallace, & Oetting, 1994; Deffenbacher, Oetting, Huff, & Thwaites, 1995) have devised and evaluated programs for general anger reduction. Although programs such as these need to be tailored to individual clients, they are often gold mines of strategies for accomplishing goals. Tailoring such generic programs to clients will be discussed in the next chapter.

## LINKING STRATEGIES TO ACTION

Although all the steps of the helping process can and should stimulate action on the part of the client, this is especially true of Step III-A, which deals with possible actions. Many clients, once they begin to see what they can do to get what they want, begin acting immediately. They don't need a formal plan. Here are a couple of examples of clients who were helped to identify strategies for implementing their goals and to act on them.

> Jeff had been in the army for about ten months. He found himself both overworked and, perhaps not paradoxically, bored. He had a couple of sessions with one of the educational counselors on the base. During these sessions Jeff began to see quite clearly that not having a high school diploma was working against him. The counselor mentioned that he could finish high school while in the army. Jeff realized that this possibility had been pointed out to him during the orientation talks, but he hadn't paid any attention to it. He had joined the army because he wasn't interested in school and couldn't find a job. Now he decided that he would get a high school diploma as soon as possible.
>
> Jeff obtained the authorization needed from his company commander to go to school. He found out what courses he needed and enrolled in time for the next school session. It didn't take him long to finish. Once he received his high school degree, he felt better about himself and found that opportunities for

more interesting jobs opened up for him in the army. Achieving his goal of getting a high school degree helped him manage the problem situation.

Jeff was one of those fortunate ones who, with a little help, quickly set a goal (the "what") and identify and implement the strategies (the "how") to accomplish it. Note, too, that his goal of getting a diploma was also a means to other goals—feeling good about himself and getting better jobs in the army.

Grace's road to problem management was quite different from Jeff's. She needed much more help.

As long as she could remember, Grace had been a fearful person. She was especially afraid of being rejected and of being a failure. As a result, she had an impoverished social life and had held a series of jobs that were safe but boring. She became so depressed that she made a halfhearted attempt at suicide, probably more an expression of anguish and a cry for help than a serious attempt to get rid of her problems by getting rid of herself.

During the resulting stay in the hospital, Grace had a few therapy sessions with one of the staff psychiatrists. The psychiatrist was supportive and helped her handle both the guilt she felt because of the suicide attempt and the depression that had led to the attempt. Just talking to someone about things she usually kept to herself seemed to help. She began to see her depression as a case of "learned helplessness." She saw quite clearly how she had let her choices be dictated by her fears. She also began to realize that she had a number of underused resources. For instance, she was intelligent and, though not good-looking, attractive in other ways. She had a fairly good sense of humor, though she seldom gave herself the opportunity to use it. She was also sensitive to others and basically caring.

After Grace was discharged from the hospital, she returned for a few outpatient sessions. She got to the point where she wanted to do something about her general fearfulness and her passivity, especially the passivity in her social life. A psychiatric social worker taught her relaxation and thought-control techniques that helped her reduce her anxiety. Once she felt less anxious, she was in a better position to do something about establishing some social relationships. With the social worker's help, she set goals of acquiring a couple of friends and becoming a member of some social group. However, she was at a loss as to how to proceed, since she thought that friendship and a fuller social life were things that should happen "naturally." She soon came to realize that many people had to work at acquiring a more satisfying social life, that for some people there was nothing automatic about it at all.

The social worker helped Grace identify various kinds of social groups that she might join. She was then helped to see which of these would best meet her needs without placing too much stress on her. She finally chose to join an arts and crafts group at a local YMCA. The group gave her an opportunity to begin developing some of her talents and to meet people without having to face demands for intimate social contact. It also gave her an opportunity to take a look at other, more socially oriented programs sponsored by the Y. In the arts and crafts program she met a couple of people she liked and who seemed to like her. She began having coffee with them once in a while and then an occasional dinner.

Grace still needed support and encouragement from her helper, but she was gradually becoming less anxious and feeling less isolated. Once in a while she would let her anxiety get the better of her. She would skip a meeting at the Y and then lie about having attended. However, as she began to let herself trust her helper more, she revealed this self-defeating game. The social worker helped her develop coping strategies for those times when anxiety seemed to be higher.

Grace's problems were more severe than Jeff's, and she did not have as many immediate resources. Therefore, she needed both more time and more attention to develop goals and strategies.

**?**

## Evaluation Questions for Step III-A

How effectively do I do the following?

- Use probes, prompts, and challenges to help clients identify possible strategies
- Help clients engage in divergent thinking with respect to strategies
- Help clients brainstorm as many ways as possible to accomplish their goals
- Use some kind of framework in helping clients be more creative in identifying strategies
- Help clients identify and begin to acquire the resources they need to accomplish their goals
- Help clients identify and develop the skills they need to accomplish their goals
- Help clients see the action implications of the strategies they identify

## CHAPTER 17

# STEP III-B: BEST-FIT STRATEGIES— WHAT STRATEGIES ARE BEST FOR ME?

# WHAT'S BEST FOR ME? THE CASE OF BUD

In the last two steps of Stage III, clients are in decision-making mode once more. After brainstorming strategies for accomplishing goals, they need to choose one or more strategies (a "package") that best fit their situation and resources and turn them into a formal plan for constructive change. Whether these steps are done with the kind of formality outlined here is not the point. Counselors, understanding the "technology" of planning, can add value by helping clients find ways of accomplishing goals (getting what they need and want) in a systematic and personalized and cost-effective way. Step III-B discusses ways of helping clients choose the strategies that are best for them. Step III-C deals with turning those strategies into some kind of step-by-step plan.

Some clients, once they are helped to develop a range of strategies to implement goals, move forward on their own; that is, they choose the best strategies, put together an action plan, and implement it. Others, however, need help in choosing strategies that best fit their situation, and so we add Step III-B to the helping process. It is useless to have clients brainstorm if they don't know what to do with all the action strategies they generate.

Consider the case of Bud, a man who was helped to discover two best-fit strategies for achieving emotional stability in his life. He achieved outcomes that surpassed anyone's wildest expectations.

One morning, Bud, then 18 years old, woke up unable to speak or move. He was taken to a hospital, where catatonic schizophrenia was diagnosed. After repeated admissions to hospitals, where he underwent both drug and electroconvulsive therapy (ECT), his diagnosis was changed to paranoid schizophrenia. He was considered incurable.

A quick overview of Bud's earlier years suggests that much of his emotional distress was caused by unmanaged life problems and the lack of human support. He was separated from his mother for four years when he was young. They were reunited in a city new to both of them, and there he suffered a great deal of harassment at school because of his "ethnic" looks and accent. There was simply too much stress and change in his life. He protected himself by withdrawing. He was flooded with feelings of loss, fear, rage, and abandonment. Even small changes became intolerable. His catatonic attack occurred in the autumn on the day of the change from daylight saving to standard time. It was the last straw.

In the hospital Bud became convinced that he and many of his fellow patients could do something about their illnesses. They did not have to be victims of themselves or of the institutions designed to help them. Reflecting on his hospital stays and the drug and ECT treatments, he later said he found his "help" so disempowering that it was no wonder that he got crazier. Somehow Bud, using his own inner resources, managed to get out of the hospital. Eventually, he got a job and was married.

One day, after a series of problems with his family and at work, Bud felt himself becoming agitated and thought he was choking to death. His doctor sent him to the hospital "for more treatment." There Bud had the good fortune to meet Sandra, a psychiatric social worker who was convinced that many of the hospital's patients were there because of lack of support before, during, and after their bouts of illness. She helped him see his need for social support, especially at times of stress. She also discovered in the inpatient counseling groups that she ran that Bud had a knack for helping others. Bud's broad goal was still emotional stability, and he wanted to do whatever was necessary to achieve it. Finding human support and helping others cope with their problems were his best strategies for achieving the stability he wanted.

Outside, Bud started a self-help group for ex-patients like himself. In the group, he was a full-fledged participant. Sandra also coached his wife on how to provide support at times of stress. As to helping others, Bud not only founded a self-help group but also turned it into a network of self-help groups for ex-patients.

This is an amazing example of a client who focused on one broad goal, emotional stability; translated it into a number of immediate, practical goals; discovered two broad strategies—finding ongoing emotional support and helping others—for accomplishing those goals; translated the strategies into practical applications; and by doing all that found the emotional stability he was looking for.

# HELPING CLIENTS CHOOSE BEST-FIT STRATEGIES

The criteria for choosing goal-accomplishing strategies are somewhat like the criteria for choosing goals outlined in Step II-B. These criteria are reviewed briefly here through a number of examples. Strategies to achieve goals should be, like goals themselves, specific, robust, prudent, realistic, sustainable, flexible, cost-effective, and in keeping with the client's values. Let's take a look at a few of these criteria as applied to choosing strategies.

**Specific strategies.** Strategies for achieving goals should be specific enough to drive behavior. In the preceding example, Bud's two broad strategies for achieving emotional stability, tapping into human support and helping others, were translated into quite specific strategies—keeping in touch with Sandra, getting help from his wife, participating in a self-help group, starting a self-help group, and founding and running a self-help organization. Contrast Bud's case with Stacy's, which is outlined next.

Stacy was admitted to a mental hospital because she had been exhibiting bizarre behavior in her neighborhood. She dressed in a slovenly way and went around admonishing the residents of the community for their "sins." Her condition was diagnosed as schizophrenia, simple type. She had been living alone for about five years, since the death of her husband. It seems that she had become more and more alienated from herself and others. In the hospital, medication helped control some of her symptoms. She stopped admonishing others and took reasonable care of herself, but she was still quite withdrawn. She was assigned to "milieu" therapy, a euphemism meaning that she was helped to follow the more or less benign routine of the hospital — a bit of work, a bit of exercise, some programmed opportunities for socializing. She remained withdrawn and usually seemed moderately depressed. No therapeutic goals had been set, and the nonspecific program to which she was assigned was totally inadequate.

So-called milieu therapy did nothing for Stacy because in no way was it specific to her needs. It was a general program that was only marginally better than drug-focused standard care. Bud's strategies, on the other hand, proved to be powerful. They not only helped him stabilize but also gave him a new perspective on life.

**Substantive strategies.** Strategies are robust to the degree that they challenge the client's resources and, when implemented, actually achieve the goal. Not only was Stacy's program too general, but it also lacked bite. Bud's strategies, on the other hand, were robust, especially the strategy of starting and running a self-help organization.

> A newly hired psychiatrist, who had been influenced by Corrigan's (1995) notion of "champions of psychiatric rehabilitation," saw immediately that Stacy needed more than either standard psychiatric or milieu-centered care. He involved her in a new comprehensive social-learning program, which included cognitive restructuring, social-skills training, and behavioral-change interventions based on incentives, shaping, modeling, and rewards. Against all odds, Stacy responded very well. She was discharged within six months and, with the help of an outpatient extension of the program, remained in the community.

For Stacy this program proved to be specific, robust, prudent, realistic, sustainable, flexible, cost-effective, and in keeping with her values. It was cost-effective in two ways. First, it was the best use of Stacy's time, energy, and psychological resources. Second, it helped her and others like her to get back into the community and stay there. It was in keeping with her values because, even though some staff members at the hospital had concluded that all she wanted was "to be left alone," Stacy did better in a community setting.

**Realistic strategies.** If clients choose strategies that are beyond their resources, they are doing themselves in. Realistic strategies are within the resources of the client, under his or her control, and unencumbered by obstacles. Bud's strategies would have appeared unrealistic to most clients and helpers. But there is a point to be made. Just as we should help clients set stretch goals whenever possible, so we should not underestimate what clients are capable of doing. In the following case, the client moves from unrealistic to realistic strategies for getting what he wants.

Desmond is in a halfway house after leaving a state mental hospital. From time to time he still has bouts of depression that incapacitate him for a few days. He wants to get a job because he thinks that a job will help him feel better about himself, become more independent, and manage his depression better. He answers job advertisements in a rather random way and is constantly getting turned down after being interviewed. He simply does not yet have the kinds of resources needed to put himself in a favorable light in job interviews. Moreover, he is not yet ready for a regular, full-time job.

On his own, Desmond does not do well in choosing strategies to achieve even modest goals.

A local university receives funds to provide outreach services to halfway houses in the metropolitan area. Part of the university program is to find companies that are willing, on a win-win basis, to work with halfway-house residents. A counselor from the program helps Desmond explore some companies that have specific programs to help people with psychiatric problems. Although some of the companies are new to the program, two have already found that some of their best workers have a variety of disabilities, including psychiatric problems. After a few interviews, Desmond gets a job in one of these companies that is within his capabilities. The entire work culture is designed to provide the kind of support he needs.

There is, of course, a difference between realism and allowing clients to sell themselves short. Robust strategies that make clients stretch for a valued goal can be most rewarding. Bud's case is an exceptional example of that.

**Strategies in keeping with the client's values.** Make sure that the strategies chosen are consistent with the client's values. Let's return to the case of the priest who had been unjustly accused of child molestation.

In preparing for the court case, the priest and his lawyer had a number of discussions. The lawyer wanted to do everything possible to destroy the credibility of the accusers. He had dug into their past and dredged up some dirt. The priest objected to these tactics. "If I let you do this," he said, "I descend to their level. I can't do that." The priest discussed this with his counselor, his superiors, and another lawyer. Then he stuck to his guns. They prepared a strong case but without the sleaze.

After the trial was over and he was acquitted, the priest said that his discussion about the lawyer's preferred tactics was one of the most difficult issues he had to face. Something in him said that since he was innocent, any means to prove his innocence was allowed. Something else told him that this was not right. The counselor helped him clarify and challenge his values but made no attempt to impose either his own or the lawyer's values on his client. Box 17-1 outlines the kinds of questions you can help clients answer as they choose best-fit strategies.

BOX 17-1
# Questions on Best-Fit Strategies

- Which strategies will be most useful in helping me get what I need and want?
- Which strategies are best for this situation?
- Which strategies best fit my resources?
- Which strategies will be most economic in the use of resources?
- Which strategies are most powerful?
- Which strategies best fit my preferred way of acting?
- Which strategies best fit my values?
- Which strategies will have the fewest unwanted consequences?

## STRATEGY SAMPLING

Some clients find it easier to choose strategies if they first sample some of the possibilities.

> Two business partners were in conflict over ownership of the firm's assets. Their goals were to see justice done, to preserve the business, and, if possible, to preserve their relationship. A colleague helped them sample some possibilities. Under her guidance they discussed with a lawyer the process and consequences of bringing their dispute to the courts, they had a meeting with a consultant-counselor who specialized in these kinds of disputes, and they visited an arbitration firm.

In this case, the sampling procedure had the added effect of giving them time to let their emotions simmer down.

Karen, the woman who, with the help of her counselor, brainstormed a wide range of strategies for disengaging from alcohol, decided to sample some of the possibilities.

> Surprised by the number of program possibilities there were to achieve the goal of getting liquor out of her life, Karen decided to sample some of them. She went to an open meeting of Alcoholics Anonymous, she attended a meeting of a women's lifestyle-issues group, she visited the hospital that had the residential treatment program, she joined up for a two-week trial physical-fitness program at a YMCA, and she had a couple of strategy meetings with her husband and children. Although none of this was done frantically, it did occupy her energies and strengthened her resolve to do something about her alcoholism.

Of course, some clients could use strategy sampling as a way of putting off action. That was certainly not the case with Bud. His attending the meeting of a self-help group after leaving the hospital was a form of strategy sampling. He was impressed by the group, but he thought that he could start a group limited to ex-patients that would focus more directly on the kinds of issues ex-patients face.

## A BALANCE-SHEET METHOD FOR CHOOSING STRATEGIES

Some form of balance sheet can be used to help clients make decisions in general. The methodology could be used for any key decision related to the helping process—whether to get help in the first place, to work on one problem rather than another, or to choose this rather than that goal. Balance sheets deal with the acceptability and unacceptability of both benefits and costs. A balance-sheet approach, applied to choosing strategies for achieving goals, poses questions such as the following:

- What are the *benefits* of choosing this strategy? for myself? for significant others?
- To what degree are these benefits acceptable? to me? to significant others?
- In what ways are these benefits unacceptable? to me? to significant others?
- What are the *costs* of choosing this strategy? for myself? for significant others?
- To what degree are these costs acceptable? to me? to significant others?
- In what ways are these costs unacceptable? to me? to significant others?

Here is an example of a client who used the balance-sheet method to assess the viability not of a goal but of strategies to achieve a goal. Karen's goal was to stop drinking. One possible strategy for accomplishing that goal was to spend a month as an inpatient at an alcoholic treatment center. This possibility appealed to her. However, since choosing this strategy would be a serious decision, the counselor, Joan, helped Karen use the balance sheet to weigh possible costs and benefits. After filling it out, Karen and Joan discussed Karen's findings. She chose to consider the pluses and minuses for herself and for her husband and children.

## Benefits of Choosing the Residential Program

- *For me.* It would help me because it would be a dramatic sign that I want to do something to change my life. It's a clean break, as it were. It would also give me time just for myself. I'd get away from all my commitments to family, relatives, friends, and work. I see it as an opportunity to do some planning. I'd have to figure out how I would act as a sober person.

• *For significant others.* I'm thinking mainly of my family here. It would give them a breather, a month without an alcoholic wife and mother around the house. I'm not saying that to put myself down. I think it would give them time to reassess family life and make some decisions about any changes they'd like to make. I think something dramatic like my going away would give them hope. They've had very little reason to hope for the last five years.

### Acceptability of benefits

• *For me.* I feel torn here. But looking at it just from the viewpoint of acceptability, I feel kind enough toward myself to give myself a month's time off. Also something in me longs for a new start in life. And it's not just time off. The program is a demanding one.

• *For significant others.* I think that my family would have no problems in letting me take a month off. I'm sure that they'd see it as a positive step from which all of us would benefit.

### Unacceptability of benefits

• *For me.* Going away for a month seems such a luxury, so self-indulgent. Also, even though taking such a dramatic step would give me an opportunity to change my current lifestyle, it would also place demands on me. My fear is that I would do fine while in the program but that I would come out and fall on my face. I guess I'm saying it would give me another chance at life, but I have misgivings about having another chance. I need some help here.

• *For significant others.* The kids are young enough to readjust to a new me. But I'm not sure how my husband would take this "benefit." He has more or less worked out a lifestyle that copes with my being drunk a lot. Though I have never left him and he has never left me, still I wonder whether he wants me back sober. Maybe this belongs under the "cost" part of this exercise. I need some help here. And, of course, I need to talk to my husband about all this. I also notice that some of my misgivings relate not to a residential program as such but to a return to a lifestyle free of alcohol. Doing this exercise helped me see that more clearly.

## Costs of Choosing the Residential Program

• *For me.* Well, there's the money. I don't mean the money just for the program, but I would be losing four weeks' wages. The major cost seems to be the commitment I have to make about a lifestyle change. And I know the residential program won't be all fun. I don't know exactly what they do there, but some of it must be demanding. Probably a lot of it.

- *For significant others.* It's a private program, and it's going to cost the family a lot of money. The services I have been providing at home will be missing for a month. It could be that I'll learn things about myself that will make it harder to live with me — though living with a drinking spouse and mother is no joke. What if I come back more demanding of them — I mean, in good ways? I need to talk this through more thoroughly.

### Acceptability of costs

- *For me.* I have no problem at all with the money or with whatever the residential program demands of me physically or psychologically. I'm willing to pay. What about the costs of the demands the program will place on me for substantial lifestyle changes? Well, in principle I'm willing to pay what that costs. I need some help here.

- *For significant others.* They will have to make financial sacrifices, but I have no reason to think that they would be unwilling. Still, I can't be making decisions for them. I see much more clearly the need to have a counseling session with my husband and children present. I think they're also willing to have a "new" person around the house, even if it means making adjustments and changing their lifestyle a bit. I want to check this out with them, but I think it would be helpful to do this with the counselor. I think they will be willing to come.

### Unacceptability of costs

- *For me.* Although I'm ready to change my lifestyle, I hate to think that I will have to accept some dumb, dull life. I think I've been drinking, at least in part, to get away from dullness; I've been living in a fantasy world, a play world a lot of the time. A stupid way of doing it, perhaps, but it's true. I have to do some life planning of some sort. I need some help here.

- *For significant others.* It strikes me that my family might have problems with a sober me if it means that I will strike out in new directions. I wonder if they want the traditional homebody wife and mother. I don't think I could stand that. All this should come out in the meeting with the counselor.

Karen concludes, "All in all, it seems like the residential program is a good idea. There is something much more substantial about it than an outpatient program. But that's also what scares me."

Karen's use of the balance sheet helps her make an initial program choice, but it also enables her to discover issues that she has not yet worked out completely. By using the balance sheet, she returns to the counselor with work to do; she does not come merely wondering what will happen next. This highlights the usefulness of exercises and other forms of structure that help clients take more responsibility for what happens both in the helping sessions and outside.

## Realism in Using the Balance Sheet

Now, a more practical and flexible approach to using the balance sheet. It is not to be used with every client to work out the pros and cons of every course of action. But you can also use parts of the balance sheet and tailor them to the kinds of action strategies the client is exploring. In fact, one of the best uses of the balance sheet is not to use it directly at all. Keep it in the back of your mind whenever clients are making decisions. Use it as a filter to listen to clients. Then turn relevant parts of it into probes to help clients focus on issues they may be overlooking. "How will this decision affect the significant people in your life?" is a probe that originates in the balance sheet. "Is there any downside to that strategy?" might help a client who is being a bit too optimistic. No formula.

# LINKING STEP III-B TO ACTION

Some clients are filled with great ideas for getting things done but never seem to do anything. They lack the discipline to evaluate their ideas, choose the best, and turn them into action. Often this kind of work seems too tedious to them, even though it is precisely what they need. Consider the following case.

> Clint came away from the doctor feeling depressed. He was told that he was in the high-risk category for heart disease and that he needed to change his lifestyle. He was cynical, a man very quick to anger, a man who did not readily trust others. Venting his suspicions and hostility did not make them go away; it only intensified them. Therefore, one critical lifestyle change was to change this pattern and develop the ability to trust others. He developed three broad goals: reducing mistrust of others' motives, reducing the frequency and intensity of such emotions as rage, anger, and irritation, and learning how to treat others with consideration. Clint read through the strategies suggested to help people pursue these broad goals (see Williams, 1989):
>
> - Keeping a hostility log to discover the patterns of cynicism and irritation in one's life
> - Finding someone to talk to about the problem, someone to trust
> - "Thought stopping," catching oneself in the act of indulging in hostile thoughts or in thoughts that lead to hostile feelings
> - Talking sense to oneself when tempted to put others down
> - Developing empathic thought patterns — that is, walking in the other's shoes
> - Learning to laugh at one's own silliness
> - Using a variety of relaxation techniques, especially to counter negative thoughts
> - Finding ways of practicing trust
> - Developing active listening skills
> - Substituting assertive for aggressive behavior
> - Getting perspective, seeing each day as one's last
> - Practicing forgiving others without being patronizing or condescending

Clint prided himself on his rationality (though his "rationality" was one of the things that got him into trouble). So, as he read down the list, he chose strategies that could form an "experiment," as he put it. He decided to talk to a counselor (for the sake of objectivity), keep a hostility log (data gathering), and use the tactics of thought stopping and talking sense to himself whenever he felt that he was letting others get under his skin. The counselor noted to himself that none of these necessarily involved changing Clint's attitudes toward others. However, he did not challenge Clint at this point. His best bet was that through "strategy sampling" Clint would learn more about his problem, that he would find that it went deeper than he thought. Clint set himself to his experiment with vigor.

Clint chose strategies that fit his values. The problem was that the values themselves needed reviewing. But Clint did act, and action gave him the opportunity to learn.

## THE SHADOW SIDE
## OF SELECTING STRATEGIES

The shadow side of decision making, discussed in Chapter 12, is certainly at work in clients' choosing strategies to implement goals. Goslin (1985) put it well.

> In defining a problem, people dislike thinking about unpleasant eventualities, have difficulty in assigning . . . values to alternative courses of action, have a tendency toward premature closure, overlook or undervalue long-range consequences, and are unduly influenced by the first formulation of the problem. In evaluating the consequences of alternatives, they attach extra weight to those risks that can be known with certainty. They are more subject to manipulation . . . when their own values are poorly thought through. . . . A major problem . . . for . . . individuals is knowing when to search for additional information relevant to decisions. (pp. 7–9)

In choosing courses of action, clients often fail to evaluate the risks involved and determine whether the risk is balanced by the probability of success. Gelatt, Varenhorst, and Carey (1972) suggested four ways in which clients may try to deal with the factors of risk and probability: wishful thinking, playing it safe, avoiding the worst outcome, and achieving some kind of balance. The first three are often pursued without reflection and therefore lie in the "shadows."

**Wishful thinking.** In this case, clients choose a course of action that might (they hope) lead to the accomplishment of a goal regardless of risk, cost, or probability. For instance, Jenny wants her ex-husband to increase the amount of support he is paying for the children. She tries to accomplish this by constantly nagging him and trying to make him feel guilty. She does not consider the risk (he might get angry and stop giving her anything), the cost (she spends a great deal of time and emotional energy arguing with him), or the probability of success (he does not react favorably to nagging). The wishful-thinking client operates blindly, engaging in some course of action without taking into account its usefulness. At its worst, this is a reckless approach. Clients who "work hard" and still "get nowhere" may be engaged in wishful thinking, persevering in using means they prefer but that are of doubtful efficacy.

**Playing it safe.** In this case, the client chooses only safe courses of action, ones that have little risk and a high degree of probability of producing at least limited success. For instance, Liam, a manager in his early 40s, is very dissatisfied with the way his boss treats him at work. His ideas are ignored, the delegation he is supposed to have is preempted, and his boss has not responded to his attempts to discuss career development. His goals are to let his boss know about his dissatisfaction and to learn what his boss thinks about him and his career possibilities. However, he fails to bring these issues up when his boss is "out of sorts." On the other hand, when things are going well, Liam doesn't want to "upset the applecart." He drops hints about his dissatisfaction, even joking about them at times. He tells others in hopes that word will filter back to his boss. During formal appraisal sessions he allows himself to be intimidated by his boss. However, in his own mind, he is doing whatever could be expected of a "reasonable" man. He does not know how safe he is playing it.

**Avoiding the worst outcome.** In this case, clients choose means that are likely to help them avoid the worst possible result. They try to minimize the maximum danger, often without identifying what that danger is. Crissy, dissatisfied with her marriage, sets a goal to be "more assertive." However, even though she has never said this either to herself or to her counselor, the maximum danger for her is to lose her partner. Therefore, her "assertiveness" is her usual pattern of compliance, with some frills. For instance, every once in a while she tells her husband that she is going out with friends and will not be around for supper. He, without her knowing it, actually enjoys these breaks. At some level of her being, she realizes that her absences are not putting him under any pressure. She continues to be assertive in this way. But she never sits down with her husband to review where they stand with each other. That might be the beginning of the end.

**Striking a balance.** In the ideal case, clients choose strategies for achieving goals that balance risks against the probability of success. This "combination" approach is the most difficult to apply, for it involves a great deal of analysis, including clarification of goals, a solid knowledge of personal values, and the ability to rank a variety of strategies according to one's values, plus the ability to predict results from a given course of action. Even more to the point, it demands challenging the shadow side of the problem, chosen goals, and the ineffectual courses of action that have been adopted in the past. Therefore, some clients have neither the skill nor the will for this approach.

**❓**

## CHAPTER 18

# Step III-C:
# Helping Clients Make Plans—What Kind of Plan Will Help Me Get What I Need and Want?

Planning in its broadest sense includes all the steps of Stage III. In Step III-C we come up with the plan itself, the *sequence* of actions — what we will do first, second, and third — that will get us what we want, our goal.

# No Plan of Action: The Case of Frank

The lack of a plan — that is, a clear step-by-step process in which strategies are used to achieve goals — keeps some clients mired in their problem situations. Consider the case of Frank, a vice president of a large West Coast corporation.

> Frank was a go-getter. He was very astute about business and had risen quickly through the ranks. Vince, the president of the company, was in the process of working out his own retirement plans. From a business point of view, Frank was the heir apparent. But there was a glitch. The president was far more than a good manager; he was a leader. He had a vision of what the company should look like five to ten years down the line. Though tough, he related well to people. Frank was quite different. He was a "hands-on" manager, meaning, in his case, that he was slow to delegate tasks to others, however competent they might be. He kept second-guessing others when he did delegate, reversed their decisions in a way that made them feel put down, listened poorly, and took a fairly short-term view of the business — "What were last week's figures like?" He was not a leader but an "operations" man.
>
> One day Vince sat down with Frank and told him that he was the heir apparent but that his promotion would not be automatic. "Frank, if it were just a question of business acumen, you could take over today. But my job, at least in my mind, demands a leader." Vince went on to explain what he meant by a leader and to point out some of the things in Frank's style that had to change.
>
> Afterward Frank saw a consultant, someone he trusted who had given him a similar opinion. They worked together for over a year, often over lunch and in hurried meetings early in the morning or late in the evening. Frank's real aim was to become president. If getting the job meant that he had to try to become the kind of leader his boss had outlined, so be it. Being very bright, he came up with some inventive strategies for moving in that direction. But he could never be pinned down to an overall program with specific milestones by which he could evaluate his progress. The consultant pushed him, but Frank was always "too busy" or would say that a formal program was "too stifling." That was odd, since formal planning was one of his strengths in the business world.
>
> Frank remained as astute as ever in his business dealings. But he merely dabbled in the strategies meant to help him achieve his goal. At the end of two years, Vince appointed someone else president of the company.

Frank had two significant blind spots that the consultant either could not or did not help him overcome. First, he thought the president's job was his, that business acumen alone would win out in the end. Second, he thought he could change his management style at the margins, when more substantial changes were called for. Frank had many good ideas, but he never put them together into a coherent program for constructive change. Rather, he jumped from one to another. He was more interested in "trying things" — that is, in activities rather than managerial-style outcomes that would have made a difference. If the consultant had said, "Come on, Frank, I know you hate doing it, but let's map out a program and find ways to get you to stick to it," maybe things would have been different. It was a question not of changing Frank's entire personality but of changing key patterns of behavior and doing so systematically. His failure to do that proved to be very expensive.

Some clients, once they have determined what they need to do to get what they want, have no difficulty deciding what needs to be done first, second, and so on. Others need help at this stage. Since some clients (and some helpers) fail to appreciate the power of a plan, it is useful to start by reviewing the advantages of planning. For a good review of the shadow side of planning, see Dorner (1996, pp. 153–183).

# How Plans Add Value to Clients' Change Programs

The focus of this chapter is formal planning. The truth is that planning goes on throughout the helping process. "Little plans," whether called such or not, are formulated and executed. They are part of the little actions that clients engage in between sessions. Therefore, every session might have a covert planning dimension. A formal plan takes strategies for accomplishing goals, divides them into workable bits, puts the bits in order, and assigns a timetable for the accomplishment of each bit. Planning, when adapted to the needs of the client, has many advantages. Here are some of them.

• *Plans help clients develop needed discipline.* Many clients get into trouble in the first place because they lack discipline. Planning places reasonable demands on clients to develop discipline. Desmond, the halfway-house resident discussed in the last chapter, needed discipline and benefited greatly from a formal job-seeking program. Indeed, ready-made programs are in themselves plans, sequences of steps designed to deliver specific outcomes.

• *Plans keep clients from being overwhelmed.* Plans help clients see goals as doable. They keep the steps toward the accomplishment of a goal "bite-size." Amazing things can be accomplished by taking bite-size steps toward substantial goals. Bud, the ex–psychiatric patient who ended up creating a network of self-help groups for ex-patients, started with the bite-size step of participating in one of those groups himself. He did not become a self-help entrepreneur overnight. It was a step-by-step process.

• *Formulating plans helps clients search for more useful ways of accomplishing goals — that is, even better strategies.* This is an example of moving back and forth among the steps of the helping process. Sy Johnson was an alcoholic. When Mr. Johnson's wife and children, working with a counselor, began to formulate a plan for coping with their reactions to his al-coholism, they realized that the strategies they had been trying were hit-or-miss. With the help of an Al-Anon self-help group, they went back to the drawing board. Mr. Johnson's drinking had introduced a great deal of disor-der into the family. Planning would help them restore order.

- *Plans provide an opportunity to evaluate the realism and ade-quacy of goals.* This is another example of the dialogue that should take place among the steps of the helping process. When Walter, a middle man-ager who had many problems in the workplace, began tracing out a plan to cope with the loss of his job and with a lawsuit filed against him by his for-mer employer, he realized that his goals of getting his job back and of filing and winning a countersuit were unrealistic. His revised goals included get-ting his former employer to withdraw the suit and of getting into better shape to search for a job by participating in a self-help group of managers who had lost their jobs.

- *Plans make clients aware of the resources they will need to imple-ment their strategies.* When Dora was helped by a counselor to formulate a plan to pull her life together after the disappearance of her younger son, she realized that she lacked the social support needed to carry out the plan. She had retreated from friends and even relatives, but now she knew she had to get back into community. Normalizing life demanded ongoing social in-volvement and support. A goal of finding the support needed to get back into community was added to her constructive-change program.

- *Formulating plans helps clients uncover unanticipated obstacles to the accomplishment of goals.* Ernesto, a U.S. soldier who had experienced a traumatic event during his stint in Bosnia, was trying to put his life back to-gether with the help of a counselor. Only when he began pulling together and trying out plans for normalizing his social life did he realize how ashamed he was of what had happened to him in the military. He felt so flawed be-cause of what had happened that it was almost impossible to involve himself intimately with others. Helping him deal with his shame became one of the most important parts of the healing process.

- *Planning can help clients manage post-decisional depression.* Sam who had decided to give up smoking, Chris who had determined to leave an abusive husband, and Sheila and Tom who had agreed to end their separation all had one thing in common—post-decisional depression, the angst that comes from wondering whether one has made the right decision. Instead of focusing on the depression, the counselor helped them formulate plans to get what they said they wanted. Immersing themselves in the plan-ning process and experimenting with parts of the plan left little room for second-guessing.

Formulating plans will not solve all our clients' problems, but it is one way of making time an ally instead of an enemy. Many clients engage in aim-less activity in their efforts to cope with problem situations. Plans help clients make the best use of their time.

# SHAPING THE PLAN: THREE CASES

Plans need "shape" to drive action. A formal plan identifies the activities or actions needed to accomplish a goal or a subgoal, puts those activities into a logical order, and sets a time frame for the accomplishment of each key step. Therefore, there are three simple questions.

- What are the concrete things that need to be done to accomplish the goal or the subgoal?
- In what sequence should these be done? What should be done first, what second, what third?
- What is the time frame? What should be done today, what tomorrow, what next month?

If clients choose goals that are complex or difficult, then it is useful to help them establish subgoals as a way of moving step-by-step toward the ultimate goal. For instance, once Bud decided to start an organization of self-help groups composed of ex-patients from mental hospitals, there were a number of subgoals that needed to be accomplished before the organization would become a reality. So his first step was to set up a test group. This provided the experience needed for further planning. A later step was to establish some kind of charter for the organization. "Charter in place" was one of the subgoals leading to his main goal.

In general, the simpler the plan the better, provided it helps clients achieve their goals. However, simplicity is not an end in itself. The question is not whether a plan or program is complicated but whether it is well shaped and designed to produce results. If complicated plans are broken down into subgoals and the strategies or activities needed to accomplish them, they are as capable of being achieved, if the time frame is realistic, as simpler ones. In schematic form, shaping looks like this:

Subprogram 1 leads to subgoal 1.

Subprogram 2 leads to subgoal 2.

Subprogram $n$ (the last in the sequence) leads to the accomplishment of the ultimate goal.

**The case of Wanda.** Take the case of Wanda, a client who set a number of goals in order to manage a complex problem situation. One of her goals was finding a job. The plan leading to this goal had a number of steps, each of which led to the accomplishment of a subgoal. The following subgoals were part of Wanda's job-finding program. They are stated as accomplishments (the past-participle approach).

Subgoal 1: Résumé written.

Subgoal 2: Kind of job wanted determined.

Subgoal 3: Job possibilities canvassed.

Subgoal 4: Best job prospects identified.

Subgoal 5: Job interviews arranged.

Subgoal 6: Job interviews completed.

Subgoal 7: Offers evaluated.

The accomplishment of these subgoals led to the accomplishment of the overall goal of Wanda's plan — that is, getting the kind of job she wants.

Wanda also had to set up a step-by-step process to accomplish each of these subgoals. For instance, the process for accomplishing the subgoal "job possibilities canvassed" included such things as reading the Help Wanted sections of the local papers, contacting friends or acquaintances who could provide leads, visiting employment agencies, reading the bulletin boards at school, and talking with someone in the job-placement office. Sometimes the sequencing of activities is important, sometimes not. In Wanda's case, it was important for her to have her résumé completed before she began to canvass job possibilities, but when it came to using different methods for identifying job possibilities, the sequence did not make any difference.

**The case of Harriet: The economics of planning.** Harriet, an undergraduate student at a small state college, wanted to become a helper. Although the college offered no formal program in counseling psychology, she identified several undergraduate courses that would help her toward her goal. One was called "Social Problem-Solving Skills"; a second, "Effective Interpersonal Communication Skills"; a third, "Developmental Psychology: The Developmental Tasks of Late Adolescence and Early Adulthood." Harriet took the courses as they came up. Unfortunately, the first course was the one in problem-solving approaches, and it included brief one-on-one practice in the skills being learned. The further into the course Harriet went, the more she realized that she would have gotten more out of the course if she had taken the communication skills program first.

Harriet had also volunteered for the peer-helper program offered by the Center for Student Services. Although those running the program were careful about whom they selected, they did not offer much training. It was a learn-as-you-go approach. Harriet realized that the developmental psychology course would have helped her in this program. Seeing that her activities needed to be better organized, she dropped out of the volunteer program and sat down with an adviser who had some background in psychology to come up with a reasonable plan. Harriet's program for constructive change would have been much more cost-effective had it been better shaped right from the beginning.

**The case of Frank revisited.** Let's see what planning might have done for Frank, the vice president who needed leadership skills. In this fantasy, Frank, like Scrooge, gets a second chance.

*What does Frank need to do?* To become a leader, Frank decides to reset his style with his subordinates by involving them more in decision making. He

BOX 18-1
# Questions on Planning

- What sequence of steps will get me to my goal?
- Which steps are most critical?
- How important is the order in which the steps take place?
- What is the best time frame for each step?
- Which steps need substeps?
- How can I build informality and flexibility into my plan?
- How do I gather the resources needed to implement the plan?

wants to listen more, set work objectives through dialogue, ask subordinates for suggestions, and delegate more. He decides to visit subordinates to see what he can do to help them, coach them when they ask for advice, give them feedback on the quality of their work, recognize their contributions, and reward them for achieving results beyond their objectives. What Frank is saying to himself through his list, then, is, "Listen more to those who report to you," "Coach those who need your advice," and so forth. He believes that he does not have to get more specific than that. The package of things he will do with each subordinate will have to be tailored to the needs of each.

*In what sequence should Frank do these things?* Frank decides that the first thing he will do is call in each subordinate and ask, "What do you need from me to get your job done? How can I add value to your work? And what management style on my part would help you most?" Their responses will help him decide which of the other activities will make sense with each subordinate. The second step is also clear. The planning cycle for the business is about to begin, and each manager needs to know what his or her objectives are. It is a perfect time to begin setting objectives. Frank therefore sends a memo to each of his direct reports, asking them to review the company's business plan and the plan for each of their functions, and to write down what they think their key managerial objectives for the coming year should be.

*What is Frank's time frame?* Frank calls in each of his subordinates immediately to discuss what they need from him. He completes his objective-setting sessions with them within three weeks. He puts off further action on delegation until he gets a better reading on their objectives and what they need from him.

This is a rough idea of what a plan for Frank might have looked like and how it might have improved his abortive effort to change his managerial style. Some clients resist planning because it does not have a "human face." It is the job of helpers to inject humanity into the planning process. Box 18-1 is a list of questions you can use to help clients think systematically about crafting a plan to get what they need and want.

# HUMANIZING THE TECHNOLOGY
## OF CONSTRUCTIVE CHANGE

Some years ago I lent a friend of mine an excellent, though somewhat detailed, book on self-development. About two weeks later he came back, threw the book on my desk, and said, "Who would go through all of that!" I retorted, "Anyone really interested in self-development." That was the righteous, not the realistic, response. Planning in the real world seldom looks like planning in textbooks. Textbooks do provide useful frameworks and principles, but they are seldom used. Most people are too impatient to do the kind of planning just outlined. One of the reasons for the dismal track record of discretionary change mentioned earlier is that even when clients do set realistic goals, they lack the discipline to develop reasonable plans. The detailed work of planning is too burdensome.

Therefore, Stages II and III of the helping process together with their six steps need a human face. If helpers skip these steps, they shortchange their clients. If they carry them out awkwardly, without giving them a human face, their clients will not become partners with them in their execution. Here, then, are some principles that can guide the constructive-change process, from Step II-A through Step III-C.

## Build a Planning Mentality into the Helping Process Right from the Start

A constructive-change mind-set should permeate the helping process right from the beginning. This is part of the hologram metaphor—the whole model should be found in each of its parts—mentioned in Chapter 2. Helpers need to see clients as change agents, not just as individuals mired in problem situations. Even while listening to a client's story, the helper needs to begin thinking of how the situation can be remedied. As mentioned earlier, significant changes can be effected within the helping sessions, but these changes need to be extended to the client's real world and complemented by further changes in the client's real-life setting.

> Cora, a battered spouse, did not want to leave her husband because of the kids. Right from the beginning, the helper saw Cora's problem situation from the point of view of the whole helping process. While she listened to Cora's story, without distorting it, she saw possible goals and strategies. Within the helping sessions, Cora learned a great deal about how battered women respond to their plight and how dysfunctional some of those responses are. She also learned how to stop blaming herself for the violence and to overcome her fears of developing more active coping strategies. At home she confronted her husband and stopped submitting to the violence in a vain attempt to avoid further abuse. She also joined a local self-help group for battered women. There she found social support and learned how to invoke both police protection and recourse to the courts. Further sessions with the counselor helped

her gradually change her identity from battered woman to survivor and, eventually, to doer.

Constructive-change scenarios like this must be in the helper's mind from the start, not as preset programs to be imposed on clients but as part of constructive-change mentality.

## Adapt the Constructive-Change Process to the Style of the Client

Setting goals, devising strategies, and making and implementing plans can be done formally or informally. There is a continuum. Some clients actually like the detailed work of devising plans; it fits their style. Gitta sought counseling as she entered the postparental period of her life. Although there were no specific problems, she saw too much emptiness as she looked into the future. The counselor helped her see the "empty nest" as a normal experience rather than a psychological or developmental problem (see Raup & Myers, 1989). After spending a bit of time discussing some of the maladaptive responses to this transitional phase of life, they embarked on a review of possible scenarios. Gitta loved brainstorming, getting into the details of the scenarios, weighing choices, setting strategies, and making formal plans. That worked for her because it was her normal style. Connor, in rebuilding his life after a serious automobile accident, very deliberately planned both a rehabilitation program and a career change. Keeping to a schedule of carefully planned actions not only helped him keep his spirits up but also helped him accomplish a succession of goals. These small triumphs buoyed his spirits and moved him, however slowly, along the rehabilitation path.

Most people, however, are not like Gitta and Connor. The distribution is skewed toward the "I hate all this detail and won't do it" end of the continuum. Yousef, a single parent with a mentally retarded son, was challenged one day by a colleague at work. "You've let your son become a ball and chain and that's not good for you or him!" his friend said. Yousef reluctantly sought counseling. He never discussed any kind of extensive change program with his helper, but with a little stimulation from her he began doing little things differently at home. When he came home from work especially tired and frustrated, he had a friend in the apartment building stop by. This helped to keep him from taking his frustrations out on his son. Instead of staying cooped up over the weekend, he found simple things to do that eased tensions, such as going to the zoo and to the art museum with a woman friend and his son. He discovered that his son enjoyed these pastimes immensely despite his limitations. In short, he discovered little ways of becoming his son's friend and not just his father. His counselor had a constructive-change mentality right from the beginning but did not try to get Yousef to engage in overly formal planning activities.

Kirschenbaum (1985) challenged the notion that planning should always provide an exact blueprint of the specific activities to be engaged in, their sequencing, and the time frame. There are three questions:

- How specific do the activities have to be?
- How rigid does the order have to be?
- How soon does each activity have to be carried out?

Kirschenbaum (p. 492) suggested that, at least in some cases, being less specific and rigid about activities, sequencing, and deadlines can "encourage people to pursue their goals by continually and flexibly choosing their activities." That is, flexibility in planning can help clients become more self-reliant and proactive. Rigid planning strategies can lead to frequent failure to achieve short-term goals. On the other hand, a slipshod approach to planning—"I will have to pull myself together one of these days"—is also self-defeating. We need only look at our own experience to see that that approach is fatal. Overall, counselors should help clients establish the specificity and the time frame that make sense for them in their situations. There are no formulas; there are only client needs and common sense. Some things need to be done now, some later. Some clients need more slack than others. Sometimes it helps to spell out the actions that need to be done in quite specific terms; at other times it is necessary only to help clients outline them in broad terms and leave the rest to their own sound judgment. The art of counseling involves helping clients find the right trade-offs, in specificity and time frame.

## Collaborate with Clients in Tailoring Generic Programs to Their Needs

There are many ready-made programs for clients with particular problems. They are often tried-and-true constructive-change programs. The 12-step approach of Alcoholics Anonymous is one of the most well known. It has been adapted to other forms of substance abuse and addiction. Counselors add value by helping clients adapt these programs to their particular needs. Consider the following case.

> After a couple of rather aimless sessions, the helper said to Ahmed, "We've talked about a lot of things, but I'm still not sure why you came in the first place." This challenged Ahmed to reveal the central issue, though he needed a great deal of help to do so. It turned out that Ahmed was sexually attracted to prepubescent children of both sexes. Although he had never engaged in pedophilic behavior, the temptation to do so was growing.
>
> The counselor adapted a New Zealand program called Kia Marama (Hudson, Marshall, Ward, Johnston, et al., 1995), a comprehensive cognitive-behavioral program for incarcerated child molesters, to Ahmed's situation. The original program included intensive work in challenging distorted attitudes, reviewing a wide range of sexual issues, seeing the world from the point of view of the victim, developing problem-solving and interpersonal-relationship skills, stress management, and relapse-prevention training. Counselor and client spent some

time assessing which parts of the program might be of most help before embarking on an intensive tailored program. In this case, of course, the objective was prevention rather than rehabilitation.

Not only did Ahmed stay out of trouble, but much of what he learned from the program applied to all dimensions of his life.

One community-based mental-health center worked extensively with people on welfare. When new legislation was passed forcing welfare recipients to get work, they searched for programs that helped people on welfare get and keep jobs. They learned a great deal from one program sponsored by a major hotel chain (see Milbank, 1996). The hotel targeted welfare recipients because it made both economic and social sense. Because of the problems with this particular population, however, their recruiters, trainers, and supervisors had to become paraprofessional helpers, though they never used that term. The people they recruited had all sorts of problems: battered women, ex-convicts, addicts, homeless people, including those who had been thrown out of shelters, and so forth. The program included helping, working with, and even doing for.

> They drive welfare trainees to work, arrange their day care, negotiate with their landlords, bicker with their case workers, buy them clothes, visit them at home, coach them in everything from banking skills to self-respect—and promise those who stick with it full-time jobs. (Milbank, 1996, A1).

They challenged their "clients'" mind-set that they are not responsible for what happens to them, enforced rules with equity, and persevered. The counselors at the mental-health center learned that some clients need wholesale involvement in the lives of people on welfare to kick-start a constructive-change process. They started a volunteer program, looking for people willing to do the kinds of things that the hotel recruiters, trainers, and supervisors did.

## Devise a Plan for the Client and Then Work with the Client on Tailoring It to His or Her Needs

The more experienced helpers become, the more they learn about the elements of program development and the more they come to know what kinds of programs work for different clients. They build up a stockpile of useful programs and know how to stitch pieces of different programs together to create new programs. And so they can use their knowledge and experience to fashion a plan for any client who lacks the skills or the temperament to pull together a plan for himself or herself. This can be done in an empowering way. For instance, the plan can be first offered as a sketch or an outline rather than as a detailed program. The helper works with the client to fill out the sketch and adapt it to his or her needs and style. Consider the following case.

Katrina, a woman who dropped out of high school but managed to get a high school equivalency diploma, was overweight and reclusive. Over the years she had restricted her activities because of her weight. Sporadic attempts at dieting had left her even heavier. Because she was chronically depressed and had little imagination, she was not able to come up with any kind of coherent plan. Once her counselor understood the dimensions of Katrina's problem situation, she pulled together an outline of a change program that included such things as blame reduction, the redefinition of beauty, increasing restricted activities (see Robinson & Bacon, 1996), other cognitive restructuring activities aimed at lessening depression, and some information and suggestions dealing with overweight drawn from health-care sources. She presented these in a simple format, adding detail only for the sake of clarity. She added further detail as Katrina got involved in the planning process and in making choices.

Although this counselor pulled together elements of a range of already existing programs, counselors are, of course, free to make up their own programs based on their expertise and experience. The point is to give clients something to work with, something to get involved in. The elaboration of the plan emerges through dialogue with the client and in the kind of detail the client can handle.

Finally, the ultimate test of the effectiveness of plans lies in the problem-managing and opportunity-developing action they drive and the results they produce. There is no such thing as a good plan in and of itself. The next and final chapter deals with results-producing action.

## ❓ Evaluation Questions for Step III-C

Helpers can ask themselves the following questions as they help clients formulate the kinds of plans that actually drive action.

- To what degree do I prize and practice planning in my own life?
- How effectively have I adopted the hologram mind-set in helping, seeing each session and each intervention in the light of the entire helping process?
- How quickly do I move to planning when I see that it is what the client needs to manage problems and develop opportunities better?
- What do I do to help clients overcome resistance to planning? How effectively do I help them identify the incentives for and the payoff of planning?
- How effectively do I help clients formulate subgoals that lead to the accomplishment of overall preferred-scenario goals?
- How practical am I in helping clients identify the activities needed to accomplish subgoals, sequence those activities, and establish realistic time frames for them?
- How well do I adapt the specificity and detail of planning to the needs of each client?
- Even at this planning step, how easily do I move back and forth among the different stages and steps of the helping model as the need arises?
- How readily do clients actually move to action because of my work with them in planning?
- How human is the technology of constructive change in my hands?
- How well do I adapt the constructive-change process to the style of the client?
- How effectively do I help clients tailor generic change programs to their specific needs?

## PART SIX

# THE ACTION ARROW: MAKING IT ALL HAPPEN

The Action Arrow of the helping model in Figure IV constitutes the "transition state" in which clients move from the current to the preferred scenario by implementing strategies and plans. However, as Ferguson (1980) noted, clients often feel at risk during transition times. They feel as if they are on a trapeze, letting go of one bar — that is, familiar but dysfunctional patterns of behavior — and grabbing hold of the other bar — that is, new and more productive patterns of behavior. Some clients, anticipating the terror of being in midair without a safety net, refuse to let go. This is one reason why it is important to encourage clients to act, to begin the transition, at least in small ways, from the first session on. If they start engaging in small problem-managing actions right from the beginning — that is, if each stage and step of the helping model is used as a driver

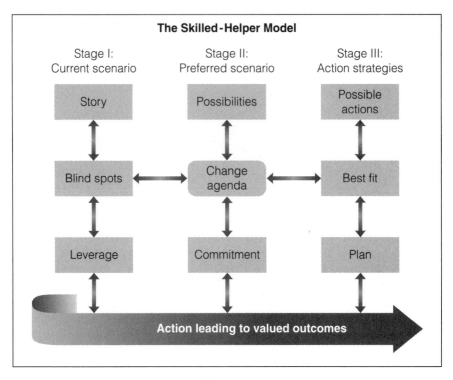

**FIGURE IV**
**The Helping Model — Action Arrow Phase**

of action — then they are less likely to be stymied by the actions called for by a more formal plan.

Some clients, once they have a clear idea of what to do to handle a problem situation, go ahead and do it, whether they have a formal plan or not. They need little or no further support and challenge from their helpers. They either find the resources they need within themselves or get support and challenge from the significant others in the social settings of their lives. Other clients choose goals and come up with strategies for implementing them but are, for whatever reason, stymied when it comes to action. Most clients fall between these two extremes.

Discipline and self-control play an important part in implementing change programs. Kirschenbaum (1987) found that many things can contribute to giving up: low initial commitment to change, weak self-efficacy, poor outcome expectations, the use of self-punishment rather than self-reward, depressive thinking, failure to cope with emotional stress, lack of consistent self-monitoring, failure to use effective habit-change techniques, giving in to social pressure, failure to cope with initial relapse, and paying attention to the wrong things — for instance, focusing on how the environment is doing me in rather than how I am failing to cope with the environment.

Kanfer and Schefft (1988, p. 58) differentiated between *decisional* self-control and *protracted* self-control. In the former, a single choice terminates a conflict. For instance, a couple makes the decision to get a divorce and goes through with it. In the latter, continued resistance to temptation is required. For instance, a client has learned how to control her anger but must continue to be on guard. Most clients need both kinds of self-control to manage their lives better. A client's choice to give up alcohol completely (decisional self-control) needs to be complemented by the ability to handle inevitable longer-term temptations. Protracted self-control calls for a preventive mentality and a certain degree of street smarts. It is easier for the client who has given up alcohol to turn down an invitation to go to a bar in the first place than to sit in a bar all evening with friends and refrain from drinking. Principles for helping clients develop and maintain protracted self-control are outlined in Chapter 19 as "tactics." Figure IV highlights the Action Arrow transition phase of the helping process. It starts from the first encounter between helper and client and continues after the formal process has been terminated.

# MAKING IT ALL HAPPEN: HELPING CLIENTS GET WHAT THEY WANT AND NEED

# HELPING CLIENTS BECOME
# EFFECTIVE TACTICIANS

In the implementation phase, strategies for accomplishing goals need to be complemented by tactics and logistics. A *strategy* is a practical plan to accomplish some objective. *Tactics* is the art of adapting a plan to the immediate situation. This includes being able to change the plan on the spot to handle unforeseen complications. *Logistics* is the art of being able to provide the resources needed for the implementation of a plan when they are needed.

> During the summer, Rebecca wanted to take an evening course in statistics so that the first semester of the following school year would be lighter. Having more time would enable her to act in one of the school plays, a high priority for her. But she didn't have the money to pay for the course, and at the university she planned to attend prepayment for summer courses was the rule. Rebecca had counted on paying for the course from her summer earnings, but she would not have the money until later. Consequently, she did some quick shopping around and found that the same course was being offered by a community college not too far from where she lived. Her tuition there was minimal, since she was a resident of the area the college served.

In this example, Rebecca keeps to her overall plan (strategy). However, she adapts the plan to an unforeseen circumstance, the demand for prepayment (tactics), by locating another resource (logistics).

Since many well-meaning and motivated clients are simply not good tacticians, counselors can add value by using the following principles to help them engage in focused and sustained goal-accomplishing action.

## Help Clients Avoid Imprudent Action

For some clients, the problem is not that they refuse to act but that they act without direction or imprudently. Rushing off to try the first "strategy" that comes to mind is often imprudent.

> Elmer injured his back and underwent a couple of operations. After the second operation he felt a little better, but then his back began troubling him again. When the doctor told him that further operations would not help, Elmer was faced with the problem of handling chronic pain. It soon became clear that his psychological state affected the level of pain. When he was anxious or depressed, the pain always seemed much worse.
> Elmer was talking this through with a counselor. One day he read about a pain clinic located in a western state. Without consulting anyone, he signed up for a six-week program. Within ten days he was back, feeling more depressed than ever. He had gone to the program with extremely high expectations because his needs were so great. The program was a holistic one that helped the participants develop a more realistic lifestyle. It included programs dealing with nutrition, stress management, problem solving, and quality of in-

terpersonal life. Group counseling was part of the program, and training was part of the group experience. For instance, the participants were trained in behavioral approaches to the management of pain.

The trouble was that Elmer had arrived at the clinic, which was located on a converted farm, with unrealistic expectations. He had not really studied the materials that the clinic had sent to him. Since he had expected to find marvels of modern medicine that would magically help him, he was extremely disappointed when he found that the program focused mainly on reducing and managing rather than eliminating pain.

Elmer's goal was to be completely free of pain, and he failed to see that that might not be possible. A more realistic goal would have centered on the reduction and management of pain. Elmer's counselor failed to help him avoid two mistakes—setting an unrealistic goal and, in desperation, acting on the first strategy that came along.

Obviously, action cannot be prudent if it is based on flawed assumptions—that is, on conditions that must be in place to assure a successful outcome but that are not in place. For instance, helpers may come up with wonderful reconciliation programs, but those programs are meaningless if the parties involved are not interested in being reconciled.

## Help Clients Develop Contingency Plans

If counselors help clients brainstorm both possibilities for a better future (goals) and strategies for achieving those goals (courses of action), then clients will have the raw materials, as it were, for developing contingency plans. Such plans answer the question, What will I do if the plan of action I choose is not working? Contingency plans help make clients more effective tacticians. The formulation of contingency plans is based on the fact that we live in an imperfect world. Goals may have to be fine-tuned or even changed. The same is true for strategies for accomplishing goals.

> Jackson, the man dying of cancer, decided to become a resident in the hospice he had visited. The hospice had an entire program in place for helping patients like Jackson die with dignity. However, although he had visited the hospice and had liked what he had seen, he could not be absolutely sure that being a resident there would work out. Fortunately, he had worked out alternative scenarios with his helper. One was living at home with some outreach services from the hospice. The other was spending his last days in a smaller hospital in a nearby town. This last alternative would not be as convenient for his family, but he would feel more comfortable there, since he hated large hospitals. He chose to stay at home as long as possible and then move to the smaller hospital, if necessary.

Contingency plans are needed especially when clients choose a high-risk program to achieve a critical goal. Having backup plans also helps clients develop more responsibility. If they see that a plan is not working, then they have to decide whether to try the contingency plan. Backup plans need not be complicated. A counselor might merely ask, "If that doesn't work, then what will you do?" As in the case of Jackson, clients can be helped to specify a contingency plan further once it is clear that the first choice is not working out.

## Help Clients Overcome Procrastination

At the other end of the spectrum are clients who keep putting action off. There are many reasons for procrastination. Take the case of Eula.

> Eula, disappointed with her relationship with her father in the family business, decided that she wanted to start her own. She thought that she could capitalize on the business skills she had picked up in school and in the family business. Her goal, then, was to establish a small software firm that created products for the family-business market.
>
> But a year went by and she still did not have any products ready for market. A counselor helped her see two things. First, her activities — researching the field, learning more about family dynamics, going to information-technology seminars, getting involved for short periods with professionals such as accountants and lawyers who did a great deal of business with family-owned firms, drawing up and redrafting business plans, and creating a brochure — did not help her move to her goal. It was as distant as ever. Second, the counselor helped Eula see that at some level of her being she was afraid of starting a new business. She ended up with a great many half-finished products. Overpreparation and half-finished products were signs of that fear. He helped her discuss and deal with her fears and identify strategies for managing them.

This was not a classic case of inertia. Eula, after all, was very active. But she avoided the most critical actions — creating and marketing products.

## Help Clients Identify Possible Obstacles to and Resources for Implementing Plans

Years ago Kurt Lewin (1969) codified common sense by developing what he called "force-field analysis." In ordinary language, this is simply a review by the client of the major obstacles to and resources for the implementation of strategies and plans. It can be used to identify obstacles and resources once plans are formulated or during implementation. Forewarned is forearmed. Clients can also be helped to identify emerging resources that can help them get what they need and want.

**Obstacles.** The identification of possible obstacles to the implementation of a program helps make clients forewarned.

> Raul and Maria were a childless couple living in a large midwestern city. They had been married for about five years and had not been able to have children. They finally decided that they would like to adopt a child, so they consulted a

counselor familiar with adoptions. The counselor helped them work out a plan of action that included helping them examine their motivation and lifestyle, contacting an agency, and preparing themselves for an interview. After the plan of action had been worked out, Raul and Maria, with the help of the counselor, identified two restraining forces: the negative feelings that often arise on the part of prospective parents when they are being scrutinized by an adoption agency and the feelings of helplessness and frustration caused by the length of time and uncertainty involved in the process.

The assumption here is that if clients are aware of some of the "wrinkles" that can accompany any given course of action, they will be less disoriented when they encounter them. Identifying possible obstacles is, at its best, a straightforward census of likely pitfalls rather than a self-defeating search for every possible thing that could go wrong.

Obstacles can come from within the clients themselves, from others, from the social settings of their lives, and from larger environmental forces. Once an obstacle is spotted, ways of coping with it can be identified. Sometimes simply being aware of a pitfall is enough to help clients mobilize their resources to handle it. At other times a more explicit coping strategy is needed. For instance, the counselor arranged a couple of role-playing sessions with Raul and Maria in which she assumed the role of the examiner at the adoption agency and took a "hard line" in her questioning. These rehearsals helped them stay calm during the actual interviews. The counselor also helped them locate a mutual-help group of parents working their way through the adoption process. The members of the group shared their hopes and frustrations and provided support for one another. In short, Raul and Maria were trained to cope with the restraining forces they might encounter on the road toward their goal.

**Possible resources.** In a more positive vein, emerging or unused resources that facilitate action can be identified.

> Nora found it extremely depressing to go to her weekly dialysis sessions. She knew that without them she would die, but she wondered whether it was worth living if she had to depend on a machine. The counselor helped her see that she was making life more difficult for herself by letting herself think such discouraging thoughts. He helped her learn how to think thoughts that would broaden her vision of the world instead of narrowing it down to herself, her pain and discomfort, and the machine. Nora was a religious person and found in the Bible a rich source of positive thinking. She initiated a new routine: The day before she visited the clinic, she began to prepare herself psychologically by reading from the Bible. Then, as she traveled to the clinic and underwent treatment, she meditated slowly on what she had read.

In this last case, the client substituted positive thinking, an underused resource, for poor-me thinking. Brainstorming resources that can counter obstacles to action can be very helpful for some clients.

# Help Clients Find the Incentives and the
# Rewards for Sustained Action

Clients avoid engaging in action programs when the incentives and the rewards for not engaging in the program outweigh the incentives and the rewards for doing so.

> Miguel, a policeman on trial for use of excessive force with a young offender, had a number of sessions with a counselor from an HMO that handled police health insurance. In the sessions the counselor learned that although this was the first time Miguel had run afoul of the law, it was in no way the first expression of a brutal streak within him. He was a bully on the beat and a despot at home, and had run-ins with strangers when he visited bars with his friends. Some of this came out during the trial.
>
> Up to the time of his arrest, he had gotten away with all of this, even though his friends had often warned him to be more cautious. His badge had become a license to do whatever he wanted. His arrest and now the trial shocked him. Before, he had seen himself as invulnerable; now, he felt very vulnerable. The thought of being a cop in prison understandably horrified him. He was found guilty, was suspended from the force for several months, and received probation on the condition that he continue to see the counselor.
>
> Beginning with his arrest, Miguel had modified his aggressive behavior a great deal, even at home. Of course, fear of the consequences of his aggressive behavior was a strong incentive to change his behavior. The next time, the courts would show no sympathy. But the counselor saw in Miguel's expressions of vulnerability the possibility of a much more decent human being, one "hiding" under the tough exterior. The counselor took a tough approach to this tough cop. He confronted Miguel for "remaining an adolescent" and for "hiding behind his badge." He called the power Miguel exercised over others "cheap power." He challenged the "decent person" to "come out from behind the screen." He told Miguel point blank that the fear he was experiencing was not enough to keep him out of trouble in the future. After probation, the fear would fade and Miguel could easily fall back into his old ways. Even worse, fear was a "weak man's" crutch. The real incentives, he said, were the rewards that come with decency. He made Miguel paint a picture of a "tough but decent" cop, family man, and friend.

The counselor was not trying to change Miguel's personality. Indeed, he didn't believe in personality transformations. But he pushed him hard to find and bring to the surface a different, more constructive set of incentives to guide his dealings with people.

The incentives and the rewards that help a client get going on a program of constructive change in the first place may not be the ones that keep the client going.

> Dwight, a man in his early 40s who was recovering from an accident at work that had left him partially paralyzed, was about to give up an arduous physical

rehabilitation program. The counselor asked him to visit the children's ward. Dwight was both shaken by the experience and amazed at the courage of many of the kids. He was especially struck by one teenager who was undergoing chemotherapy. "He seems so positive about everything," Dwight said. The counselor told him that the boy was tempted to give up, too. Dwight and the boy saw each other frequently. Dwight put up with the pain. The boy hung in there. Three months later the boy died. Dwight's response, besides grief, was, "I can't give up now; that would really be letting him down."

Dwight's partnership with the teenager proved to be an excellent incentive. In fact, he began to see all sorts of growth possibilities in different kinds of partnering.

Constructive-change activities that are not rewarded tend over time to lose their vigor, decrease, and even disappear. This process is called extinction. It was happening with Luigi.

Luigi, a middle-aged man, had been in and out of mental hospitals a number of times. He discovered that one of the best ways of staying out was to use some of his excess energy helping others. He had not returned to the hospital once during the three years he worked at a soup kitchen. However, finding himself becoming more and more manic over the past six months and fearing that he would be rehospitalized, he sought the help of a counselor.

Luigi's discussions with the counselor led to some interesting findings. He discovered that whereas in the beginning he had worked at the soup kitchen because he wanted to, he was now working there because he thought he should. He felt guilty about leaving and also thought that doing so would lead to a relapse. In sum, he had not lost his interest in helping others, but his current work was no longer interesting or challenging. As a result of his sessions with the counselor, Luigi began to work for a group that provided housing for the homeless and the elderly. He poured his energy into his new work and no longer felt manic.

The lesson here is that incentives cannot be put in place and then be taken for granted. They need tending.

# Help Clients Develop Action-Focused Self-Contracts and Agreements

Earlier we discussed self-contracts as a way of helping clients commit themselves to what they want, their goals. Self-contracts are also useful in helping them both initiate and sustain problem-managing action. For instance, Feller (1984) developed a "job-search agreement" to help job seekers persist in their search. The following agreement — clients were to respond "true" to all the following statements and then act on these "truths" — requires clients to commit themselves not only to job-seeking behavior but also to sound psychological practices that promote the right mentality for job-seeking behavior.

I agree that no matter how many times I enter the job market, or the level of skills, experiences, or academic success I have, the following appear TRUE:

1. It takes only one YES to get a job; the number of no's does not affect my next interview.
2. The open market lists about 20% of the jobs presently open to me.
3. About 80% of the job openings are located by talking to people.
4. The more people who know my skills and know that I'm looking for a job, the more I increase the probability that they'll tell me about a job lead.
5. The more specifically I can tell people about the problems I can solve or outcomes I can attain, rather than describe the jobs I've had, the more jobs they may think I qualify for.

I agree that regardless of how much I need a job, the following appear TRUE:

6. If I cut expenses and do more things for myself, I reduce my money problems.
7. The more I remain positive, the more people will be interested in me and my job skills.
8. If I relax and exercise daily, my attitude and health will appear attractive to potential employers.
9. The more I do positive things and the more I talk with enthusiastic people, the more I will gain the attention of new contacts and potential employers.
10. Even if things don't go as I would like them to, I choose my own thoughts, feelings, and behaviors each day.

It is easy to see how similar "agreements" could act as drivers of goal-accomplishing actions in many different kinds of problem situations. Self-contracts and agreements with others are especially helpful for more difficult aspects of action programs; they help focus clients' energies.

In the following example, several parties had to commit themselves to the provisions of the contract.

A boy in the seventh grade was causing a great deal of disturbance by his out-breaks in class. The usual kinds of punishment did not seem to work. After the teacher discussed the situation with the school counselor, the counselor called a meeting of all the stakeholders — the boy, his parents, the teacher, and the principal. The counselor offered a simple contract. When the boy disrupted the class, one and only one thing would happen: He would go home. Once the teacher indicated that his behavior was disruptive, he was to go to the principal's office and check out without receiving any kind of lecture. He was to go immediately home and check in with whichever parent was at home, again without receiving any further punishment. The next day he was to return to school. All agreed to the contract, though both principal and parents said they would find it difficult not to add to the punishment.

The first month, the boy spent a fair number of days or partial days at home. The second month, however, he missed only two partial days, and the third month only one. The truth is that he really wanted to be in school with his classmates. That's where the action was. And so he paid the price of self-control to get what he wanted. The contract proved an effective tool of behavioral change.

The counselor had suspected that the boy found socializing with his classmates rewarding. But now he had to pay for the privilege of socializing. Reasonable behavior in the classroom was not too high a price.

## GETTING ALONG WITHOUT A HELPER: DEVELOPING SOCIAL NETWORKS FOR SUPPORTIVE CHALLENGE

In most cases helping is a short-term process. But even in longer-term therapy, clients must eventually get on with life without their helpers. Ideally, the counseling process not only helps clients deal with specific problem situations and unused opportunities, but also, as outlined in Chapter 1, equips them with the working knowledge and skills needed to manage those situations more effectively on their own.

Since adherence to constructive-change programs is often difficult, social support and challenge in their everyday lives can help them move to action, persevere in action programs, and both consolidate and maintain gains. When it comes to social support and challenge, there are a number of possible scenarios at the implementation stage and beyond.

- The formal sessions end, and it is up to the client to continue to pursue goals identified in the helping sessions or to maintain gains made there. That is, clients take the responsibility for implementation on their own and run with it.
- Clients continue to see a helper regularly in the implementation phase.
- Clients see a helper occasionally either "on demand" or in scheduled stop-and-check sessions.
- Clients join some kind of self-help group together with one-to-one counseling sessions, which are eventually eliminated.
- Clients develop social relationships that provide both ongoing support and challenge for the changes they are making in their lives.

Support was discussed earlier and, since the literature tends to focus on caring support, a few words about caring challenge in everyday life are in order.

**Challenging relationships.** It was suggested earlier that support without challenge can be hollow and that challenge without support can be abrasive. Ideally, the people in the lives of clients provide a judicious mixture of support and challenge.

Harry, a man in his early 50s, was suddenly stricken with a disease that called for immediate and drastic surgery. He came through the operation quite well, even getting out of the hospital in record time. For the first few weeks he seemed, within reason, to be his old self. However, he had problems with the drugs he had to take following the operation. He became quite sick and took on many of the mannerisms of a chronic invalid. Even after the right mix of drugs was found, he persisted in invalid-like behavior. Whereas right after the operation he had "walked tall," he now began to shuffle. He also talked constantly about his symptoms and generally used his "state" to excuse himself from normal activities.

At first Harry's friends were in a quandary. They realized the seriousness of the operation and tried to put themselves in his place. They provided all sorts of support. But gradually they realized that he was adopting a style that would alienate others and keep him out of the mainstream of life. Support was essential, but it was not enough. They used a variety of ways to challenge his behavior: mocking his "invalid" movements, engaging in serious one-to-one talks, turning a deaf ear to his discussion of symptoms, and routinely including him in their plans.

Harry did not always react graciously to his friends' challenges, but in his better moments he admitted that he was fortunate to have such friends. Counselors can help clients, as they attempt to change their behavior, find people willing to provide a judicious mixture of support and challenge.

**Feedback from significant others.** Gilbert (1978, p. 175), in his book on human competence, claimed that "improved information has more potential than anything else I can think of for creating more competence in the day-to-day management of performance." Feedback is one way of providing both support and challenge. If clients are to be successful in implementing their action plans, they need adequate information about how well they are performing. Sometimes they know themselves; other times they need a more objective view. The purpose of feedback is not to pass judgment on the performance of clients but rather to provide guidance, support, and challenge. There are two kinds of feedback.

- Through *confirmatory* feedback, significant others such as helpers, relatives, friends, and colleagues let clients know that they are on course — that is, moving successfully through the steps of an action program toward a goal.
- Through *corrective* feedback, significant others let clients know that they have wandered off course and what they need to do to get back on.

Corrective feedback, whether from helpers or people in the client's everyday life, should incorporate the following principles:

- It should be given in the spirit of caring.
- It should be confirmatory, corrective, or both.

- It should be brief and to the point.
- It should focus on the client's behaviors rather than on more elusive personality characteristics.
- It should be given in moderate doses. Overwhelming the client defeats the purpose of the entire exercise.
- The client should be invited not only to comment on the feedback but also to expand on it.
- The client should be helped to discover alternative ways of doing things.

One of the main problems with feedback is finding people in the client's day-to-day life who see the client in action enough to make it meaningful, who care enough to give it, and who have the skills to provide it constructively.

**An amazing case of getting along without a helper.** As indicated earlier, many client problems are coped with and managed, not solved. Consider the following very real case of a woman who certainly did not choose not to change. Quite the contrary. Her case is a good example of a no-formula approach to developing and implementing a program for constructive change.

> Vickey readily admits that she has never fully "conquered" her illness. Some 20 years ago, she was diagnosed as manic-depressive. The picture looked something like this: She would spend about six weeks on a high; then the crash would come, and for about six weeks she'd be in the pits. After that she'd be normal for about eight weeks. This cycle meant many trips to the hospital. Some seven years into her illness, during a period in which she was in and out of the hospital, she made a decision. "I'm not going back into the hospital again. I will so manage my life that hospitalization will never be necessary." This nonnegotiable goal was her manifesto.
>
> Starting with this declaration of intent, Vickey moved on, in terms of Step II-B, to spell out what she wanted: (1) She would channel the energy of her "highs"; (2) she would consistently manage or at least endure the depression and agony of her "lows"; (3) she would not disrupt the lives of others by her behavior; (4) she would not make self-defeating decisions when either high or low. Vickey, with some help from a rather nontraditional counselor, began to do things to turn those goals into reality. She used her broad goals to provide direction for everything she did.
>
> Vickey learned as much as she could about her illness, including cues about crisis times and how to deal with both highs and lows. To manage her highs, she learned to channel her excess energy into useful — or at least nondestructive — activity. Some of her strategies for controlling her highs centered on the telephone. She knew instinctively that controlling her illness meant not just managing problems but also developing opportunities. During her free time, she would spend long hours on the phone with a host of friends, being careful not to overburden any one person. Phone marathons became part of her lifestyle. She made the point that a big phone bill was infinitely better than a stay in the hospital. She called the telephone her "safety valve." She even set up her own phone-answering business and worked very hard to make it work.

BOX 19-1

# Questions on Implementing Plans

- Now that I have a plan, how do I move into action?
- What kind of self-starter am I? How can I improve?
- What obstacles lie in my way? Which are critical?
- How can I manage these obstacles?
- How do I keep my efforts from flagging?
- What do I do when I feel like giving up?
- What kind of support will help me to keep going?

At the time of her highs, she would do whatever she had to do to tire herself out and get some sleep, for she had learned that sleep was essential if she was to stay out of the hospital. This included working longer shifts at the business. She developed a cadre of supportive people, including her husband. She took special care not to overburden him. She made occasional use of a drop-in crisis center but preferred avoiding any course of action that reminded her of the hospital.

It must be noted that the central issue in this case is the client's decision to stay out of the hospital. Her determination drove everything else. This case also exemplifies the spirit of action that ideally characterizes the implementation stage of the helping process. She did not let inertia get the best of her. She did not let entropy do her in. Here is a woman who, with occasional help from a counselor, took charge of her life. She set some simple goals and devised a set of simple strategies for accomplishing them. And she never went back into the hospital. Some will say that she was not "cured" by this process. Her goal was not to be cured but to lead as normal a life as possible in the real world. Some would say that her approach lacked elegance. It certainly did not lack results. Box 19-1 outlines the kinds of questions you can help clients ask themselves as they implement change programs.

# THE SHADOW SIDE
## OF IMPLEMENTING CHANGE

There are many reasons why clients fail to act in their own behalf. Three are discussed here: helpers who do not have an action mentality, client inertia, and client entropy.

# Helpers as Agents

Driscoll (1984, pp. 91–97) discussed the temptation of helpers to respond to the passivity of their clients with a kind of passivity of their own, a "sorry, it's up to you" stance. This, he claimed, is a mistake.

> A client who refuses to accept responsibility thereby invites the therapist to take over. In remaining passive, the therapist foils the invitation, thus forcing the client to take some initiative or to endure the silence. A passive stance is therefore a means to avoid accepting the wrong sorts of responsibility. It is generally ineffective, however, as a long-run approach. Passivity by a therapist leaves the client feeling unsupported and thus further impairs the already fragile therapeutic alliance. Troubled clients, furthermore, are not merely unwilling but generally and in important ways unable to take appropriate responsibility. A passive countermove is therefore counterproductive, for neither therapist nor client generates solutions, and both are stranded together in a muddle of entangling inactivity. (p. 91)

To help others act, helpers must be agents and doers in the helping process, not mere listeners and responders. The best helpers are active in the helping sessions. They keep looking for ways to enter the worlds of their clients, to get them to become more active in the sessions, to get them to own more of it, to help them see the need for action — action in their heads and action outside their heads in their everyday lives. And they do all this while espousing the client-centered values outlined in Chapter 3. Although they don't push reluctant clients too hard, thus turning reluctance into resistance, neither do they sit around waiting for reluctant clients to act.

# Client Inertia: Reluctance to Get Started

Inertia is the human tendency to put off problem-managing action. With respect to inertia, I often say to clients, "The action program you've come up with seems to be a sound one. The main reason that sound action programs don't work, however, is that they are never tried. Don't be surprised if you feel reluctant to act or are tempted to put off the first steps. This is quite natural. Ask yourself what you can do to get by that initial barrier." The sources of inertia are many, ranging from pure sloth to paralyzing fear. Understanding what inertia is like is easy. We need only look at our own behavior. The list of ways in which we avoid taking responsibility is endless. We'll examine several of them here: passivity, learned helplessness, disabling self-talk, and getting trapped in vicious circles.

**Passivity.**  One of the most important ingredients in the generation and perpetuation of the "psychopathology of the average" is passivity, the failure of people to take responsibility for themselves in one or more developmental areas of life or in various life situations that call for action. Passivity takes many forms: doing nothing—that is, not responding to problems and options; uncritically accepting the goals and solutions suggested by others; acting aimlessly; and becoming paralyzed or even violent—that is, shutting down or blowing up (see Schiff, 1975).

> When Zelda and Jerzy first noticed small signs that things were not going right in their relationship, they did nothing. They noticed certain incidents, mused on them for a while, and then forgot about them. They lacked the communication skills to engage each other immediately and to explore what was happening. Zelda and Jerzy had both learned to remain passive before the little crises of life, not realizing how much their passivity would ultimately contribute to their downfall. Endless unmanaged problems led to major blowups until they decided to end their marriage.

Passivity in dealing with little things can prove very costly. The little things have a way of turning into big things.

**Learned helplessness.** Seligman's (1975) concept of "learned helplessness" and its relationship to depression has received a great deal of attention since he first introduced it. Some clients learn to believe from an early age that there is nothing they can do about certain life situations. There are degrees in feelings of helplessness—from mild forms of "I'm not up to this" to feelings of total helplessness coupled with deep depression. Learned helplessness, then, is a step beyond mere passivity.

Bennett and Bennett (1984) saw the positive side of helplessness. If the problems clients face are indeed out of their control, then it is not helpful for them to have an illusory sense of control, unjustly assign themselves responsibility, and indulge in excessive expectations. Somewhat paradoxically, they found that challenging clients' tendency to blame themselves for everything actually fostered realistic hope and change. The trick is helping clients learn what is and what is not in their control. A man with a physical disability may not be able to do anything about the disability itself, but he does have some control over how he views his disability and the power to pursue certain life goals despite it. The opposite of helplessness is "learned optimism" (Seligman, 1991) and resourcefulness. If helplessness can be learned, so can resourcefulness. Indeed, increased resourcefulness is one of the principal goals of successful helping.

**Disabling self-talk.** Challenging dysfunctional self-talk on the part of clients was discussed earlier. Clients often talk themselves out of things, thus talking themselves into passivity. They say to themselves such things as "I can't do it," "I can't cope," "I don't have what it takes to engage in that program; it's too hard," and "It won't work." Such self-defeating conversations with themselves get people into trouble in the first place and then prevent them from getting out. Helpers can add great value by helping clients challenge the kind of self-talk that interferes with action.

**Vicious circles.** Pyszczynski and Greenberg (1987) developed a theory about self-defeating behavior and depression. They said that people whose actions fail to get them what they want can easily lose a sense of self-worth and become mired in a vicious circle of guilt and depression.

> Consequently, the individual falls into a pattern of virtually constant self-focus, resulting in intensified negative affect, self-derogation, further negative outcomes, and a depressive self-focusing style. Eventually, these factors lead to a negative self-image, which may take on value by providing an explanation for the individual's plight and by helping the individual avoid further disappointments. The depressive self-focusing style then maintains and exacerbates the depressive disorder. (p. 122)

It does sound depressing. One client, Amanda, fits this theory perfectly. She had aspirations of moving up the career ladder where she worked. She was very enthusiastic and dedicated, but she was unaware of the "gentleman's club" politics of the company in which she worked and didn't know how to "work the system." She kept doing the things that she thought should get her ahead. They didn't. Finally, she got down on herself, began making mistakes in the things that she usually did well, and made things worse by constantly talking about how she "was stuck," thus alienating her friends. By the time she saw a counselor, she felt defeated and depressed. She was about to give up. The counselor focused on the entire "circle" — low self-esteem producing passivity producing even lower self-esteem — and not just the self-esteem part. Instead of just trying to help her change her inner world of disabling self-talk, he also helped her intervene in her life to become a better problem solver. Small successes in problem solving led to the start of a "benign" circle — success producing greater self-esteem leading to greater efforts to succeed.

**Disorganization.** Tico lived out of his car. No one knew exactly where he spent the night. The car was chaos, and so was his life. He was always going to get his career, family relations, and love life in order, but he never did. Living in disorganization was his way of putting off life decisions. Ferguson (1987) painted a picture that may well remind us of ourselves, at least at times.

> When we saddle ourselves with innumerable little hassles and problems, they distract us from considering the possibility that we may have chosen the wrong job, the wrong profession, or the wrong mate. If we are drowning in unfinished housework, it becomes much easier to ignore the fact that we have become estranged from family life. Putting off an important project — painting a picture, writing a book, drawing up a business plan — is a way of protecting ourselves from the possibility that the result may not be quite as successful as we had hoped. Setting up our lives to insure a significant level of disorganization allows us to continue to think of ourselves as inadequate or partially-adequate people who don't have to take on the real challenges of adult behavior. (p. 46)

Many things can be behind this unwillingness to get our lives in order, like defending ourselves against a fear of succeeding.

Driscoll (1984, pp. 112–117) has provided us with a great deal of insight into this problem. He described inertia as a form of control. He says that if we tell some clients to jump into the driver's seat, they will compliantly do so—at least, until the journey gets too rough. The most effective strategy, he claimed, is to show clients that they have been in the driver's seat right along: "Our task as therapists is not to talk our clients into taking control of their lives, but to confirm the fact that they already are and always will be." That is, inertia, in the form of staying disorganized, is itself a form of control. The client is actually successful, sometimes against great odds, at remaining disorganized and thus preserving inertia.

## Entropy: The Tendency of Things to Fall Apart

Entropy is the tendency to give up action that has been initiated. Kirschenbaum (1987), in a review of the research literature, uses the term "self-regulatory failure." Programs for constructive change, even those that start strong, often dwindle and disappear. All of us have experienced the problems involved in trying to implement programs. We make plans, and they seem realistic to us. We start the steps of a program with a good deal of enthusiasm. However, we soon run into tedium, obstacles, and complications. What seemed so easy in the planning stage now seems quite difficult. We become discouraged, flounder, recover, flounder again, and finally give up, rationalizing to ourselves that we did not want to accomplish those goals anyway.

Phillips (1987, p. 650) identified what he called the "ubiquitous decay curve" in both helping and in medical-delivery situations. Attrition, noncompliance, and relapse are the name of the game. A married couple trying to reinvent their marriage might eventually say to themselves, "We had no idea that it would be so hard to change ingrained ways of interacting with each other. Is it worth the effort?" Their motivation is on the wane. Wise helpers know that the decay curve is part of life and help clients deal with it. With respect to entropy, a helper might say, "Even sound action programs begun with the best of intentions tend to fall apart over time, so don't be surprised when your initial enthusiasm seems to wane a bit. That's only natural. Rather, ask yourself what you need to do to keep yourself at the task."

Brownell and her associates (1986) provided a useful caution. They drew a fine line between preparing clients for mistakes and giving them "permission" to make mistakes by implying that they are inevitable. They also made a distinction between "lapse" and "relapse." A slip or a mistake in an action program (a lapse) need not lead to a relapse—that is, giving up the program entirely. Consider Graham, a man who has been trying to change what others see as his "angry interpersonal style." Using a variety of self-monitoring and self-control techniques, he has made great progress in changing his style. On occasion, he loses his temper, but never in any extreme way. He makes mistakes, but he does not let an occasional lapse end up in relapse.

# Choosing Not to Change

Some clients who seem to do well in analyzing problems, developing goals, and even identifying reasonable strategies and plans end up by saying — in effect, if not directly — something like this: "Even though I've explored my problems and understand why things are going wrong — that is, I understand myself and my behavior better, and I realize what I need to do to change — right now I don't want to pay the price called for by action. The price of more effective living is too high."

The question of human motivation seems almost as enigmatic now as it must have been at the dawning of the history of the human race. So often we seem to choose our own misery. Worse, we choose to stew in it rather than endure the relatively short-lived pain of behavioral change. Helpers can and should challenge clients to search for incentives and rewards for managing their lives more effectively. They should also help clients understand the consequences of not changing. But in the end it is the client's choice.

The shadow side of change stands in stark contrast to the case of Vickey. Savvy helpers are not magicians, but they understand the shadow side of change, learn to see signs of it in each individual case, and, in keeping with the values outlined in Chapter 3, do whatever they can to challenge clients to deal with the shadows.

# ❓ Evaluation Questions for the Action Arrow

How well do I do the following as I try to help this client make the transition to action?

- Understand how widespread both inertia and entropy are and how they are affecting this client
- Help clients become effective tacticians
- Help clients avoid both procrastination and imprudent action
- Help clients develop contingency plans
- Help clients discover and manage obstacles to action
- Help clients discover resources that will enable them to begin acting, to persist, and to accomplish their goals
- Help clients find the incentives and the rewards they need to persevere in action
- Help clients acquire the skills they need to act and to sustain goal-accomplishing action
- Help clients develop a social support and challenge system in their day-to-day lives
- Prepare clients to get along without a helper
- Come to grips with what kind of agent of change I am in my own life
- Face up to the fact that not every client wants to change

# PRACTICAL APPENDICES: FURTHER USEFUL MATERIAL

Since the appendices contain materials relevant to specific chapters, they are listed by chapter number. If a chapter number is missing, there is no relevant appendix material for that chapter.

## CHAPTER 1

### The Self-Help Movement

Since self-help groups are one of most popular forms of helping, at least in the United States, a few words about such groups are in order. We live in a society that has been at times somewhat suspicious of and confrontational toward professional help. Further, there is a growing realization that help must begin at home — that is, with the person with the problem. The spirit of welfare reform is in the air. Thus, the self-help movement (Gartner & Riessman, 1984; *Harvard Mental Health Letter*, 1993a, 1993b; Humphreys & Rappaport, 1994; Lavoie, Gidron, & Borkman, 1995; Riessman, 1985, 1990; Riessman & Carroll, 1995), which has always made sense, makes even more sense today as a partial answer to many social problems.

Riessman and Carroll describe the self-help paradigm as a strategic response to a world where problems are expanding, resources are shrinking, and expert solutions are wanting. Members of self-help groups, they note, are not just receivers but also providers of help. Self-help converts dependent receivers of help into active helpers. Ten million alcoholics can be seen as a resource rather than just a problem. Eleven million diabetics can monitor themselves individually and cope collectively in groups. Partnerships between professional helpers and self-help groups may increase the success rate of helping and address the financial squeeze of the helping professions that comes from managed care.

**Self-help groups.** The self-help movement has been flourishing (Katz & Bender, 1990) in the United States. Across the country, groups have been established to help people cope with almost every conceivable problem. In

1992 a study conducted by Princeton's Center for the Study of American Religion showed that there are some three million small groups meeting regularly in the United States (see Lattin, 1992). Some of the groups focus on spiritual self-help and constitute what the center's Robert Wuthnow has called a kind of "do-it-yourself" religion. Others provide support for all the problems that beset humankind. Admittedly, it is difficult to see where spiritual support ends and psychological support begins.

Some 25 years ago, Hurvitz (1970, 1974) claimed that these groups were more effective and efficient than other forms of helping: "It is likely that more people have been and are being helped by [self-help groups] than have been and are being helped by all types of professionally trained psychotherapists combined, with far less theorizing and analyzing and for much less money" (1970, p. 48). Wuthnow's study certainly provides support for that hypothesis today. The size and robustness of this movement—the average group lasts some five years—might surprise many professionals because most groups are quiet, small, and local. The wide range of self-help groups includes Alcoholics Anonymous, groups that help women cope with the aftermath of a mastectomy operation, Weight Watchers, and groups of people who are HIV-positive but have not yet developed AIDS.

As might be expected, the movement has its critics (see Riessman, 1990; Rosen, 1993). Common criticisms are that it has too "religious" a flavor, that self-help groups are themselves addictive, and that the movement is too psychologically focused and therefore distracts people from larger issues such as social criticism and needed structural changes in society.

**The self-help bias toward action.** Although research suggests that self-help groups differ greatly in their approaches to helping (McFadden, Seidman, & Rappaport, 1992), the pragmatic, action-oriented, democratic ethos of self-help groups described in an issue of the *Self-Help Reporter* (Fall, 1985, p. 5) are congruent with the values approach taken in this book:

- A noncompetitive, cooperative orientation
- An anti-elite, anti-bureaucratic focus
- An experiential emphasis—people who have the problem know a great deal about it from the "inside," from experiencing it
- A pragmatic "do what you can do," "one day at a time," and "you can't solve everything at once" approach
- A shared, often rotating leadership
- Being helped by helping others
- A refusal to see helping as a commodity to be bought and sold
- An accent on using one's own power in taking control over one's own life
- A strong optimism regarding the ability to change

- A belief that small steps are important
- A critical stance toward professionalism, which is often pretentious, purist, distant, and mystifying
- Simplicity and informality
- The centrality of helping — knowing how to receive help, give help, and help oneself
- An avoidance of self-victimization
- A focus on the group, with "getting back into community" as essential

Those who join such groups are more motivated to take an active role in changing their lives. Self-help groups also use peer pressure to get passive group members to act. In the past, professional helpers tended to ignore self-help groups — perhaps as a reaction to the movement's suspicion of professionals. Now more and more professionals are taking self-help groups seriously. Given the inexhaustability of human need, there is a place for professionals, for the self-help movement, and for partnerships between the two.

Of course, any given individual or self-help group can deviate from the self-responsibility and action focus just outlined. Some people become obsessed with their own disabilities and with the self-help group itself — another form of addiction — and that undermines the capacity for personal action (see Kaminer, 1992). On the other hand, there are many examples of individuals and groups that have used the self-help forum as a basis for social and political action (see *Self-Help Reporter,* Summer, 1992, p. 1).

# CHAPTER 2

## A Future-Centered Approach to Helping

Since helping too often gets bogged down in Stage I, here is a nonlinear, future-oriented version of the helping model. The focus is on a possibility-rich future rather than on a past that cannot be changed or a present that is problematic.

**1. Telling the story.** First, help clients tell their stories. But it may be a mistake to get the client to reveal all the details of the problem situation all at once at the beginning of the helping encounter. Sometimes a relatively quick look at the highlights of the story is enough to begin with.

Jay, a single man, age 25, tells a story of driving while drinking and of an auto accident in which he was responsible for the death of a friend as well as his own severe back injuries. He has told the story over and over again to many different people, as if telling it enough will make what happened go away. When Jay met the counselor, he had already told his story a number of times to different people. Retelling the whole thing in great detail was not going to help. A summary was enough.

**2. Possibilities for a better future: A sense of hope.** Help clients take an initial cut at exploring what they want. Describing what they want—preferred-scenario possibilities—can help clients develop a sense of hope. They identify a better future early in the helping process, and that future pervades the rest of the process.

Jay outlines a range of possibilities. He outlines a future in which he is more responsible for his actions, he is not hounded by regret and guilt, he has made peace with the family of his friend, he is physically rehabilitated within the limits of his injuries, and he is coping with the physical and psychological consequences of the accident. This is not the time to set clear-cut behavioral goals: The client is being helped to outline a picture, not draft a blueprint. The purpose of this quick move to the preferred scenario is to help the client focus on a better future rather than a problematic and unchangeable past.

**3. Return to Stage I.** Much of the work of Stage I can be done more effectively in the light of possibilities for a better future.

- It is easier for clients to spot the real issues and to develop them in more detail. They also see more clearly the gaps between where they are and where they want to be.

- When clients outline what they really want, it is easier for them to identify and manage blind spots and develop new perspectives.

- When clients begin to sketch possibilities for a better future, it is easier for them to see which problems and which unused opportunities have leverage for them. They know better what to work on.

Even a brief visit to Stage II—what things would look like if the problem situation were being managed a bit better—can act like a spotlight that illuminates the client's story and gives it a different cast. The current scenario is now seen as a gap or a set of gaps. A gap is not the same as a problem. A gap implies the need for action to bridge the gap.

In our example, Jay is helped to explore some of the gaps between his current and his preferred scenarios and to develop new perspectives. For instance, he comes to see his overpowering regret and guilt as partially self-inflicted. When he engages in self-recrimination, he is not in the right frame of mind either to make peace with the family of his friend or to get seriously involved in a physical rehabilitation program. Managing the current flood of self-defeating emotions is a point of leverage. Until now he has not seen this constant self-chastisement as actually standing in the way of what he wants. He has unwittingly decided that self-chastisement is "good" and that more is even better. At this point the exploration of the problem situation is no longer an end in itself but a tool to be used to fashion the future.

**4. Return to Stage II.** Help clients use what they have learned in their return to Stage I to complete the process of determining what they want and what they are willing to do to get what they want.

Through a dialogue between what is and what could be, Jay begins to formulate a set of goals for himself. He cannot change the past, but he can learn from it. The energy he has been putting into self-recrimination can be put into self-reformation. He begins to explore more specifically what a self-responsible Jay would look like: not a person hounded by guilt, but not the irresponsible "free spirit" that lives on the edge of disaster, either. He sees a person whose energy is not quashed but channeled. These ideas begin to be translated into specific goals — for instance, the goals of the physical rehabilitation program he is involved in.

**5. *Moving on to Stage III: Strategies, plans, and actions.*** Help clients formulate and implement strategies and plans to accomplish goals.

Jay is helped to develop and implement strategies for controlling the flood of emotions he has let paralyze him. He uses his guilt as a stimulus to develop a way of meeting and making peace with the family of his friend. As he develops new perspectives, he begins to formulate strategies for making his emotions his allies instead of his enemies.

The purpose of this brief example is to illustrate a different, more upbeat movement in the helping process, what might be called a Stage II – centered approach to helping. It demonstrates that the helping process, at its best, is not a slavish, linear working through of the three stages.

Listen to the marriage counselor friend of mine who has adopted this "back and forth" approach to his work with couples:

> When a couple comes, they don't need to spend a great deal of time talking about what is going wrong. They know what's going wrong because they're living it every day. They are only too willing to tell their stories in great detail, but they tell entirely different stories. It's as if they were not even married to each other. When telling their stories, they begin to fight the way they fight at home. I don't find this very useful. And so, early on I make them focus on the future. I ask them questions like, "What do you want, individually and together? What would your marriage look like if it were a little bit better, not perfect, but just a little bit better?" I have them brainstorm separately and then share the possibilities. Instead of fighting, they are actually working together. If they get stuck, I ask them to think about other couples they know, couples that seem to have decent, if not perfect, marriages. "What's their marriage like? What parts of it would you like to see incorporated into your own marriage?" What I try to do is to help them use their imaginations to create a vision of a marriage that is different from the one they are experiencing. It's not that I never let them talk about the present. Rather, I try to help them talk about the present by having them engage in a dialogue between a better future and the unacceptable present.

This is not a quick lesson in how to do marriage counseling, but it does give the flavor of a more future-oriented approach.

# Using the Model as a "Browser": The Search for Best Practice

The claim in this book is this: Problem management and opportunity development is one of the principal processes—perhaps *the* principal process—underlying all successful counseling and psychotherapy.

**A bewildering array of models and methods.** Even though there is only a handful of "major brands" of psychotherapy—psychoanalytic, Adlerian, existential, person-centered, Gestalt, interpersonal, behavioral, cognitive, systemic, culture-sensitive, and integrative (see Corey, 1996; Prochaska & Norcrosss, 1994)—there are literally hundreds of different approaches. And each year both new approaches and variations of older approaches are added. In sum, novice helpers are faced with a bewildering number of schools, systems, methods, and techniques, all of which are proposed with equal seriousness and all of which claim to lead to success.

**Eclecticism.** Furthermore, many advanced helpers, even when they choose one specific school or approach to helping, often borrow methods and techniques from other approaches. Other helpers, without declaring allegiance to any particular school, stitch together their own approaches to helping. This borrowing and stitching is called "eclecticism" (Jensen, Bergin, & Greaves, 1990; Lazarus, Beutler, & Norcross, 1992; Prochaska & Norcross, 1994). In one study, some 40% of helpers said that eclecticism was their primary approach to helping (Milan, Montgomery, & Rogers, 1994). Effective eclecticism, however, must be more than a random borrowing of ideas and techniques from here and there. There must be some integrating framework to give coherence to the entire process; that is, to be effective, eclecticism must be systematic.

**Problem-management as underlying process.** However, when any school, model, or eclectic mixture is successful, it is so precisely because it helps clients (1) identify and explore problem situations and unused opportunities, (2) determine what they need and want, (3) discover ways of getting what they need and want, and (4) translate their learnings into problem-managing action. *That is, the problem-management and opportunity-development process outlined earlier underlies or is embedded in all approaches to helping,* since all approaches deal with *constructive client change.* Therefore, the model outlined in these pages together with the basic skills and techniques that make it work is a very useful starting point for novices no matter what school or approach or eclectic system they may ultimately choose or devise.

**The "browser" approach.** The helping model in this book can also be used as a tool — a "browser," to use an Internet term — for mining, organizing, and evaluating concepts and techniques that work, no matter what their origin.

1. **Mining.** First, helpers can use the problem-management model to mine any given school or approach, "digging out" whatever is useful without having to accept everything else that is offered. The stages and steps of the model serve as tools for identifying methods and techniques that will serve the needs of clients.

2. **Organizing.** Second, since the problem-management model is organized by stages and steps, it can be used to organize the methods and techniques that have been mined from the rich literature on helping. For instance, a number of contemporary therapies have elaborated excellent techniques for helping clients identify blind spots and develop new perspectives on the problem situations they face. As we shall see, these techniques can be organized in Step I-B, the "new perspectives" step of the problem-management model.

3. **Evaluating.** Since the problem-management model is pragmatic and focuses on outcomes of helping, it can be used to evaluate the vast number of helping techniques that are constantly being devised. The model enables helpers to ask in what way a technique or method contributes to the "bottom line" — that is, to outcomes that serve the needs of clients.

The problem-management and opportunity-development model can serve this function because it is an open-systems model, not a closed school. Although it takes a stand on how counselors may help their clients, it is open to being corroborated, complemented, and challenged by any other approach, model, or school of helping. The needs of clients, not the egos of model builders, must remain central to the helping process. Our clients deserve "best practice," whatever its source.

# Becoming a Skilled Helper:
## The Basic Steps of the Training Process

As the title of this book suggests, helpers-to-be need not just a helping model but also a range of techniques and skills to make the model work. They include basic and advanced communication skills, establishing working relationships with clients, helping clients challenge themselves, problem clarification, goal setting, development of an action plan, the implementation of the plan, and ongoing evaluation. These are creative and humane ways of being with clients in their efforts to develop a program of constructive change. The only way to acquire these skills is by learning them experientially, practicing them, and using them until they become second nature. The following are standard steps in a skills-training program.

**1. Cognitive understanding.** Develop a cognitive understanding of a particular helping method or the skill of delivering it. You can do this by reading the text and listening to lectures.

**2. Clarification.** Clarify what you have read or heard. This can be done through instructor-led questioning and discussion.

*The desired outcome of steps 1 and 2 is* **cognitive clarity.**

**3. Modeling.** Watch experienced instructors model the skill or method in question. This can be done "live" or through films and videotapes.

**4. Written exercises.** Do the exercises in *Exericses in Helping Skills*, a manual that accompanies this book, that are related to the skill or method you are learning. The purpose of this initial use of the method or skill is to demonstrate to yourself that you understand the helping method or skill enough to begin to practice it. The exercises in that manual are a way of practicing the skills and methods "in private" before practicing them with your fellow trainees. They provide a behavioral link between the introduction to a skill or method that takes place in the first four steps of this training format and actual practice in a group.

*The desired outcome of steps 3 and 4 is* **behavioral clarity.**

**5. Practice.** Move into smaller groups to practice the skill or method in question with your fellow trainees.

**6. Feedback.** During these practice sessions evaluate your own performance and get feedback from a trainer and from your fellow trainees. This feedback serves to confirm what you are doing right and to correct what you are doing wrong. The use of video to provide feedback is also helpful. Steps 3, 4, 5, and 6 constitute the heart of experiential learning.

*The desired outcome of steps 5 and 6 is* **initial competence**
*in using the model and in the skills that make it work.*

**7. Evaluating the learning experience.** From time to time stop and reflect on the training process itself. Take the opportunity to express how you feel about the learning program and how you feel about your own progress. Whereas steps 1 through 6 deal with the task of learning the helping model and the methods and skills that make it work, step 7 deals with group maintenance — that is, managing the needs of individual trainees. Doing this kind of group maintenance work helps establish a learning community.

**8. Supervised practice with actual clients.** Finally, when it is deemed that you are ready, apply what you have learned to actual clients. Supervision is an extremely important part of the learning process. Indeed, effective helpers never stop learning about themselves, their clients, and the helping process itself.

*The desired outcome of steps 7 and 8 is **initial mastery.***

The program in which you are enrolled may cover only a few of these steps. Comprehensive programs for training professional helpers, however, should eventually include all steps. Of course, the best helpers continue to learn throughout their careers — from their own experiences, from the experiences of their colleagues, from reading, from seminars, from formal and informal research, and, perhaps most important, from their interactions with their clients. Full mastery, therefore, is a journey rather than a destination.

# CHAPTER 15
## The Balance-Sheet Methodology

Since choosing goals and strategies to accomplish them often requires choosing from among a number of alternatives, counselors need ways of helping clients do so. Decision making, as we have already seen, can be a messy process. Part of the messiness of decision making is the failure on the part of clients to consider the consequences of the decisions they make during the helping process. For instance, a counselor who helped the homeless found that some of her clients rejected out of hand the notion of any kind of shelter. They prized their freedom. However, some of them did not weigh the consequences of their decision. This counselor challenged them to do so. The balance sheet, illustrated in Figure A-1, is a decision-making aid counselors can use to help clients through difficult choices (see The Shadow Side of Decision Making in Chapter 12). It is a way of helping clients examine the consequences of their choices with regards to both utility — "Will this goal help me move beyond the problems that I am currently struggling with?" — and acceptability — "Can I live with this choice?" Ambivalence — "Should I or shouldn't I" — often paralyzes clients. The balance sheet often helps clients move beyond ambivalence.

The balance sheet can be used to help clients choose both goals and strategies for accomplishing goals. As you can readily see, the balance sheet is rather thorough. Therefore, it need never be used in its entirety, nor should it be used to help clients evaluate every decision they make. As with other helping methods, clients should be helped to use it only when it will add value.

| If I choose this course of action: | | |
|---|---|---|
| **The self** | | |
| Gains for self: | Acceptable to me because: | Not acceptable to me because: |
| Losses for self: | Acceptable to me because: | Not acceptable to me because: |
| **Significant others** | | |
| Gains for significant others: | Acceptable to me because: | Not acceptable to me because: |
| Losses for significant others: | Acceptable to me because: | Not acceptable to me because: |
| **Social setting** | | |
| Gains for social setting: | Acceptable to me because: | Not acceptable to me because: |
| Losses for social setting: | Acceptable to me because: | Not acceptable to me because: |

FIGURE A-1
The Decision Balance Sheet

# REFERENCES

ABRAMSON, P. R., Cloud, M. Y., Keese, N., & Keese, R. (1994, Spring). How much is too much? Dependency in a psychotherapeutic relationship. *American Journal of Psychotherapy*, 48(2), 294–301.

ACKOFF, R. (1974). *Redesigning the future*. New York: Wiley.

ALBEE, G. W., & K. D. Ryan-Finn. (1993, November/December). An overview of primary prevention. *Journal of Counseling and Development*, 115–123.

ALLISON, K. W., Echemendia, R. J., Crawford, I., & LaVome Robinson, W. (1996). Predicting cultural competence: Implications for practice and training. *Professional Psychology: Research and Practice*, 27(4), 386–393.

AMERICAN PSYCHOLOGIST. (1992, December). Ethical principles of psychologists and code of conduct.

AMERICAN PSYCHOLOGIST. (1996, October). Special Issue: Outcome assessment of psychotherapy.

ANDERSON, T., & Leitner, L. M. (1996). Symptomatology and the use of affect constructs to influence value and behavior constructs. *Journal of Counseling Psychology*, 43(1), 77–83.

ARGYRIS, C. (1982). *Reasoning, learning, and action*. San Francisco: Jossey-Bass.

ARNKOFF, D. B. (1995). Two examples of strains in the therapeutic alliance in an integrative cognitive therapy. *Psychotherapy in Practice*, 1(1), 33–46.

ATKINSON, D. R., Worthington, R. L., Dana, D. M., & Good, G. E. (1991). Etiology beliefs, preferences for counseling orientations, and counseling effectiveness. *Journal of Counseling Psychology*, 38(3), 258–264.

BAILEY, K. G., Wood, H. E., & Nava, G. R. (1992). What do clients want? Role of psychological kinship in professional helping. *Journal of Psychotherapy Integration*, 2(2), 125–147.

BALDWIN, B. A. (1980). Styles of crisis intervention: Toward a convergent model. *Journal of Professional Psychology*, 11, 113–120.

BANDURA, A. (1977). Self-efficacy: Toward a unifying theory of behavioral change. *Psychological Review*, 84, 191–215.

BANDURA, A. (1980). Gauging the relationship between self-efficacy judgment and action. *Cognitive Therapy and Research*, 4, 263–268.

BANDURA, A. (1982). Self-efficacy mechanism in human agency. *American Psychologist*, 37, 122–147.

BANDURA, A. (1986). *Social foundations of thought and action: A social cognitive theory*. Englewood Cliffs, NJ: Prentice Hall.

BANDURA, A. (1989). Human agency in social cognitive theory. *American Psychologist*, 44, 1175–1184.

BANDURA, A. (1990). Foreword to E. A. Locke, & G. P. Latham (1990), *A theory of goal setting and task performance*. Englewood Cliffs, NJ: Prentice Hall.

BANDURA, A. (1991). Human agency: The rhetoric and the reality. *American Psychologist, 46,* 157–161.

BANKOFF, E. A. (1994). The social network of the psychotherapy patient. *Psychotherapy, 31*(3), 503–514.

Basic Behavioral Science Task Force of the National Advisory Mental Health Council (1996). Basic behavioral science research for mental health — Family processes and social networks. *American Psychologist, 51*(6), 622–630.

BEIER, E. G., & Young, D. M. (1984). *The silent language of psychotherapy: Social reinforcement of unconscious processes* (2nd ed.). New York: Aldine.

BENBENISHTY, R., & Schul, Y. (1987). Client-therapist congruence of expectations over the course of therapy. *British Journal of Clinical Psychology, 26*(1), 17–24.

BENNETT, M. I., & Bennett, M. B. (1984). The uses of hopelessness. *American Journal of Psychiatry, 141,* 559–562.

BERENSON, B. G., & Mitchell, K. M. (1974). *Confrontation: For better or worse.* Amherst, MA: Human Resource Development Press.

BERGER, D. M. (1989). Developing the story in psychotherapy. *American Journal of Psychotherapy, 43*(2), 248–259.

BERGIN, A. E. (1991). Values and religious issues in psychotherapy and mental health. *American Psychologist, 46*(4), 394–403.

BERGIN, A. E., & Garfield, S. L. (1994). *Handbook of psychotherapy and behavior change* (4th ed.). New York: Wiley.

BERNARD, M. E. (Ed.). (1991). *Using rational-emotive therapy effectively.* New York: Plenum.

BERNARD, M. E., & DiGiuseppe, R. (Eds.). (1989). *Inside rational-emotive therapy.* San Diego: Academic Press.

BERNE, E. (1964). *Games people play.* New York: Grove Press.

BERNHEIM, K. F. (1989). Psychologists and the families of the severely mentally ill: The role of family consultation. *American Psychologist, 44,* 561–564.

BERNSTEIN, R. (1994). *Dictatorship of virtue: Multiculturalism and the battle for America's future.* New York: Knopf.

BERSOFF, D. N. (1995). *Ethical conflicts in psychology.* Hyattsville, MD: American Psychological Association.

BEUTLER, L. E., & Bergan, J. (1991). Values change in counseling and psychotherapy: A search for scientific credibility. *Journal of Counseling Psychology, 38*(1), 16–24.

BINDER, C. (1990, September). Closing the confidence gap. *Training,* 49–56.

BISCHOFF, M. M., & Tracey, T. J. G. (1995). Client resistance as predicted by therapist behavior: A study of sequential dependence. *Journal of Counseling Psychology, 42*(4), 487–495.

BORDIN, E. S. (1979). The generalizability of the psychoanalytic concept of the working alliance. *Psychotherapy: Theory, Research and Practice, 16,* 252–260.

BORGEN, F. H. (1992). Expanding scientific paradigms in counseling psychology. In S. D. Brown & R. W. Lent (Eds.), *Handbook of counseling psychology.* New York: Wiley.

BRAMMER, L. (1973). *The helping relationship: Process and skills.* Englewood Cliffs, NJ: Prentice Hall.

BRONFENBRENNER, U. (1977). Toward an experimental ecology of human development. *American Psychologist,* 513–531.

BROSKOWSKI, A. T. (1995). The evolution of health care. *Professional Psychology: Research and Practice, 26,* 156–162.

BROWNELL, K. D., Marlatt, G. A., Lichtenstein, E., & Wilson, G. T. (1986). Understanding and preventing relapse. *American Psychologist, 41,* 765–782.

BRYANT, B. K. (1994). *Counseling for radial understanding.* Alexandria, VA: American Counseling Association.

CANTER, M. B., Bennett, B. E., Jones, S. E., & Nagy, T. F. (1994). *Ethics for psychologists: A commentary on the APA ethics code.* Hyattsville, MD: American Psychological Association.

CARKHUFF, R. R. (1969a). *Helping and human relations: Vol. 1. Selection and training.* New York: Holt, Rinehart & Winston.

CARKHUFF, R. R. (1969b). *Helping and human relations: Vol. 2. Practice and research.* New York: Holt, Rinehart & Winston.

CARKHUFF, R. R. (1971). Training as a preferred mode of treatment. *Journal of Counseling Psychology, 18,* 123–131.

CARKHUFF, R. R. (1987). *The art of helping* (6th ed.). Amherst, MA: Human Resource Development Press.

CARKHUFF, R. R., & Anthony, W. A. (1979). *The skills of helping: An introduction to counseling.* Amherst, MA: Human Resource Development Press.

CARTER, J. A. (1996). Measuring transference: Can we identify what we have not defined? *Journal of Counseling Psychology, 43*(3), 257–258.

CERVONE, D., and Scott, W. D. (1995). Self-efficacy theory and behavioral change: Foundations, conceptual issues, and therapeutic implications. In W. O'Donohue & L. Krasner (Eds.), *Theories of behavior therapy: Exploring behavior change* (pp. 349–383). Washington, DC: American Psychological Association.

CHANG, E. C. (1996). Cultural differences in optimism, pessimism, and coping: Predictors of subsequent adjustment in Asian American and Caucasian American college students. *Journal of Counseling Psychology, 43*(1), 113–123.

CLAIBORN, C. D. (1982). Interpretation and change in counseling. *Journal of Counseling Psychology, 29,* 439–453.

CLAIBORN, C. D., Berberoglu, L. S., Nerison, R. M., & Somberg, D. R. (1994). The client's perspective: Ethical judgments and perceptions of therapist practices. *Professional Psychology: Research and Practice, 25*(3), 268–274.

CLARK, A. J. (1991). The identification and modification of defense mechanisms in counseling. *Journal of Counseling and Development, 69,* 231–236.

COLE, H. P., & Sarnoff, D. (1980). Creativity and counseling. *Personnel and Guidance Journal, 59,* 140–146.

COLE, J. D., Watt, N. F., West, S. G., Hawkins, J. D., Asarnow, J. R., Markman, H. J., Ramey, S. L., Shure, M. B., & Long, B. (1993). The science of prevention: A conceptual framework and some directions for a national research program. *American Psychologist, 48,* 1013–1022.

CONOLEY, C. W., Padula, M. A., Payton, D. S., & Daniels, J. A. (1994). Predictors of client implementation of counselor recommendations: Match with problem, difficulty level, and building on client strengths. *Journal of Counseling Psychology, 41*(1), 3–7.

CONSUMER REPORTS. (1994). Annual questionnaire.

CONSUMER REPORTS. (1995, November). Mental health: Does therapy help? 734–739.

COREY, G. (1996). *Theory and practice of counseling and psychotherapy* (5th ed.). Pacific Grove, CA: Brooks/Cole.

COREY, G., Corey, M. S., & Callanan, P. (1993). *Issues and ethics in the helping professions* (4th ed.). Pacific Grove, CA: Brooks/Cole. Chapter 2, "The Counselor as a Person and as a Professional," includes questions and discussion of issues arising from demands of the counseling profession and needs of its practitioners.

CORRIGAN, P. W. (1995). Wanted: Champions of psychiatric rehabilitation. *American Psychologist, 50*(7), 514–521.

COSIER, R. A., & Schwenk, C. R. (1990). Agreement and thinking alike: Ingredients for poor decisions. *Academy of Management Executive, 4*(1), 69–74.

COVEY, S. R. (1989). *The seven habits of highly effective people*. New York: Simon & Schuster (Fireside edition, 1990).

COWEN, E. L. (1982). Help is where you find it. *American Psychologist, 37,* 385–395.

CROSS, J. G., & Guyer, M. J. (1980). *Social traps*. Ann Arbor: University of Michigan Press.

CUMMINGS, N. A. (1979). Turning bread into stones: Our modern antimiracle. *American Psychologist, 34,* 1119–1129.

DAS, A. K. (1995, Fall). Rethinking multicultural counseling: Implications for counselor education. *Journal of Counseling and Development, 74,* 45–52.

DAUSER, P. J., Hedstrom, S. M., & Croteau, J. M. (1995). Effects of disclosure of comprehensive pretherapy information on clients at a university counseling center. *Professional Psychology: Research and Practice, 26*(2), 190–195.

DE BONO, E. (1992). *Serious creativity: Using the power of lateral thinking to create new ideas*. New York: Harper Business.

DEFFENBACHER, J. L., Oetting, E. R., Huff, M. E., & Thwaites, G. A. (1995). Fifteen-month follow-up of social skills and cognitive-relaxation approaches to general anger reduction. *Journal of Counseling Psychology, 42*(3), 400–405.

DEFFENBACHER, J. L., Thwaites, G. A., Wallace, T. L., & Oetting, E. R. (1994). Social skills and cognitive-relaxation approaches to general anger reduction. *Journal of Counseling Psychology, 41,* 386–396.

DEUTSCH, M. (1954). Field theory in social psychology. In G. Lindzey (Ed.), *The handbook of social psychology* (Vol. 1). Cambridge, MA: Addison-Wesley.

DIMOND, R. E., Havens, R. A., & Jones, A. C. (1978). A conceptual framework for the practice of prescriptive eclecticism in psychotherapy. *American Psychologist, 33,* 239–248.

DORN, F. J. (1984). *Counseling as applied social psychology: An introduction to the social influence model*. Springfield, IL: Chas. C Thomas.

DORN, F. J. (Ed.). (1986). *The social influence process in counseling and psychotherapy*. Springfield, IL: Chas. C Thomas.

DORNER, D. (1996). *The logic of failure: Why things go wrong and what we can do to make them right*. New York: Holt.

DRISCOLL, R. (1984). *Pragmatic psychotherapy*. New York: Van Nostrand Reinhold.

DUAN, C., & Hill, C. E. (1996). The current state of empathy research. *Journal of Counseling Psychology, 43*(3), 261–274.

DURLAK, J. A. (1979). Comparative effectiveness of paraprofessional and professional helpers. *Psychological Bulletin, 86,* 80–92.

DWORKIN, S. H., & Kerr, B. A. (1987). Comparison of interventions for women experiencing body image problems. *Journal of Counseling Psychology, 34*(2), 136–140.

EDWARDS, C. E., & Murdock, N. L. (1994). Characteristics of therapist self-disclosure in the counseling process. *Journal of Counseling and Development, 72,* 384–389.

EGAN, G. (1970). *Encounter: Group processes for interpersonal growth*. Pacific Grove, CA: Brooks/Cole.

EISENBERG, W., & Strayer, J. (Eds.). (1987). *Empathy and its development*. New York: Cambridge University Press.

EKMAN, P. (1992). *Telling lies: Clues to deceit in the marketplace, politics, and marriage*. New York: Norton.

EKMAN, P. (1993). Facial expression and emotion. *American Psychologist, 48,* 384–392.

EKMAN, P., & Davidson, R. J. (1994). *The nature of emotion*. New York: Oxford University Press.

ELIAS, M. J., & Clabby, J. F. (1992). Building social problem-solving skills. San Francisco: Jossey-Bass.

ELLIOTT, R. (1985). Helpful and nonhelpful events in brief counseling interviews: An empirical taxonomy. *Journal of Counseling Psychology, 32,* 307–322.

ELLIS, A. (1984). Must most psychotherapists remain as incompetent as they are now? In J. Hariman (Ed.), *Does psychotherapy really help people?* Springfield, IL.: Chas. C Thomas.

ELLIS, A. (1985). *Overcoming resistance: Rational-emotive therapy with difficult clients*. New York: Springer.

ELLIS, A. (1987a). The evolution of rational-emotive therapy (RET) and cognitive behavior therapy (CBT). In J. K. Zeig (Ed.), *The evolution of psychotherapy*. New York: Brunner/Mazel.

ELLIS, A. (1987b). Integrative developments in rational-emotive therapy (RET). *Journal of Integrative and Eclectic Psychotherapy, 6,* 470–479.

ELLIS, A. (1991). The revised ABCs of rational-emotive therapy (RET). *Journal of Rational-Emotive and Cognitive-Behavior Therapy, 9,* 139–172.

ELLIS, A., & DRYDEN, W. (1987). *The practice of rational-emotive therapy*. New York: Springer.

ELSON, M. L., & Neufeldt, S. A. (1996, Summer). Building on an empirical foundation: Strategies to enhance good practice. *Journal of Counseling and Development, 74,* 609–615.

ETZIONI, A. (1989, July-August). Humble decision making. *Harvard Business Review,* 120–126.

EYSENCK, H. J. (1984). The battle over psychotherapeutic effectiveness. In J. Hariman (Ed.), *Does psychotherapy really help people?* Springfield, IL: Chas. C Thomas.

EYSENCK, H. J. (1994). The outcome problem in psychotherapy: What have we learned? *Behavior Research and Therapy, 32*(5), 477–495.

FARRELLY, F., & Brandsma, J. (1974). *Provocative therapy*. Cupertino, CA: Meta Publications.

FELLER, R. (1984). *Job-search agreements*. Monolith, Colorado State University, Fort Collins.

FERGUSON, M. (1980). *The aquarian conspiracy: Personal and social transformation in the 1980s*. Los Angeles: J. P. Tarcher.

FERGUSON, T. (1987, January-February). Agreements with yourself. *Medical Self-Care,* 44–47.

FESTINGER, S. (1957). *A theory of cognitive dissonance*. New York: Harper & Row.

FISH, J. M. (1995, Spring). Does problem behavior just happen? Does it matter? *Behavior and Social Issues, 5*(1), 3–12.

FISHER, R., & Ury, W. (1981). *Getting to yes: Negotiating agreement without giving in*. Boston: Houghton Mifflin.

FOWERS, B. J., & Richardson, F. C. (1996, Summer). Why is multiculturalism good? *American Psychologist, 51*(6), 609–621.

FOX, R. E. (1995). The rape of psychotherapy. *Professional Psychology: Research and Practice, 26,* 147–155.

FRANCES, A., Clarkin, J., & Perry, S. (1984). *Differential therapeutics in psychiatry.* New York: Brunner/Mazel.

FRASER, J. S. (1996). All that glitters is not always gold: Medical offset effects and managed behavioral health care. *Professional Psychology: Research and Practice, 27,* 335–344.

FREIRE, P. (1970). *Pedagogy of the oppressed.* New York: Seabury.

FREMONT, S. K., & Anderson, W. (1986). What client behaviors make counselors angry? An exploratory study. *Journal of Counseling and Development, 65,* 67–70.

FRIEDLANDER, M. L., & Schwartz, G. S. (1985). Toward a theory of strategic self-presentation in counseling and psychotherapy. *Journal of Counseling Psychology, 32,* 483–501.

GALASSI, J. P., & Bruch, M. A. (1992). Counseling with social interaction problems: Assertion and social anxiety. In S. D. Brown & R. W. Lent (Eds.), *Handbook of counseling psychology.* New York: Wiley.

GARTNER, A., & Riessman, F. (Eds.). (1984). *The self-help revolution.* New York: Human Sciences Press.

GASTON, L., Goldfried, M. R., Greenberg, L. S., Horvath, A. O., Raue, P. J., & Watson, J. (1995). The therapeutic alliance in psychodynamic, cognitive-behavioral, and experimental therapies. *Journal of Psychotherapy Integration, 5*(1), 1–26.

GATI, I., Krausz, M., & Osipow, S. H. (1996). A taxonomy of difficulties in career decision making. *Journal of Counseling Psychology, 43*(4), 510–526.

GELATT, H. B. (1989). Positive uncertainty: A new decision-making framework for counseling. *Journal of Counseling Psychology, 36,* 252–256.

GELATT, H. B., Varenhorst, B., & Carey, R. (1972). *Deciding: A leader's guide.* Princeton, NJ: College Entrance Examination Board.

GELSO, C. J., & Carter, J. A. (1994). Components of the psychotherapy relationship: Their interaction and unfolding during treatment. *Journal of Counseling Psychology, 41*(3), 296–306.

GEORGES, J. C. (1988, April). Why soft-skills training doesn't take. *Training,* 40–47.

GIBSON, W. T., & Pope, K. S. (1993). The ethics of counseling: A national survey of certified counselors. *Journal of Counseling and Development, 71,* 330–336.

GILBERT, T. F. (1978). *Human competence: Engineering worthy performance.* New York: McGraw-Hill.

GOODCHILDS, J. D. (1991). *Psychological perspectives on human diversity in America.* Washington, DC: American Psychological Association.

GOODYEAR, R. K., & Shumate, J. L. (1996). Perceived effects of therapist self-disclosure of attraction to clients. *Professional Psychology: Research and Practice, 27*(6), 613–616.

GOSLIN, D. A. (1985). Decision making and the social fabric. *Society, 22*(2), 7–11.

GRACE, M., Kivlighan, D. M., Jr., & Kunce, J. (1995, Summer). The effect of non-verbal skills training on counselor trainee nonverbal sensitivity and responsiveness and on session impact and working alliance ratings. *Journal of Counseling and Development, 73,* 547–552.

GREENBERG, L. S. (1986). Change process research. *Journal of Consulting and Clinical Psychology, 54,* 4–9.

GREENBERG, L. S. (1994). What is "real" in the relationship? Comment on Gelso and Carter (1994). *Journal of Counseling Psychology, 41*(3), 307–309.

GREENSON, R. R. (1967). *The technique and practice of psychoanalysis.* New York: International Universities Press.

GUISINGER, S., & Blatt, S. J. (1994). Individuality and relatedness: Evolution of a fundamental dialectic. *American Psychologist, 49*(2), 104–111.

HALEY, J. (1976). *Problem solving therapy.* San Francisco: Jossey-Bass.

HALL, E. T. (1977). *Beyond culture.* Garden City, NJ: Anchor Press.

HALLECK, S. L. (1988). Which patients are responsible for their illnesses? *American Journal of Psychotherapy, 42,* 338–353.

HANDELSMAN, M. M., & Galvin, M. D. (1988). Facilitating informed consent for outpatient psychotherapy: A suggested written format. *Professional Psychology: Research and Practice, 19*(2), 223–225.

HANNA, F. J. (1994). A dialectic of experience: A radical empiricist approach to conflicting theories in psychotherapy. *Psychotherapy, 31,* 124–136.

HANNA, F. J., & Ottens, A. J. (1995). The role of wisdom in psychotherapy. *Journal of Psychotherapy Integration, 5,* 195–219.

HARE-MUSTIN, R., & Marecek, J. (1986). Autonomy and gender: Some questions for therapists. *Psychotherapy, 23,* 205–212.

*HARVARD MENTAL HEALTH LETTER.* (1993a, March). Self-help groups — Part I.

*HARVARD MENTAL HEALTH LETTER.* (1993b, April). Self-help groups — Part II.

HATTIE, J. A., Sharpley, C. E., & Rogers, H. J. (1984). Comparative effectiveness of professional and paraprofessional helpers. *Psychological Bulletin, 95,* 534–541.

HAYS, P. A. (1996, Spring). Addressing the complexities of culture and gender in counseling. *Journal of Counseling and Development, 74,* 332–338.

HEADLEE, R., & Kalogjera, I. J. (1988). The psychotherapy of choice. *American Journal of Psychotherapy, 42*(4), 532–542.

HEINSSEN, R. K. (1994, June). *Therapeutic contracting with schizophrenic patients: A collaborative approach to cognitive-behavioral treatment.* Paper presented at the 21st International Symposium for the Psychotherapy of Schizophrenia, Washington, DC.

HEINSSEN, R. K., Levendusky, P. G., & Hunter, R. H. (1995). Client as colleague: Therapeutic contracting with the seriously mentally ill. *American Psychologist, 50*(7), 522–532.

HELLER, K. (1993, November/December). Prevention activities for older adults: Social structures and personal competencies that maintain useful social roles. *Journal of Counseling and Development,* 124–130.

HELMS, J. E. (1994). How multiculturalism obscures racial factors in the therapy process: Comment on Ridley et al. (1994), Sodowsky et al. (1994), Ottavi et al. (1994), and Thompson et al. (1994). *Journal of Counseling Psychology, 41*(2), 162–165.

HENDRICK, S. S. (1988). Counselor self-disclosure. *Journal of Counseling and Development, 66,* 419–424.

HENDRICK, S. S. (1990). A client perspective on counselor disclosure. *Journal of Counseling and Development, 69,* 184.

HENGGELER, S. W., Schoenwald, S. K., & Pickrel, S. G. (1995, Fall). Multisystemic therapy: Bridging the gap between university and community-based treatment. *Journal of Consulting and Clinical Psychology, 63*(5), 709–717.

HEPPNER, P. P. (1989). Identifying the complexities within clients' thinking and decision making. *Journal of Counseling Psychology, 36,* 257–259.

HEPPNER, P. P., & Claiborn, C. D. (1989). Social influence research in counseling: A review and critique (Monograph). *Journal of Counseling Psychology, 36,* 365–387.

HEPPNER, P. P., & Frazier, P. A. (1992). Social psychological processes in psychotherapy: Extrapolating basic research to counseling psychology. In S. D. Brown & R. W. Lent (Eds.), *Handbook of counseling psychology.* New York: Wiley.

HIGHLEN, P. S., & Hill, C. E. (1984). Factors affecting client change in counseling. In S. D. Brown & R. W. Lent (Eds.), *Handbook of counseling psychology* (pp. 334–396). New York: Wiley.

HILL, C. E. (1994). What is the therapeutic relationship? A reaction to Sexton and Whiston. *The Counseling Psychologist, 22*(1), 90–97.

HILL, C. E., Nutt-Williams, E., Heaton, K. J., Thompson, B. J., & Rhodes, R. H. (1996). Therapist retrospective recall of impasses in long-term psychotherapy: A qualitative analysis. *Journal of Counseling Psychology, 43*(2), 207–217.

HILL, C. E., Thompson, B. J., Cogar, M. C., & Denmann III, D. W. (1993). Beneath the surface of long-term therapy: Therapist and client report of their own and each other's covert processes. *Journal of Counseling Psychology, 40*(3), 278–287.

HILLS, M. D. (1984). *Improving the learning of parents' communication skills by providing for the discovery of personal meaning.* Doctoral dissertation, University of Victoria, British Columbia, Canada.

HORVATH, A. O., & Symonds, B. D. (1991). Relation between working alliance and outcome in psychotherapy: A meta-analysis. *Journal of Counseling Psychology, 38,* 139–149.

HOWARD, G. S. (1991). Culture tales: A narrative approach to thinking, cross-cultural psychology, and psychotherapy. *American Psychologist, 46,* 187–197.

HOWELL, W. S. (1982). *The empathic communicator.* Belmont, CA: Wadsworth.

HOYT, W. T. (1996). Antecedents and effects of perceived therapist credibility: A meta-analysis. *Journal of Counseling Psychology, 43*(4), 430–447.

HUBER, C. H., & Baruth, L. G. (1989). *Rational-emotive family therapy.* New York: Springer.

HUDSON, S. M., Marshall, W. L., Ward, T., Johnston, P. W., et al. (1995). Kia Marama: A cognitive-behavioral program for incarcerated child molesters. *Behavior Change, 12*(2), 69–80.

HUMPHREYS, K. (1996). Clinical psychologists as therapists: History, future, and alternatives. *American Psychologist, 51,* 190–197.

HUMPHREYS, K., & Rappaport, J. (1994). Researching self-help mutual aid groups and organizations: Many roads, one journey. *Applied and Preventive Psychology, 3.*

HUNTER, R. H. (1995). Benefits of competency-based treatment programs. *American Psychologist, 50,* 509–513.

HURVITZ, N. (1970). Peer self-help psychotherapy groups and their implications for psychotherapy. *Psychotherapy: Theory, Research, and Practice, 7,* 41–49.

HURVITZ, N. (1974). Similarities and differences between conventional psychotherapy and peer self-help psychotherapy groups. In P. S. Roman & H. M. Trice (Eds.), *The sociology of psychotherapy.* New York: Jason Aronson.

ISHIYAMA, F. I. (1990). A Japanese perspective on client inaction: Removing attitudinal blocks through Morita therapy. *Journal of Counseling and Development, 68,* 566–570.

IVEY, A. E. (1994). *Intentional interviewing and counseling* (3rd ed.). Pacific Grove, CA: Brooks/Cole.

IVEY, A. E., Ivey, M. B., & Simek-Morgan, L. (1993). *Counseling and psychotherapy: A multicultural perspective* (3rd ed.). Needham Heights, MA: Allyn & Bacon.

IVEY, A. E., Ivey, M. B., & Simek-Morgan, L. (1997). *Counseling and psychotherapy: A multicultural perspective* (4th ed.). Needham Heights, MA: Allyn & Bacon.

JANIS, I. L., & Mann, L. (1977). *Decision making: A psychological analysis of conflict, choice, and commitment.* New York: Free Press.

JANOSIK, E. H. (Ed.). (1984). *Crisis counseling: A contemporary approach.* Belmont, CA: Wadsworth.

JENSEN, J. P., Bergin, A. E., & Greaves, D. W. (1990). The meaning of eclecticism: New survey and analysis of components. *Professional Psychology: Research and Practice, 21,* 124–130.

JONES, A. S., & Gelso, C. J. (1988). Differential effects of style of interpretation: Another look. *Journal of Counseling Psychology, 35*(4), 363–369.

KAGAN, J. (1996). Three pleasing ideas. *American Psychologist, 51*(9), 901–908.

KAGAN, N. (1973). Can technology help us toward reliability in influencing human interaction? *Educational Technology, 13,* 44–51.

KAHN, M. (1990). *Between therapist and client.* New York: W. H. Freeman.

KAMINER, W. (1992). *I'm Dysfunctional, You're Dysfunctional.* Reading, MA: Addison-Wesley.

KANFER, F. H., & Schefft, B. K. (1988). *Guiding therapeutic change.* Champaign, IL: Research Press.

KAROLY, P. (1995). Self-control theory. In W. O'Donohue & L. Krasner (Eds.), *Theories of behavior therapy: Exploring behavior change* (pp. 259–285). Washington, DC: American Psychological Association.

KATZ, A. H., & Bender, E. I. (1990). *Helping one another: Self-help groups in a changing world.* Oakland, CA: Third Party Publishing Company.

KAUFMAN, G. (1989). *The psychology of shame.* New York: Springer.

KAYE, H. (1992). *Decision power.* Upper Saddle River, NJ: Prentice Hall.

KEITH-SPIEGEL, P. (1994, November). The 1992 ethics code: Boon or bane? *Professional Psychology: Research and Practice, 25,* 315–316.

KELLY, E. W., Jr. (1994). *Relationship-centered counseling: An integration of art and science.* New York: Springer.

KENDALL, P. C. (1992). Healthy thinking. *Behavior Therapy, 23*(1), 1–11.

KERR, B. & Erb, C. (1991). Career counseling with academically talented students: Effects of a value-based intervention. *Journal of Counseling Psychology, 38*(3), 309–314.

KIERULFF, S. (1988). Sheep in the midst of wolves: Person-responsibility therapy with criminals. *Professional Psychology: Research and Practice, 19,* 436–440.

KIESLER, D. J. (1988). *Therapeutic metacommunication.* Palo Alto, CA: Consulting Psychologists Press.

KIRSCHENBAUM, D. S. (1985). Proximity and specificity of planning: A position paper. *Cognitive Therapy and Research, 9,* 489–506.

KIRSCHENBAUM, D. S. (1987). Self-regulatory failure: A review with clinical implications. *Clinical Psychological Review, 7,* 77–104.

KIVLIGHAN, D. M., Jr. (1990). Relation between counselors' use of intentions and clients' perception of working alliance. *Journal of Counseling Psychology, 37,* 27–32.

KIVLIGHAN, D. M., Jr., & Schmitz, P. J. (1992). Counselor technical activity in cases with improving work alliances and continuing-poor working alliances. *Journal of Counseling Psychology, 39,* 32–38.

KNAPP, M. L. (1978). *Nonverbal communication in human interaction* (2nd ed.). New York: Holt, Rinehart & Winston.

KOHUT, H. (1978). The psychoanalyst in the community of scholars. In P. H. Ornstein (Ed.), *The search for self: Selected writings of H. Kohut.* New York: International Universities Press.

KOTTLER, J. A. (1992). *Compassionate therapy: Working with difficult clients.* San Francisco: Jossey-Bass.

KOTTLER, J. A. (1993). *On being a therapist* (Rev. ed.). San Francisco: Jossey-Bass.

KOTTLER, J. A. (1997). *Finding your way as a counselor.* Alexandria, VA: American Counseling Association.

KOTTLER, J. A., & Blau, D. A. (1989). *The imperfect therapist: Learning from failure in therapeutic practice.* San Francisco: Jossey-Bass.

LAMBERT, M. J., & Cattani-Thompson, K. (1996, Summer). Current findings regarding the effectiveness of counseling: Implications for practice. *Journal of Counseling and Development, 74,* 601–608.

LANDRETH, G. L. (1984). Encountering Carl Rogers: His views on facilitating groups. *Personnel and Guidance Journal, 62,* 323–326.

LANDSMAN, M. S. (1994). Needed: Metaphors for the prevention model of mental health. *American Psychologist, 49,* 1086–1087.

LANG, P. J. (1995, May). The emotion probe: Studies of motivation and attention. *American Psychologist, 50*(5), 372–385.

LATTIN, D. (1992, November 21). Challenging organized religion: Spiritual small-group movement. *San Francisco Chronicle,* pp. A1, A9.

LAVOIE, F., Gidron, B., & Borkman, T. (Eds.) (1995). *Self-help and mutual aid groups: International and multicultural perspectives.* Binghamton, NY: Haworth Press.

LAZARUS, A. A. (1976). *Multimodal behavior therapy.* New York: Springer.

LAZARUS, A. A. (1981). *The practice of multimodal therapy.* New York: McGraw-Hill.

LAZARUS, A. A. (1993, Fall). Tailoring the therapeutic relationship, or being an authentic chameleon. *Psychotherapy, 30,* 404–407.

LAZARUS, A. A., Beutler, L. E., & Norcross, J. C. (1992). The future of technical eclecticism. *Psychotherapy, 29*(1), 11–20.

LEAHEY, M., & Wallace, E. (1988). Strategic groups: One perspective on integrating strategic and group therapies. *Journal for Specialists in Group Work, 13,* 209–217.

LEE, C. C. (1995). *Counseling for diversity: A guide for school counselors and related professionals.* Needham Heights, MA: Allyn and Bacon.

LEE, C. C. (1997). *Multicultural issues in counseling: New approaches to diversity.* Alexandria, VA: American Counseling Association.

LEIGH, I. W., Corbett, C. A., Gutman, V., & Morere, D. A. (1996). Providing psychological services to deaf individuals: A response to new perceptions of diversity. *Professional Psychology: Research and Practice, 27*(4), 364–371.

LEVIN, L. S., & Shepherd, I. L. (1974). The role of the therapist in Gestalt therapy. *The Counseling Psychologist, 4,* 27–30.

LEWIN, K. (1969). Quasi-stationary social equilibria and the problem of permanent change. In W. G. Bennis, K. D. Benne, & R. Chin (Eds.), *The planning of change*. New York: Holt, Rinehart & Winston.

LIEBERMAN, L. R. (1997). Psychologists as psychotherapists. *American Psychologist, 52,* 181.

LIGHTSEY, O. R. Jr. (1996, October). What leads to wellness? The role of psychological resources in well-being. *The Counseling Psychologist, 24,* 589–735.

LIPSEY, M. W., & Wilson, D. B. (1993). The efficacy of psychological, educational, and behavioral treatment: Confirmation from meta-analysis. *American Psychologist, 48*(12), 1181–1209.

LLEWELYN, S. P. (1988). Psychological therapy as viewed by clients and therapists. *British Journal of Clinical Psychology, 27*(3), 223–237.

LOCKE, E. A., & Latham, G. P. (1984). *Goal setting: A motivational technique that works*. Englewood Cliffs, NJ: Prentice Hall.

LOCKE, E. A., & Latham, G. P. (1990). *A theory of goal setting and task performance*. Englewood Cliffs, NJ: Prentice Hall.

LOWENSTEIN, L. (1993). Treatment through traumatic confrontation approaches: The story of S. *Education Today, 43*(3), 198–201.

LUBORSKY, L. (1993, Fall). The promise of new psychosocial treatments or the inevitability of nonsignificant differences — A poll of the experts. *Psychotherapy and Rehabilitation Research Bulletin* (2), 6–8.

LUBORSKY, L., Crits-Christoph, P., McLellan, A. T., Woody, G., Piper, W., Liberman, B., Imber, S., & Pilkonis, P. (1986). The nonspecific hypothesis of therapeutic effectiveness: A current assessment. *American Journal of Orthopsychiatry, 56,* 501–512.

LYNCH, R. T., & Gussel, L. (1996). Disclosure and self-advocacy regarding disability-related needs: Strategies to maximize integration in postsecondary education. *Journal of Counseling and Development, 74,* 352–357.

LYND, H. M. (1958). *On shame and the search for identity*. New York: Science Editions.

MacDONALD, G. (1996). Inferences in therapy: Process and hazards. *Professional Psychology: Research and Practice, 27*(6), 600–603.

MADDUX, J. E. (Ed.). (1995). *Self-efficacy, adaptation, and adjustment: Theory, research, and application*. New York: Plenum

MAGER, R. F. (1992, April). No self-efficacy, no performance. *Training,* 32–36.

MAHALIK, J. R. (1994). Development of the client resistance scale. *Journal of Counseling Psychology, 41*(1), 58–68.

MAHONEY, M. J. (1991). *Human change processes*. New York: Basic Books.

MAHONEY, M. J., & Patterson, K. M. (1992). Changing theories of change: Recent developments in counseling. In S. D. Brown & R. W. Lent (Eds.), *Handbook of counseling psychology*. New York: Wiley.

MAHRER, A. R. (1993, Fall). The experiential relationship: Is it all-purpose or is it tailored to the individual client? *Psychotherapy, 30*(3), 413–416.

MALLINCKRODT, B. (1996). Change in working alliance, social support, and psychological symptoms in brief therapy. *Journal of Counseling Psychology, 43*(4), 448–455.

MARCH, J. G. (1982, November-December). Theories of choice and making decisions. *Society,* 29–39.

MARKUS, H., & Nurius, P. (1986). Possible selves. *American Psychologist, 41,* 954–969.

MARTIN, J. (1994). *The construction and understanding of psychotherapeutic change.* New York: Teachers College Press.

MASH, E. J., & Hunsley, J. (1993). Assessment considerations in the identification of failing psychotherapy: Bringing the negatives out of the darkroom. *Psychological Assessment, 5*(3), 292–301.

MASLOW, A. H. (1968). *Toward a psychology of being* (2nd ed.). New York: Van Nostrand Reinhold.

MASSON, J. F. (1988). *Against therapy: Emotional tyranny and the myth of psychological healing.* New York: Atheneum.

MATHEWS, B. (1988). The role of therapist self-disclosure in psychotherapy: A survey of therapists. *American Journal of Psychotherapy, 42*(4), 521–531.

McFADDEN, J. (1993). *Transcultural counseling.* Alexandria, VA: American Counseling Association.

McFADDEN, J. (1996, Spring). A transcultural perspective: Reaction to C. H. Patterson's "Multicultural counseling: From diversity to universality." *Journal of Counseling and Development, 74,* 232–235.

McFADDEN, L., Seidman, E., & Rappaport, J. (1992). A comparison of espoused theories of self- and mutual-help groups: Implications for mental health professionals. *Professional Psychology: Research and Practice, 23,* 515–520.

McMILLEN, C., Zuravin, S., & Rideout, G. (1995). Perceived benefit from child sexual abuse. *Journal of Consulting and Clinical Psychology, 63*(6), 1037–1043.

McNEILL, B., & Stolenberg, C. D. (1989). Reconceptualizing social influence in counseling: The elaboration likelihood model. *Journal of Counseling Psychology, 36*(1), 24–33.

McWHIRTER, E. H. (1996). *Counseling for empowerment.* Alexandria, VA: American Counseling Association.

MEARA, N. M., Schmidt, L. D., & Day, J. D. (1996). Principles and virtues: A foundation for ethical decisions, policies, and character. *The Counseling Psychologist, 24*(1), 4–77.

MEHRABIAN, A. (1971). *Silent messages.* Belmont, CA: Wadsworth.

MEHRABIAN, A. (1972). *Nonverbal communication.* Chicago: Aldine-Atherton.

MEHRABIAN, A., & Reed, H. (1969). Factors influencing judgments of psychopathology. *Psychological Reports, 24,* 323–330.

MILAN, M. A., Montgomery, R. W., & Rogers, E. C. (1994). Theoretical orientation revolution in clinical psychology: Fact or fiction? *Professional Psychology: Research and Practice, 4,* 398–402.

MILBANK, D. (1996, October 31). Hiring welfare people, hotel chain finds, is tough but rewarding. *Wall Street Journal,* pp. A1, A14.

MILLER, G. A., Galanter, E., & Pribram, K. H. (1960). *Plans and the structure of behavior.* New York: Holt, Rinehart & Winston.

MILLER, I. J. (1996). Managed care is harmful to outpatient mental health services: A call for accountability. *Professional Psychology: Research and Practice, 27,* 349–363.

MILLER, L. M. (1984). *American spirit: Visions of a new corporate culture.* New York: Morrow.

MILLER, W. C. (1986). *The creative edge: Fostering innovation where you work.* Reading, MA: Addison-Wesley.

MOHR, D. C. (1995, Spring). Negative outcome in psychotherapy: A critical review. *Clinical Psychology: Science and Practice, 2*(1), 1–27.

MORRIS, T. (1994). *True success: A new philosophy of excellence*. New York: Grosset/ Putman.

MULTON, K. D., Brown, S. D., & Lent, R. W. (1991). Relation of self-efficacy beliefs to academic outcomes: A meta-analytic investigation. *Journal of Counseling Psychology, 38*(1), 30–38.

MULTON, K. D., Patton, M. J., & Kivlighan, D. M., Jr. (1996a). Development of the Missouri Identifying Transference scale. *Journal of Counseling Psychology, 43*(3), 243–252.

MULTON, K. D., Patton, M. J., & Kivlighan, D. M., Jr. (1996b). Counselor recognition of transference reactions: Reply to Mallinckrodt (1996) and Carter (1996). *Journal of Counseling Psychology, 43*(3), 259–260.

MURPHY, K. C., & Strong, S. R. (1972). Some effects of similarity self-disclosure. *Journal of Counseling Psychology, 19*, 121–124.

NEIMEYER, G. (Ed.). (1993). *Constructivist assessment, a casebook*. Newbury Park, CA: Sage Publications.

NEIMEYER, R. A., & Mahoney, M. J. (1995). *Constructivism in psychotherapy*. Hyattsville, MD: American Psychological Association.

NORCROSS, J. C., & Wogan, M. (1987). Values in psychotherapy: A survey of practitioners' beliefs. *Professional Psychology: Research and Practice, 18*(1), 5–7.

O'HANLON, W. H., & Weiner-Davis, M. (1989). *In search of solutions: A new direction in psychotherapy*. New York: Norton.

O'LEARY, A. (1985). Self-efficacy and health. *Behavior Research and Therapy, 23*, 437–451.

ORLINSKY, D. E., & Howard, K. I. (1987, Spring). A generic model of psychotherapy. *Journal of Integrative and Eclectic Psychotherapy, 6*, 6–27.

OTANI, A. (1989). Client resistance in counseling: Its theoretical rationale and taxonomic classification. *Journal of Counseling and Development, 67*, 458–461.

PATRICELLI, R. E., & Lee, F. C. (1996). Employer-based innovations in behavioral health benefits. *Professional Psychology: Research and Practice, 27*, 325–334.

PATTERSON, C. H. (1985). *The therapeutic relationship: Foundations for an eclectic psychotherapy*. Pacific Grove, CA: Brooks/Cole.

PATTERSON, C. H. (1996, Spring). Multicultural counseling: From diversity to universality. *Journal of Counseling and Development, 74*, 227–231.

PAYNE, E. C., Robbins, S. B., & Dougherty, L. (1991). Goal directedness and older-adult adjustment. *Journal of Counseling Psychology, 38*(3), 302–308.

PEDERSEN, P. (1994). *A handbook for developing multicultural awareness* (2nd ed.). Alexandria, VA: American Counseling Association.

PEDERSEN, P. (1996, Spring). The importance of both similarities and differences in multicultural counseling: Reaction to C. H. Patterson. *Journal of Counseling and Development, 74*, 236–237.

PEDERSEN, P., & Ivey, A. E. (1993). *Culture-centered counseling and interviewing skills: A practical guide*. Westport, NC: Praeger.

PENNEBAKER, J. W. (1995a). *Emotion, disclosure, and health*. Washington, DC: American Psychological Association

PENNEBAKER, J. W. (1995b). Emotion, disclosure, and health: An overview. In J. W. Pennebaker (Ed.), *Emotion, Disclosure, and Health*, 3–10.

PERRY, M. J., & Albee, G. W. (1994). On "the science of prevention." *American Psychologist, 49*, 1087–1088.

PETERSON, C., Seligman, M. E. P., & Vaillant, G. E. (1988). Pessimistic explanatory style as a risk factor for physical illness: A thirty-five-year longitudinal study. *Journal of Personality and Social Psychology, 55,* 23–27.

PHILLIPS, E. L. (1987). The ubiquitous decay curve: Service delivery similarities in psychotherapy, medicine, and addiction. *Professional Psychology: Research and Practice, 18,* 650–652.

PONTEROTTO, J. G., Casas, J. M., Suzuki, L. A., & Alexander, C. M. (Eds.) (1995). *Handbook of multicultural counseling.* Thousand Oaks, CA: Sage Publications.

POPE, K. S., & Vasquez, M. J. T. (1991). *Ethics in psychotherapy and counseling: A practical guide for psychologists.* San Francisco: Jossey-Bass.

PROCHASKA, J. O., & Norcross, J. C. (1994). *Systems of psychotherapy: A transtheoretical analysis* (3rd ed.). Pacific Grove, CA: Brooks/Cole.

PROCTOR, E. K., & Rosen, A. (1983). Structure in therapy: A conceptual analysis. *Psychotherapy: Theory, Research, and Practice, 20,* 202–207.

*PROFESSIONAL PSYCHOLOGY: RESEARCH AND PRACTICE.* (1994, November, Vol. 25). Special Section: The 1992 ethics code: Boon or bane?

PYSZCZYNSKI, T., & Greenberg, J. (1987). Self-regulatory preservation and the depressive self-focusing style: A self-awareness theory of depression. *Psychological Bulletin, 102,* 122–138.

RAUP, J. L., & Myers, J. E. (1989). The empty nest syndrome: Myth or reality? *Journal of Counseling & Development, 68,* 180–183.

REANDEAU, S. G., & Wampold, B. E. (1991). Relationship of power and involvement to working alliance: A multiple-case sequential analysis of brief therapy. *Journal of Counseling Psychology, 38,* 107–114.

RENNIE, D. L. (1994). Clients' deference in psychotherapy. *Journal of Counseling Psychology, 41*(4), 427–437.

RIDLEY, C. R., Mendoza, D. W., Kanitz, B. E., Angermeier, L., & Zenk, R. (1994). Cultural sensitivity in multicultural counseling: A perceptual schema model. *Journal of Counseling Psychology, 41*(2), 125–136.

RIESSMAN, F. (1985). New dimensions in self-help. *Social Policy, 15*(3), 2–4.

RIESSMAN, F. (1990, Summer/Fall). Bashing self-help. *Self-Help Reporter,* 1–2.

RIESSMAN, F., & Carroll, D. (1995). *Redefining self-help.* San Francisco: Jossey-Bass.

ROBERTSHAW, J. E., Mecca, S. J., & Rerick, M. N. (1978). *Problem-solving: A systems approach.* New York: Petrocelli Books.

ROBINSON, B. E., & Bacon J. G. (1996). The "If only I were thin . . ." treatment program: Decreasing the stigmatizing effects of fatness. *Professional Psychology: Research and Practice, 27*(2), 175–183.

ROBITSCHEK, C. G., & McCarthy, P. A. (1991). Prevalence of counselor self-reliance in the therapeutic dyad. *Journal of Counseling and Development, 69,* 218–221.

ROGERS, C. R. (1951). *Client-centered therapy.* Boston: Houghton Mifflin.

ROGERS, C. R. (1957). The necessary and sufficient conditions of therapeutic personality change. *Journal of Consulting Psychology, 21,* 95–103.

ROGERS, C. R. (1965). *Client-centered therapy: Its current practice, implications and theory.* Boston: Houghton, Mifflin.

ROGERS, C. R. (1980). *A way of being.* Boston: Houghton Mifflin.

ROGERS, C. R., Perls, F., & Ellis, A. (1965). *Three approaches to psychotherapy 1* [Film]. Orange, CA: Psychological Films, Inc.

ROGERS, C. R., Shostrom, E., & Lazarus, A. (1977). *Three approaches to psychotherapy 2* [Film]. Orange, CA: Psychological Films, Inc.

ROSEN, G. M. (1993). Self-help or hype? Comments on psychology's failure to advance self-care. *Professional Psychology: Research and Practice, 24*(3), 340–345.

ROSEN, S., & Tesser, A. (1970). On the reluctance to communicate undesirable information: The MUM effect. *Sociometry, 33,* 253–263.

ROSEN, S., & Tesser, A. (1971). Fear of negative evaluation and the reluctance to transmit bad news. *Proceedings of the 79th Annual Convention of the American Psychological Association, 6,* 301–302.

ROSENBERG, S. (1996). Health maintenance organization penetration and general hospital psychiatric services: Expenditure and utilization trends. *Professional Psychology: Research and Practice, 27,* 345–348.

ROSSI, P. H., & Wright, J. D. (1984). Evaluation research: An assessment. *Annual Review of Sociology, 10,* 331–352.

RUSSELL, J. A. (1995). Facial expressions of emotion: What lies beyond minimal universality? *Psychological Bulletin, 118*(3), 379–391.

RUSSO, J. E., & Schoemaker, P. J. H. (1992, Winter). Managing overconfidence. *Sloan Management Review,* 7–17.

SAFRAN, J. D., & Muran, J. C. (1995a). Resolving therapeutic alliance ruptures: Diversity and integration. *In Session: Psychotherapy in Practice, 1*(1), 81–92.

SAFRAN, J. D., & Muran, J. C. (Eds.). (1995b, Spring, Vol. 1). The therapeutic alliance [Special issue]. *In Session: Psychotherapy in Practice.*

SCHEIN, E. H. (1990, Spring). A general philosophy of helping: Process consultation. *Sloan Management Review,* 57–64.

SCHIFF, J. L. (1975). *Cathexis reader: Transactional analysis treatment of psychosis.* New York: Harper & Row.

SCHMIDT, F. L. (1992). What do data really mean? Research findings, meta-analysis, and cumulative knowledge in psychology. *American Psychologist, 47*(10), 1173–1181.

SCHMIDT, F. L., & Hunter, J. E. (1993). Tacit knowledge, practical intelligence, general mental ability, and job knowledge. *Current Directions in Psychological Science, 1,* 8–9.

SCHOEMAKER, P. J. H., & Russo, J. E. (1990). *Decision traps.* New York: Doubleday.

SCOTT, N. E., & Borodovsky, L. G. (1990). Effective use of cultural role taking. *Professional Psychology: Research and Practice, 21*(3), 167–170.

*SELF-HELP REPORTER.* (1985, Vol. 7, No. 2). The self-help ethos.

*SELF-HELP REPORTER.* (1992, Summer). From self-help to social action.

SELIGMAN, M. (1991). *Learned optimism.* New York: Knopf.

SELIGMAN, M. (1994). *What you can change and what you can't.* New York: Knopf.

SELIGMAN, M. (1995). The effectiveness of psychotherapy: The consumer reports study. *American Psychologist, 50*(12), 965–974.

SELIGMAN, M. E. P. (1975). *Helplessness: On depression, development, and death.* San Francisco: W. H. Freeman.

SEXTON, T. L., & Whiston, S. C. (1994). The status of the counseling relationship: An empirical review, theoretical implications, and research directions. *The Counseling Psychologist, 22*(1), 6–78.

SHADISH, W. R., & Sweeney, R. B. (1991). Mediators and moderators in meta-analysis: There's a reason we don't let dodo birds tell us which psychotherapies should have prizes. *Journal of Consulting and Clinical Psychology, 59*(6), 883–893.

SHAPIRO, D., & Shapiro, D. (1982). Meta-analysis of comparative therapy outcome studies: A replication and refinement. *Psychological Bulletin, 92,* 581–604.

SIEGMAN, A. W., & Feldstein, S. (Eds.). (1987). *Nonverbal behavior and communication* (2nd ed.). Hillsdale, NJ: Erlbaum.

SIMON, J. C. (1988). Criteria for therapist self-disclosure. *American Journal of Psychotherapy, 42*(3), 404–415.

SMABY, M., & Tamminen, A. W. (1979). Can we help belligerent counselees? *Personnel and Guidance Journal, 57,* 506–512.

SMITH, M., Glass, G., & Miller, T. (1980). *The benefit of psychotherapy.* Baltimore: Johns Hopkins University Press.

SNYDER, C. R. (1984, September). Excuses, excuses. *Psychology Today,* 50–55.

SNYDER, C. R., & Higgins, R. L. (1988). Excuses: Their effective role in the negotiation of reality. *Psychological Bulletin, 104,* 23–35.

SNYDER, C. R., Higgins, R. L., & Stucky, R. J. (1983). *Excuses: Masquerades in search of grace.* New York: Wiley.

SOMBERG, D. R., Stone, G. L., & Claiborn, C. D. (1993). Informed consent: Therapists' beliefs and practices. *Professional Psychology: Research and Practice, 24*(2), 153–159.

STEENBARGER, B. N., & Smith, H. B. (1996, November/December). Assessing the quality of counseling services. *Journal of Counseling and Development,* 145–150.

STERNBERG, R. J. (1990). Wisdom and its relations to intelligence and creativity. In R. J. Sternberg (Ed.), *Wisdom: Its nature, origins, and development* (pp. 124–159). New York: Cambridge University Press.

STERNBERG R. J., Wagner, R. K., Williams, W. M., & Horvath, J. A. (1995, Winter). Testing common sense. *American Psychologist, 50,* 912–927.

STILES, W. B. (1994, Fall). Drugs, recipes, babies, bathwater, and psychotherapy process-outcome relations. *Journal of Consulting and Clinical Psychology, 62*(5), 955–959.

STILES, W. B., & Shapiro, D. A. (1994, Fall). Disabuse of the drug metaphor: Psychotherapy process-outcome correlations. *Journal of Consulting and Clinical Psychology, 62*(5), 942–948.

STONE, G. L. (1995). Culture and counseling. *Counseling Psychologist, 23,* 3–121.

STRICKER, G. (1995). Failures in psychotherapy. *Journal of Psychotherapy Integration, 5*(2), 91–93.

STRICKER, G., & Fisher, A. (Eds.). (1990). *Self-disclosure in the therapeutic relationship.* New York: Plenum.

STROH, P., & Miller, W. W. (1993, May). HR professionals should thrive on paradox. *Personal Journal,* 132–135.

STRONG, S. R. (1968). Counseling: An interpersonal influence process. *Journal of Counseling Psychology, 15,* 215–224.

STRONG, S. R. (1991). Social influence and change in therapeutic relationships. In C. R. Snyder & D. R. Forsyth (Eds.), *Handbook of social and clinical psychology: The health perspective* (pp. 540–562). New York: Pergamon Press.

STRONG, S. R., & Claiborn, C. D. (1982). *Change through interaction: Social psychological processes of counseling and psychotherapy.* New York: Wiley.

STRONG, S., Yoder, B., & Corcoran, J. (1995). Counseling: A social process for encouraging personal powers. *The Counseling Psychologist, 23*(2), 374–384.

STRUPP, H. H., Hadley, S. W., & Gomes-Schwartz, B. (1977). *Psychotherapy for better or worse: The problem of negative effects.* New York: Jason Aronson.

SUE, D. W. (1990). Culture-specific strategies in counseling: A conceptual framework. *Professional Psychology: Research and Practice, 21*(6), 424–433.

SUE, D. W., Arredondo, P., & McDavis, R. J. (1992). Multicultural counseling competencies and standards: A call to the profession. *Journal of Counseling and Development, 70,* 477–486.

SUE, D. W., & Sue, D. (1990). *Counseling the culturally different: Theory and practice* (2nd ed.). New York, Wiley.

SULLIVAN, T., Martin, W., Jr., & Handelsman, M. (1993). Practical benefits of an informed-consent procedure: An empirical investigation. *Professional Psychology: Research and Practice, 24*(2), 160–163.

SYKES, C. J. (1992). *A nation of victims.* St. Martin's.

TAUSSIG, I. M. (1987). Comparative responses of Mexican Americans and Anglo-Americans to early goal setting in a public mental health clinic. *Journal of Counseling Psychology, 34,* 214–217.

TESSER, A., & Rosen, S. (1972). Similarity of objective fate as a determinant of the reluctance to transmit unpleasant information: The MUM effect. *Journal of Personality and Social Psychology, 23,* 46–53.

TESSER, A., Rosen, S., & Batchelor, T. (1972). On the reluctance to communicate bad news (the MUM effect): A role play extension. *Journal of Personality, 40,* 88–103.

TESSER, A., Rosen, S., & Tesser, M. (1971). On the reluctance to communicate undesirable messages (the MUM effect): A field study. *Psychological Reports, 29,* 651–654.

TOMES, H. (1994). Minority recruitment, retention is a priority. *APA Monitor, 25,* 35.

TRACEY, T. J. (1991). The structure of control and influence in counseling and psychotherapy: A comparison of several definitions and measures. *Journal of Counseling Psychology, 38,* 265–278.

TREVINO, J. G. (1996, April). Worldview and change in cross-cultural counseling. *Counseling Psychologist, 24*(2), 198–215.

TRYON, G. S., & Kane, A. S. (1993). Relationship of working alliance to mutual and unilateral termination. *Journal of Counseling Psychology, 40*(1), 33–36.

TYLER, F. B., Pargament, K. I., & Gatz, M. (1983). The resource collaborator role: A model for interactions involving psychologists. *American Psychologist, 38,* 388–398.

VACHON, D. O., & Agresti, A. A. (1992). A training proposal to help mental health professionals clarify and manage implicit values in the counseling process. *Professional Psychology: Research and Practice, 23*(6), 509–514.

VARGAS, L. A., & Willis, D. J. (1994). New directions in the treatment of ethnic minority children and adolescents. *Journal of Clinical Child Psychology, 23,* 2–4.

WACHTEL, P. L. (1989, August 6). Isn't insight everything? [Book review]. *New York Times,* p. 18.

WAEHLER, C. A., & Lenox, R. (1994). A concurrent (versus stage) model for conceptualizing and representing the counseling process. *Journal of Counseling and Development, 73,* 17–22.

WAHLSTEIN, D. (1991). Nonverbal behavior and self-presentation. *Psychological Bulletin, 110*(3), 587–595.

WATKINS, C. E., Jr. (1990). The effects of counselor self-disclosure: A research review. *The Counseling Psychologist, 18*(3), 477–500.

WATKINS, C. E., Jr., & Schneider, L. J. (1989). Self-involving versus self-disclosing counselor statements during an initial interview. *Journal of Counseling and Development, 67,* 345–349.

WATSON, D. L., & Tharp, R. G. (1993). *Self-directed behavior* (6th ed.). Pacific Grove, CA: Brooks/Cole.

WEICK, K. E. (1979). *The social psychology of organizing* (2nd ed.). Reading, MA: Addison-Wesley.

WEINBERGER, J. (1995, Spring). Common factors aren't so common: The common factors dilemma. *Clinical Psychology: Science and Practice, 2*(1), 45–69.

WEINER, M. F. (1983). *Therapist disclosure: The use of self in psychotherapy* (2nd ed.). Baltimore: University Park Press.

WEINRACH, S. G. (1989). Guidelines for clients of private practitioners: Committing the structure to print. *Journal of Counseling and Development, 67,* 299–300.

WEINRACH, S. G. (1995). Rational emotive behavior therapy: A tough-minded therapy for a tender-minded profession. *Journal of Counseling and Development, 73,* 296–300.

WEINRACH, S. G. (1996). Nine experts describe the essence of rational emotive therapy while standing on one foot. *Journal of Counseling and Development, 74,* 326–331.

WEINRACH, S. G., & Thomas, K. R. (1996, Summer). The counseling profession's commitment to diversity-sensitive counseling: A critical reassessment. *Journal of Counseling and Development, 74,* 472–477.

WEISZ, J. R., Donenberg, G. R., Han, S. S., & Weiss, B. (1995, Fall). Bridging the gap between laboratory and clinic in child and adolescent psychotherapy. *Journal of Consulting and Clinical Psychology, 63*(5), 688–701.

WEISZ, J. R., Rothbaum, F. M., & Blackburn, T. C. (1984). Standing out and standing in: The psychology of control in America and Japan. *American Psychologist, 39,* 955–969.

WELLENKAMP, J. (1995). Cultural similarities and differences regarding emotional disclosure: Some examples from Indonesia and the Pacific. In J. W. Pennebaker (Ed.), *Emotion, Disclosure, and Health,* 293–311.

WESTERMAN, M. A. (Ed.). (1989). Putting insight to work [Special section]. *Journal of Integrative and Eclectic Psychotherapy, 8,* 195–250.

WHEELER, D. D., & Janis, I. L. (1980). *A practical guide for making decisions.* New York: Free Press.

WHYTE, G. (1991). Decision failures: Why they occur and how to prevent them. *Academy of Management Executive, 5*(3), 23–31.

WILLIAMS, R. (1989, January-February). The trusting heart. *Psychology Today,* 36–42.

WINBORN, B. (1977). Honest labeling and other procedures for the protection of consumers of counseling. *Personnel and Guidance Journal, 56,* 206–209.

WOODY, R. H. (1991). *Quality care in mental health.* San Francisco: Jossey-Bass.

WORSLEY, A. (1981). In the eye of the beholder: Social and personal characteristics of teenagers and their impressions of themselves and fat and slim people. *British Journal of Medical Psychology, 54,* 231–242.

YALOM, I. D. (1989). *Love's executioner and other tales of psychotherapy.* Scranton, PA: Basic Books.

YANKELOVICH, D. (1992, October 5). How public opinion really works. *Fortune,* 102–108.

YUTRZENKA, B. (1995). Making a case for training in ethnic and cultural diversity in increasing treatment efficacy. *Journal of Consulting and Clinical Psychology, 63,* 197–206.

ZANE, N. W. S., Sue, S., Hu, L., & Kwon, J. (1991). Asian-American assertion: A social learning analysis of cultural differences. *Journal of Counseling Psychology, 38*(1), 63–70.

ZAYAS, L. H., Torres, L. R., Malcom, J., & DesRosiers, F. S. (1996). Clinicians' definitions of ethnically sensitive therapy. *Professional Psychology: Research and Practice, 27*(1), 78–82.

# NAME INDEX

Abramson, P. R., 51
Ackoff, R., 204
Agresti, A. A., 42
Alexander, C. M., 46
Allison, K. W., 46
Anderson, T., 86
Anderson, W., 138
Angermeier, L., 46
Anthony, W. A., 180
Argyris, C., 151
Arnkoff, D. B., 40
Arrendondo, P., 46
Atkinson, D. R., 167

Bacon, J. G., 310
Bailey, K. G., 40
Baldwin, B. A., 210
Bandura, A., 191, 195, 265
Baruth, L. G., 153
Basic Behavioral Science Task Force of the National Advisory Mental Health Council, 279
Batchelor, T., 166
Beier, E. G., 164, 165
Benbenishty, R., 57
Bender, E. I., 335
Bennett, B. E., 56
Bennett, M. B., 330
Bennett, M. I., 330
Berberoglu, L. S., 56
Berenson, B. G., 190, 191
Bergan, J., 42
Berger, D. M., 170
Bergin, A. E., 13, 42, 340
Bernard, M. E., 147, 153
Berne, E., 164
Bernstein, R., 46
Bersoff, D. N., 56
Beutler, L. E., 42, 340
Binder, C., 113
Bischoff, M. M., 138
Blatt, S. J., 46
Blau, D. A., 12, 167
Bordin, E. S., 41
Borgen, F. H., 151

Borkman, T., 335
Borodovsky, L. G., 94
Brammer, L., 108
Brandsma, J., 52
Bronfenbrenner, U., 265
Brown, S. D., 195
Brownell, K. D., 332
Bruch, M. A., 195
Bryant, B. K., 46

Callanan, P., 56
Canter, M. B., 56
Carey, R., 296
Carkhuff, R. R., 60, 110, 159, 180, 281
Carroll, D., 335
Carter, J. A., 40
Casas, J. M., 46
Cattani-Thompson, K., 9
Cervone, D., 195
Chang, E. C., 46
Clabby, J. F., 53
Claiborn, C. D., 51, 53, 56, 57, 147, 178
Clark, A. J., 138
Clarkin, J., 208
Cloud, M. Y., 51
Cogar, M. C., 79
Cole, H. P., 227
Conoley, C. W., 185
Consumer Reports, 10
Corbett, C. A., 46
Corcoran, J., 51
Corey, G., 56, 340
Corey, M. S., 56
Cosier, R. A., 203
Covey, S. R., 83, 132
Cowen, E. L., 11
Crawford, I., 46
Cross, J. G., 17
Croteau, J. M., 57
Cummings, N. A., 185

Daniels, J. A., 185
Das, A. K., 46

Dauser, P. J., 57
Day, J. D., 56
De Bono, E., 228
Deffenbacher, J. L., 282
Denmann, D. W., III, 79
DesRosiers, F. S., 46
Deutsch, M., 126
DiGiuseppe, R., 153
Dimond, R. E., 140
Donenberg, G. R., 11
Dorn, F. J., 51, 147
Dorner, D., 301
Dougherty, L., 223
Driscoll, R., 7, 55, 89, 140, 158, 275, 329, 332
Dryden, W., 153
Duan, C., 73
Durlak, J. A., 11
Dworkin, S. H., 96

Echemendia, R. J., 46
Edwards, C. E., 178
Egan, G., 139, 148
Eisenberg, W., 73
Ekman, P., 70
Elias, M. J., 53
Elliott, R., 151
Ellis, A., 12, 138, 153, 261
Erb, C., 42
Etzioni, A., 203
Eysenck, H. J., 11

Farrelly, F., 52
Feldstein, S., 70
Feller, R., 323
Ferguson, M., 313
Ferguson, T., 331
Festinger, S., 165
Fish, J. M., 126
Fisher, A., 178
Fisher, R., 45
Fowers, B. J., 46
Frances, A., 208
Frazier, P. A., 51
Freire, P., 52

365

# SUBJECT INDEX

TO THE OWNER OF THIS BOOK:

We hope that you have found *The Skilled Helper,* Sixth Edition, useful. So that this book can be improved in a future edition, would you take the time to complete this sheet and return it? Thank you.

School and address: ———————————————————————————————

Department: ——————————————————————————————————

Instructor's name: ——————————————————————————————

1. What I like most about this book is: ————————————————————

————————————————————————————————————————

————————————————————————————————————————

2. What I like least about this book is: ————————————————————

————————————————————————————————————————

————————————————————————————————————————

3. My general reaction to this book is: —————————————————————

————————————————————————————————————————

4. The name of the course in which I used this book is: ————————————

————————————————————————————————————————

5. Were all of the chapters of the book assigned for you to read? ——————

    If not, which ones weren't? ——————————————————————————

6. In the space below, or on a separate sheet of paper, please write specific suggestions for improving this book and anything else you'd care to share about your experience in using the book.

————————————————————————————————————————

————————————————————————————————————————

————————————————————————————————————————

————————————————————————————————————————

————————————————————————————————————————

Optional:

Your name: _____ Date: _____

May Brooks/Cole quote you, either in promotion for *The Skilled Helper,* Sixth Edition, or in future publishing ventures?

Yes: _____ No: _____

Sincerely,

*Gerard Egan*

FOLD HERE

- - - - - - - - - - - - - - - - - - - - - - - - - - - - - - - - - - - - - -

‖‖ ‖‖

## BUSINESS REPLY MAIL
FIRST CLASS      PERMIT NO. 358      PACIFIC GROVE, CA

POSTAGE WILL BE PAID BY ADDRESSEE

ATT: *Gerard Egan* _____

**Brooks/Cole Publishing Company**
**511 Forest Lodge Road**
**Pacific Grove, California 93950-9968**

‖‖‖‖‖‖‖‖‖‖‖‖‖‖‖‖‖‖‖‖‖‖‖‖‖‖‖‖‖‖‖

- - - - - - - - - - - - - - - - - - - - - - - - - - - - - - - - - - - - - -

FOLD HERE